CLIMATE CHANGE

CLIMATE CHANGE: PAST, PRESENT AND FUTURE

Marie-Antoinette Mélières
Université Grenoble Alpes, France

and

Chloé Maréchal
Université Claude Bernard Lyon 1, France

Translated by

Erik Geissler
Université Grenoble Alpes, France

and **Catherine Cox**

WILEY Blackwell

Registered office: John Wiley & Sons, Ltd, The Atrium, Southern Gate, Chichester, West Sussex, PO19 8SQ, UK

Editorial offices: 9600 Garsington Road, Oxford, OX4 2DQ, UK
The Atrium, Southern Gate, Chichester, West Sussex, PO19 8SQ, UK
111 River Street, Hoboken, NJ 07030-5774, USA

For details of our global editorial offices, for customer services and for information about how to apply for permission to reuse the copyright material in this book please see our website at www.wiley.com/wiley-blackwell.

Library of Congress Cataloging-in-Publication Data
Mélières, Marie-Antoinette.
  [Climat et société. English]
  Climate change : past, present, and future / Marie-Antoinette Mélières and Chloé Maréchal ; translated by Erik Geissler and Catherine Cox.
    pages cm
  Originally published: Grenoble : CRDP de l' Académie de Grenoble, 2010.
  Includes index.
  ISBN 978-1-118-70852-1 (cloth) – ISBN 978-1-118-70851-4 (pbk.)   1. Climatic changes.
2. Biodiversity.   3. Greenhouse effect, Atmospheric.   4. Global warming.   5. Climate change
mitigation.   I. Maréchal, Chloé.   II. Title.
  QC902.9.M4513 2015
  551.6–dc23
                                                                                           2014039734

A catalogue record for this book is available from the British Library.

About the cover: Banded iceberg drifting a couple of miles off the French station, Dumont d'Urville (Terre Adélie), Antarctica. White stripes are normal glacier ice resulting from snow compaction including air bubbles. The coloured stripes are the result of the melting and refreezing processes within the layered structure of the iceberg. Due to enhanced melting at the contact with the ocean, the immersed part rapidly reduces leading to a tipping over as revealed by the now tilted stripes. © Emmanuel Le Meur, 2008/9.

Set in 11/14 pt Adobe Garamond Pro by Aptara

1   2015

Contents

Foreword

Discovering complexity: a springboard for action

The present moment of the present century is singular. Climate change threatens humanity with an irreversible and unwanted change in the biosphere. This book by Marie-Antoinette Mélières and Chloé Maréchal presents the scientific background for a comprehension of what is so vitally at stake.

The depth and range of knowledge deployed, over the scales of time and space, enable us to grasp the physical and biological processes that drive the Earth's present climate and understand how it has changed through the ages.

This timely and up-to-date work incorporates the most recent scientific advances, including the 2013 IPCC report (*The Physical Science Basis*). Its precise and jargon-free explanations are supported by figures and illustrations of excellent quality. It offers clear answers to the questions that are most frequently raised in the public domain. Principally addressed to students and teachers, it will also prove to be a mine of information for journalists, as well as for scientists of all disciplines.

Climate and biodiversity are the basis of life, and indeed are life itself. They are universal assets that belong to all humanity. States, as well as individuals, must reflect on and invent new relationships through which these assets are accessed and governed. The reflection can develop and progress only if scientific information that is rigorous and free of ideology is available. Our task is to devise and bring about more than a paradigm shift, a change of era that has no precedent in the history of mankind. It is our own human species that is at the heart of the problem and, since climate knows no frontiers, we all now have our backs to the wall.

My hope is that this book will contribute to expanding our field of vision. Three components of the same reality are too frequently separated: society, biodiversity and climate change. They must be reconnected. Once we acknowledge our own responsibility, all possibilities become open. Not a single person, whatever his or her status, is immune to the risks. But in full awareness of them, we must grasp this opportunity to rethink how we lead and organize our lives. It is imperative to rediscover the means of living and evolving in harmony with the living tissue of the Earth and its climate.

Let us not sacrifice the future for the present. Let us instead seize the chance in an all-encompassing vision and together open Chapter 2 of our history.

Nicolas Hulot
Président de la Fondation Nicolas Hulot pour la nature et l'Homme
President of the Nicolas Hulot Foundation for Nature and Mankind
www.fnh.org

Acknowledgements

This book is a completely revised and updated edition of the original French version *Climat et Société – Climats passés, passage de l'homme, climat futur: repères essentiels*, published in 2010 by the Centre de Recherche Pédagogique de Grenoble. We are particularly grateful to R. Briatte and S. Duchaffaut, who were the first to recognize the need for the book and whose good-tempered persistence finally led to its publication.

We especially thank P. Gibbard, whose suggestion it was to publish this work in English, and R. Bradley and G. Brasseur, who gave their backing to this enterprise. The support of H. Letreut and J.-C. Hourcade, which was crucial in seeking financial assistance for the translation, is gratefully acknowledged. Our thanks go to the following organizations for their financial support: Centre National du Livre, Rhône-Alpe Region, Ferthé Foundation and Nicolas Hulot Foundation.

We are indebted to E. Geissler and C. Cox for this careful translation, an undertaking that required close collaboration and permanent commitment. In addition to his role as translator, E. Geissler, with unfailing availability, brought the critical and attentive regard of an eminent physicist.

We also thank N. Roberts and the anonymous referees for their reading and helpful comments on the manuscript.

We extend our gratitude to all our colleagues who contributed to the final document through their reading and constructive criticisms of several parts; in particular, V. Badeau, G.Beaugrand, B. Barnier, P. Braconnot, J.-P. Brugal, A. Cazenave, J.-L. Dufresne, M. Fontugne, M. Magny, V. Masson-Delmotte, H. Mercier, J.-J. Moisselin, J. Orr, D. Paillard, J.-R. Petit, S. Planton, E. Rignot, C. Robinet, J. Severinghaus, C. Vincent and C. Waelbroeck.

We also gratefully acknowledge our many colleagues for their valuable collaboration, for their discussions and for the information that they brought to the contents of this book, notably E. Bard, W. Barthlott, O. Boucher, J. Blondel, J. Chappelaz, P. Ciais, N. Coltice, D. Couvet, M. De Angelis, G. Delaygues, V. Devictor, G. Escarguel, X. Fettweis, B. Francou, S. Gillihan, W. Haeberli, A. Hauchecorne, G. Jone, J. Jouzel, E. Lemeur, M.-F. Loutre, L. Mercalli, R. Ricklefs , D. Salas, M. Sato, and P. Seguin.

Many scientists and organizations – in particular, the IPCC (Intergovernmental Panel on Climate Change) – generously provided figures with authorization for reproduction, which have been of immense value in illustrating the examples in the book. The list is long. We are deeply indebted to them.

Finally, we wish to express our gratitude to R. and H. Cox for their support and advice, and especially to M. Chenevier for his unfailing help and encouragement during the writing of this book.

We extend our sincere thanks to the editorial staff of Wiley for their competence and patience in overseeing the birth of this book.

It is dedicated to the youth of today and to the children that will follow them. It is also dedicated to Gabin and Helia.

About the companion website

This book is accompanied by a companion website:

www.wiley.com\go\melieres\climatechange

The website includes:
- PDFs of figures from the book for downloading
- Powerpoints of tables from the book for downloading

Introduction

Aims

The purpose of this book is to offer a unified view of the climate challenge that confronts modern society. The subject matter spans many disciplines, ranging from materials science to the social sciences. Its complexity is immense. Here we adopt a pragmatic approach, marking out the principal reference points that circumscribe the whole subject. These consist essentially of the basic mechanisms that govern climate equilibrium, the various causes of natural changes in the climate, the history of the climate and its troubled past, the changes that are taking place now and the role played by mankind.

The second aim of the book is to trace the relationship between climate change and changes in the biosphere. For this reason, we focus on vegetation, on which the existence and survival of the animal kingdom depends. Only the principal traits are outlined, but this is sufficient for the reader to grasp the enormous impact that the traditional 'business as usual' scenario could have by the end of the present century.

How can these topics be addressed? At the outset, it should be made clear that many of the facts and numbers in this book are taken from the two most recent Intergovernmental Panel for Climate Change (IPCC) reports, published in 2007 and 2013. This group was created in 1988 by the World Meteorological Organization (WMO) and the United Nations Environment Programme (UNEP). The IPCC is an intergovernmental organization that is open to all member countries of the United Nations and the WMO (194 countries).

Climate Change: Past, Present and Future, First Edition. Marie-Antoinette Mélières and Chloé Maréchal.
© 2015 John Wiley & Sons, Ltd. Published 2015 by John Wiley & Sons, Ltd.
Companion website: www.wiley.com\go\melieres\climatechange

The situation in brief . . .

Climates on Earth vary greatly from one place to another, be it in the temperature, the amount of precipitation or the strength of the prevailing winds. It might therefore seem incongruous to speak of an 'average temperature' to describe our planet. In fact, the global climate system that reigns from the poles to the Equator is defined, and sustained, by continual consumption of the energy available at the surface of the Earth, which is constantly renewed. Temperature is the physical parameter that expresses the average amount of available energy. That is why changes in all the climates on Earth are defined first by changes in the mean temperature on Earth. *Part of this book will therefore focus on the notion of temperature.*

The changes in climate that occurred before the middle of the 19th century are known from reconstructions based on various proxies. From the middle of the 19th century on, measurements of temperature had become sufficiently numerous and widespread for reliable estimates to be made of the mean temperature at the surface of the Earth. These measurements demonstrate that the temperature is rising; in other words, the *average* climate is warming on the scale of the planet. It is now known that, over the 20th century, the average global warming attained about +1°C. It has been especially pronounced in the three most recent decades.

Over the same time period, from the mid-19th century, with the progress of industrialization, increasing amounts of carbon dioxide (CO_2) have been injected into the atmosphere, along with other gases (CH_4, N_2O etc.). The effect of these emissions has been to modify the natural exchange cycle of the gases between the Earth's surface and the atmosphere, and to increase their concentration in the atmosphere. These gases are greenhouse gases (GHG). Their presence in the atmosphere reinforces the natural greenhouse effect, making the planet warmer.

. . . and the questions that arise

The *first question* is as follows. *Is the recent warming in the 20th century the result of human activity, especially in the last few decades, during which GHG concentrations have greatly increased? Or, on the contrary, is this contribution much smaller than other natural causes?*

If mankind is indeed the cause, then a *second question* arises, about the future. If GHG emissions from human activity continue on a basis of economic and industrial development similar to that of the present, then in the course of the 21st century the general warming of the 20th century will continue, but at a much greater pace. So, our second question has to do with the climate change that follows in the wake of the future warming, and its consequences. *What will the Earth's climate be in this situation? How will temperature and precipitation be affected in the various regions?*

Here, the *third question* arises. *How does this warming compare with natural changes that the Earth has already known in the past?* Why is the scientific community, by almost general consensus, sending out such a strong alarm signal? If there is to be pronounced warming in future decades, what need we be afraid of? What will the consequences be? To answer these questions we must measure past climate changes by the yardstick of life on Earth. It is the cost to living organisms that is at stake – life in all its forms, from the base of the pyramid to its apex, the world of plants and the world of animals alike, a world in which mankind depends intimately on these different forms of life. It is in this context that disruption caused by a large and rapid climate change takes on its full meaning.

Finally, if future conditions threaten to become too disruptive for the present equilibrium, a *fourth question* arises: *to what extent must mankind change its lifestyle?* This delicate question implies a decision on the limits of degradation that we can tolerate. To avoid climate perturbations with consequences for life that are too drastic, a hierarchy must be established among the various economic scenarios.

The different parts of this book

To answer the above questions, the following points must be examined in turn:

- First, we must understand the basis of climate equilibrium: how the climate engine 'works', and how to estimate the orders of magnitude of the different parameters involved, either natural or those involving human agency (*Parts I and II*). This will enable us to discern the different factors that drive the climate changes that the planet continually undergoes (*Part III*).
- Next follows a rapid overview of past climate changes and their causes, over different timescales. The aim is to understand the context of the present warming, and what changes may be anticipated in the future (*Part IV*).
- The climate warming of the recent decades is then described, together with its worldwide implications and its consequences for our environment. The cause of this change is discussed (*Part V*).
- Lastly, we review different possible future economic scenarios, and discuss computer simulations of the climate changes during the 21st century. In the light of our knowledge of what has happened in the past, this can help us to select the scenarios of the future that deserve preference (*Part VI*).

PART I
THE CLIMATE ENGINE OF THE EARTH: ENERGY

Chapter 1
Why are there many different climates on Earth?

At any one place, the climate is defined essentially by the prevailing temperature and by the rainfall. These two quantities, both in their annual averages and in their seasonal variations, are distributed unevenly over the surface of the planet. The result is a mosaic of extremely varied climates. *Why is this? What are the factors that produce such a wide range of temperatures that water exists in abundance in all its three phases (gas, liquid and solid), and that our planet occupies a unique place in the solar system? What factors govern the distribution of temperature and rainfall?* The primary driving force is the annual amount of energy arriving at the surface of the Earth, and its seasonal distribution. The guiding principle of this energy distribution will become apparent as we introduce the various relevant parameters.

1. First, we enquire into the source of energy that continuously supplies the Earth's surface and sets the operating range of temperature. *This is the Sun alone*, all the other sources of energy being incomparably weaker. The flow of energy is defined primarily by the Sun's radiation and is a function of its *activity*. The amount of solar energy received on the Earth also depends on the *distance between the Earth and the Sun*. The position of the Earth in the solar system is thus the first key factor that, unlike our neighbouring planets, enables it to host life in abundance.
2. The second characteristic of planet Earth is its *atmosphere*, which by its composition modifies the flow of energy arriving at the surface. The greenhouse gases (GHGs) in our atmosphere play a leading role in this flow, increasing the energy available at the surface of the planet and raising its average temperature.

Climate Change: Past, Present and Future, First Edition. Marie-Antoinette Mélières and Chloé Maréchal.
© 2015 John Wiley & Sons, Ltd. Published 2015 by John Wiley & Sons, Ltd.
Companion website: www.wiley.com\go\melieres\climatechange

3. Since the Earth is practically *spherical*, the solar flux falling on its surface is spread very unevenly over the different latitudes. At higher latitudes, the Sun's rays become increasingly tilted with respect to the surface and, on moving from the Equator to the poles, less and less energy is received per square metre (Part I, Note 1). This property defines the first major characteristic of climates on Earth: temperature decreases from the Equator to the polar regions.

4. The temperature difference between the Equator and the poles is nonetheless *attenuated* by the universal principle that heat propagates from hot regions to cold regions. Heat is transferred from the tropics towards higher latitudes by three transport mechanisms: atmospheric circulation, ocean circulation and the water cycle.

5. Owing to the tilt of the axis of rotation of the Earth in the ecliptic (the plane in which the Earth moves around the Sun during the year), the slope of the Sun's rays, and hence the energy delivered to each point on the Earth, oscillates throughout the year. This gives rise to the different *seasons*, as described in Box 1.1.

6. Finally, since the *orbit* of the Earth around the Sun is slightly elliptical rather than perfectly circular, the Earth–Sun distance varies over the course of the year. This is accompanied by variations in the amount of energy received during the year, but these variations are much smaller than those that give rise to the seasons. Over thousands of years, however, their slow changes have a major impact.

BOX 1.1 THE SEASONS

In the course of a year, the Earth travels in an almost circular orbit around the Sun, in a plane called the ecliptic plane (Fig. B1.1). The axis of rotation of the Earth (the polar axis) is at present tilted at 23°27' with respect to the normal to this plane. At any given place, therefore, the angle of the Sun's rays at zenith (i.e. the angle of the rays with respect to the normal to the Earth's surface) varies throughout the year, with accompanying changes in the amount of sunlight each day. At latitude 45°N, for example, this angle varies between 21°33' (summer solstice) and 68°27' (winter solstice). At the Equator, it varies in the range ± 23°27'. The amount

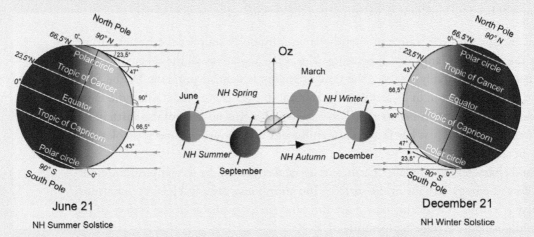

Fig. B1.1 The position of the Earth at different seasons. The 23°27' tilt (obliquity) of the Earth's axis with respect to the normal to the ecliptic plane defines the Tropic of Cancer (23°27'N) and the Tropic of Capricorn (23°27'S). At those latitudes, the rays from the Sun at zenith fall perpendicularly on the surface of the Earth at the June and December solstices, respectively. The obliquity also defines the polar circles (66.33°N and 66.33°S). Between the pole and the polar circle, the day lasts for 24 hours at summer solstice.

of solar energy received, also called the *solar irradiance* (in W/m^2), varies during the year, bringing in its train the succession of seasons. In each hemisphere, the year is marked by four dates that define the beginning of each season: the two solstices, summer and winter, when the irradiance is at a maximum (summer solstice) and then at a minimum (winter solstice) at latitudes situated between the tropics and the poles, and the two equinoxes (spring and autumn) when the day and night have the same duration.

At the equinox of 20 March – the spring equinox in the Northern Hemisphere (NH) and autumn in the Southern Hemisphere (SH) – the Sun's rays strike the Earth vertically at the Equator and tangentially at the poles. Day and night have the same length (12 hours) over the whole planet, from the North Pole to the South Pole. Then, in the NH, for example, between 20 March and 21 June, the daily irradiance gradually increases in latitudes situated above the Tropic of Cancer (23°27'N), reaching a maximum on 21 June, the longest day of the year. On this date, the Sun's rays at solar zenith fall vertically on the surface of the Earth at the Tropic of Cancer, and the irradiance is at a maximum. This is the summer solstice in the NH, which marks the beginning of summer. North of the Polar Circle (67°33'N), daylight lasts for 24 hours. Then, after 21 June, from the Tropic of Cancer to the North Pole, the daily irradiance decreases, together with the length of the day. At the 22 September equinox (the autumn equinox in the NH), day and night are once again of the same length over the whole planet and, at solar midday, the Sun's rays fall perpendicularly on the Earth's surface at the Equator. After that, the irradiance continues to decrease further until 21 December (the NH winter solstice), on which date the length of the day is at a minimum. Beyond the NH polar circle, the night lasts for 24 hours. In the SH, the situation is reversed.

These different mechanisms, which are listed in Table 1.1, generate differences in temperature and average rainfall that enable a great variety of climates to flourish on Earth. From frozen regions or scorching deserts to very wet zones, which may be either cool or warm, each zone is a home to *suitably adapted forms of life*.

TABLE 1.1 SUCCESSIVE STEPS IN THE DISTRIBUTION OF AVAILABLE ENERGY AT GROUND LEVEL ON THE EARTH'S SURFACE. CONTROLLING FACTORS ARE SHOWN IN RED. (a) THE AVERAGE ENERGY AVAILABLE AT THE SURFACE OF THE EARTH DEPENDS ON: (I) THE ACTIVITY OF THE SUN; (II) THE DISTANCE OF THE EARTH FROM THE SUN; AND (III) THE ENERGY TRANSFER THROUGH THE ATMOSPHERE. IN THIS TRANSFER, THE COMPOSITION OF THE ATMOSPHERE PLAYS A MAJOR ROLE. (b) THE AMOUNT OF SOLAR ENERGY RECEIVED AT THE SURFACE VARIES WITH LATITUDE AND SEASON. THE CONTROLLING FACTORS ARE: (I) THE SPHERICAL SHAPE OF THE PLANET; (II) THE ENERGY TRANSFER FROM THE EQUATOR TO THE POLES BY OCEAN AND ATMOSPHERIC CIRCULATION, TOGETHER WITH THE WATER CYCLE; AND (III) THE TILT OF THE POLAR AXIS, WHICH GOVERNS THE SEASONS.

(a) The average energy available at ground level on the Earth

Sun	⟶	Earth (top of atmosphere)	⟶	Earth (ground level)
Average flux radiated		*Average flux received*		*Energy available at surface*
Solar activity		*Earth–Sun distance*		*Atmosphere*

(b) The distribution of energy over the Earth's surface

Geographical distribution at top of atmosphere	⟶	Geographical distribution at ground level	⟶	Seasonal distribution
Sphericity of Earth		*Atmospheric circulation*		*Tilt of polar axis on the ecliptic*
		Ocean circulation		
		Water cycle		

In Parts I, II and III of this book, we focus on the mechanisms that give rise to the average climate at the surface of our planet. In this first step, the orders of magnitude in the climate are identified. We also briefly discuss the circulation of the atmosphere and of the oceans, since they play a major role in defining the pattern of rainfall over the planet and also in transferring heat from the Equator to the poles. But first we present an overview of the various climates that prevail on Earth, and their relationship to life forms. This will put into perspective what is at stake in climate change and what its implications are for society.

Chapter 2
Different climates . . . such diversity of life

2.1 The different climates on Earth

The major climate zones are distinguished by their annual mean values of temperature and precipitation, modulated by seasonal variations. As these averages fluctuate from one year to another, about 30 years are needed to establish a meaningful average value. The distribution of these climate zones over the Earth's surface is, roughly speaking, a function of latitude. This zonal distribution, however, is modulated by two factors: the topographical relief, and the relative positions of the land masses and oceans. The zonal distribution affects the climate on continents via the thermal regulation of the ocean, the effect of which is greatest in coastal regions.

The classification for continents established by Köppen at the beginning of the last century is used as a reference. This climate distribution is shown in Fig. 2.1, updated by Kottek et al. (2006):

- Temperature: in climate types A and B, the annual temperature is above 18°C, E climates have a monthly average temperature that is always below 10°C, D climates have only 3 months above 10°C and, finally, C climates are intermediate.

Climate Change: Past, Present and Future, First Edition. Marie-Antoinette Mélières and Chloé Maréchal.
© 2015 John Wiley & Sons, Ltd. Published 2015 by John Wiley & Sons, Ltd.
Companion website: www.wiley.com\go\melieres\climatechange

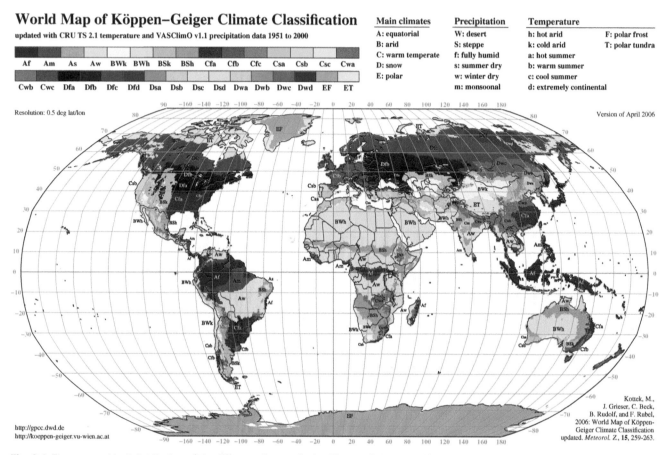

World Map of Köppen–Geiger Climate Classification

updated with CRU TS 2.1 temperature and VASClimO v1.1 precipitation data 1951 to 2000

Main climates	Precipitation	Temperature	
A: equatorial	W: desert	h: hot arid	F: polar frost
B: arid	S: steppe	k: cold arid	T: polar tundra
C: warm temperate	f: fully humid	a: hot summer	
D: snow	s: summer dry	b: warm summer	
E: polar	w: winter dry	c: cool summer	
	m: monsoonal	d: extremely continental	

Af Am As Aw BWk BWh BSk BSh Cfa Cfb Cfc Csa Csb Csc Cwa

Cwb Cwc Dfa Dfb Dfc Dfd Dsa Dsb Dsc Dsd Dwa Dwb Dwc Dwd EF ET

Resolution: 0.5 deg lat/lon

Version of April 2006

Kottek, M.,
J. Grieser, C. Beck,
B. Rudolf, and F. Rubel,
2006: World Map of Köppen-
Geiger Climate Classification
updated. *Meteorol. Z.*, **15**, 259-263.

http://gpcc.dwd.de
http://koeppen-geiger.vu-wien.ac.at

Fig. 2.1 The geographical distribution of the different climates in the Köppen–Geiger classification.

Source: Kottek et al. (2006). Reproduced by permission of F. Rubel/Institute for Veterinary Public Health, Vienna, Austria.

- Precipitation during the year: the existence or not of a dry period (A climates can thus be subdivided into climates without a dry season, A$_f$, or with a dry season, A$_w$ etc.). The various denominations are specified in the caption to Fig. 2.1.

To simplify, the amount of rainfall in tropical regions defines the boundary between hot wet climates (A) and hot dry climates (B). The hot deserts of both hemispheres are in B climates. Moderate climates (C), varying from oceanic to continental, prevail at medium latitudes. Then, at increasingly high latitudes, boreal (D) and polar (E) climates (characterized respectively by tundra and ice) set in. Subclasses are distinguished according to the amount of water deposited during the seasons.

The development of life on the continents is to a great extent dependent on the physical conditions that are defined by the different climates, by virtue of the vegetation and the fauna that climate supports. The importance of climate for the development of life in all its forms is fundamental. To place this in context, we briefly recall how life organizes itself around these various climates.

2.2 Climates, biomes and biodiversity

2.2.1 DISTRIBUTIONS

The greatest mass of living matter on Earth is vegetation. That of animals is several orders of magnitude smaller. Its distribution can be estimated from the amount of vegetation per unit area. In oceans, an indirect measure of this quantity is provided by the amount of chlorophyll, a molecule that is fundamental to the structure of vegetable matter. Measurements of chlorophyll are made continuously over the whole of the Earth's surface by the NASA SeaWIFS satellite. For the continents, satellites can estimate the quantity of vegetation from the Normalized Difference Vegetation Index (NDVI), which describes the vegetation coverage on the ground. In the map in Fig. 2.2, these two types of data from satellite measurements are combined to yield a composite picture of the Earth's coverage. Thus, *in spite of very different climates, the stock of vegetation of certain high-latitude areas appears to be comparable with that of equatorial areas.*

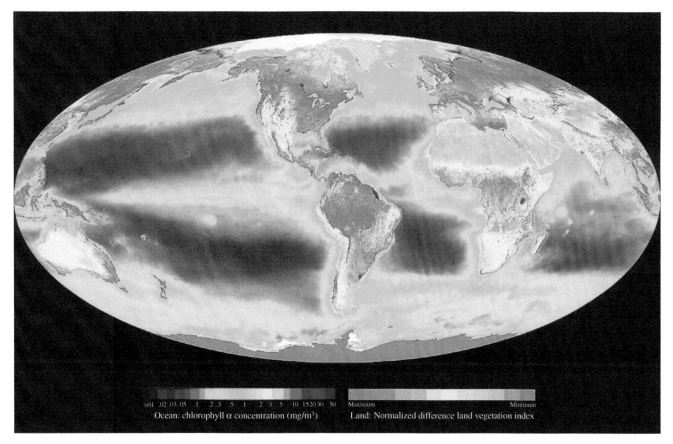

Fig. 2.2 A composite map showing average photosynthetic activity from September 1997 to August 2000. This activity reflects the amount of synthesized plant biomass. (a) Ocean: chlorophyl concentration (mg/m³). (b) Continents: the scale is the Normalized Difference Vegetation Index (NDVI) for plant coverage of the soil.

Sources: SeaWiFS Project, NASA/Goddard Space Flight Center and ORBIMAGE.

Another method of assessing the activity of the biosphere is to estimate the amount of vegetation synthesized during the year (Net Primary Production). This corresponds to the flux of synthesized organic matter, as opposed to the total stock illustrated in Fig. 2.2. The map in Fig. 2.3 displays this flux for the continents and oceans on the same scale. It demonstrates strikingly that production in the oceans is substantially smaller than on the continents. The figure also illustrates the major characteristics of the climate zones, showing that the growth of vegetation is particularly low in the subtropics, both on the continents and in the oceans. This second map enables us to compare the quantity of living matter generated annually at high and low latitudes. *In spite of the very different climates, this quantity appears to be similar in high-latitude areas and in the majority of low-latitude areas.*

But what can we say about the diversity of the life that is thus generated? What is the complexity of these interacting living systems? A first approach to this complexity is to be found in the *biome*

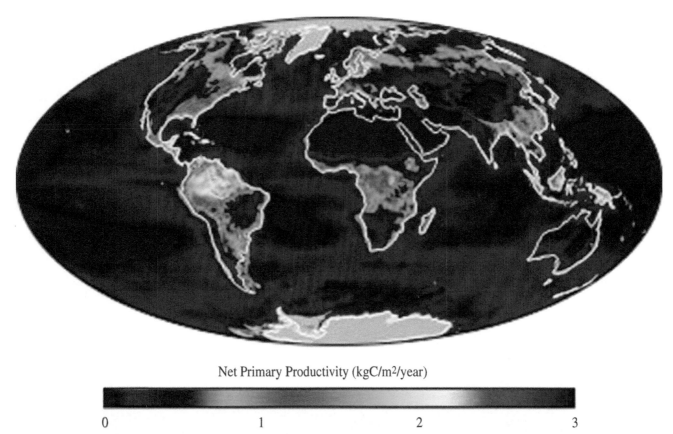

Net Primary Productivity (kgC/m²/year)

0   1   2   3

Fig. 2.3 The distribution of vegetation synthesized annually per square metre (the global average of the net productivity of vegetation for 2002), measured by the rate at which plants absorb carbon from the atmosphere. Note the difference between continents and oceans: production is much lower in the latter. The highest primary productivity is in the rainforest areas of South America, Africa and South-East Asia.

Source: NASA.

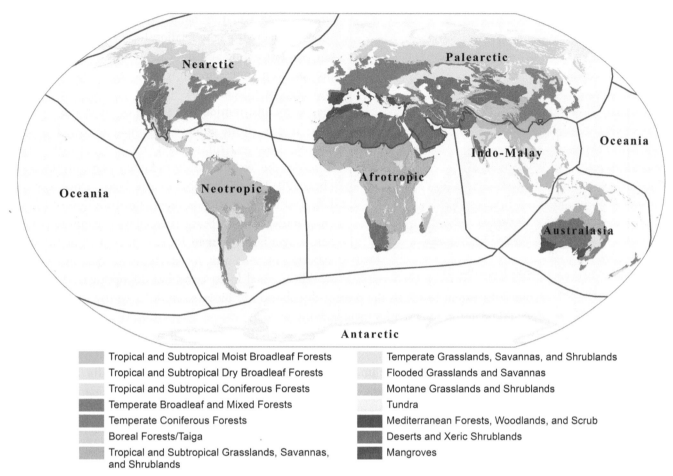

Fig. 2.4 The distribution of the different biomes on the continents. Since species adapt to the temperature and water resources of the region, this distribution closely resembles that of the major climate zones (Fig. 2.1).

Sources: Map data from Olson et al. (2001); map by L. Spurrier, WWF (2013). Reprinted with permission of the WWF.

classification. A biome is a set of ecosystems that are characteristic of a biogeographical area. It is named according to the predominant vegetation and animal species that live and have adapted there. On the continents, this set of ecosystems responds to the current prevailing climate as well as to the resources and characteristics of the soil, which in turn are partly related to the past history of the climate. The first bioclimatic classifications appeared in the 1950s, with that of Holdridge. Nowadays, several classifications exist. Not all classification systems are equivalent, as the defining criteria tend to be selected according to the objectives of the specific country or organization. Here we employ a classification published by Olson et al. (2001).

Figure 2.4 shows the distribution of the different biomes on the continents, from the Equator to the arctic zones. Since living systems must adapt to the prevailing temperatures and to the available water resources, the distribution of the biomes on Earth must be broadly

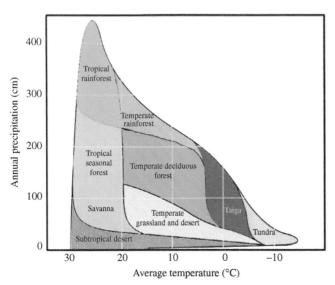

Fig. 2.5 The distribution of biomes in terms of annual temperature and precipitation.
Source: Modified from Whittaker (1975) by B. Ricklefs/CC-BY-SA-3.0.

similar to that of the major climate zones. Figure 2.5 shows a diagram of the relationship between the different types of biome and the climate features in terms of precipitation (rain and snow) and temperature, which was first established by Whittaker (1975). The biome map is the distribution that would occur in the natural state; that is, in the absence of human activity. Human activity, however, substantially modifies the distribution. At the website http://old.unep-wcmc.org/media-library/2011/09/27/cc69b7bf/currorig.jpg, a map can be found of the approximate original distribution of forest cover under current (postglacial) climatic conditions and before significant human impact, as well as of the distribution of remaining forest. Approximately half of the world's original forest cover has disappeared. This illustrates the extent to which the present distribution of biomes can differ widely from what it would be in the absence of human activity.

Figures 2.2 and 2.3 could give the impression that continental areas are comparable, both in existing stocks of vegetation and in net primary production, despite being at different latitudes and having different climates (e.g. equatorial and high-latitude areas). Thus, in spite of very different climates, the *productivity of certain equatorial areas appears to be comparable with that of high-latitude areas. But are these areas truly comparable in wealth, expressed in number of species? Do these different biomes harbour a similar profusion of life or, like the climate zones, does their wealth, or biodiversity, vary progressively from the Equator to the poles?*

2.2.2 BIODIVERSITY

Different methods exist for characterizing biodiversity. For the purposes of illustration, we restrict ourselves to the world of vascularized plants (i.e. the great majority of plants). Since the number of species is one of the means of characterizing biodiversity, in Fig. 2.6 we show the distribution of the number of vascular species of plants indexed by geographical area (Barthlott et al. 1996, 2007).

It clearly appears that this distribution is strongly correlated with the climate zones, the equatorial zone being by far the richest. A marked decrease is observed from the Equator to the poles. The number of species, which reaches several thousand in the tropics, decreases to about a thousand towards the mid-latitudes, and then falls to a few hundred on approaching the polar regions. The only exceptions to this almost regular decrease are two bands of desert that gird both hemispheres in the tropics: there, biodiversity collapses to a mere few hundred species, and is comparable to that of the polar deserts. The influence of the past on this distribution of biodiversity on the continents is by no means negligible, since continental drift, long-term climate evolution and recent climate changes in the Quaternary (with alternating glacial and interglacial climates) contributed to the heritage. We shall see later, in greater detail, the impact on the vegetation that the major climate cycles of the Quaternary have had in the past 2 million years.

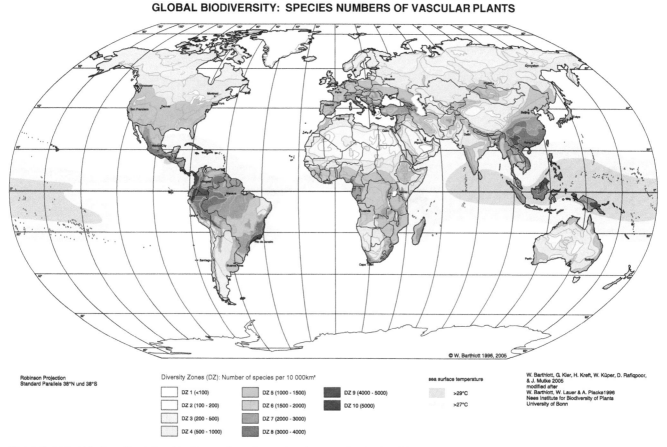

GLOBAL BIODIVERSITY: SPECIES NUMBERS OF VASCULAR PLANTS

Fig. 2.6 Global biodiversity, defined by the number of vascular plant species in the different geographical regions. Continental biodiversity is closely related to climate zones (the equatorial zone being by far the richest), and decreases on moving from the Equator to the poles. Diversity is particularly high in the mountainous regions of the tropics. Updated 2013.

Source: Barthlott et al. (1997). Reproduced with permission of W. Barthlott, Nees-Institut für Biodiversität der Pflanzen der Universitat, Bonn, Germany.

2.3 Climate and society

This chapter on life cannot be closed without mentioning human beings, whose place among the living organisms on Earth is exceptional. Like any other form of life, humans tend to settle in places that offer both favourable living conditions and the availability of food. For the vast majority of Earth dwellers, the availability of food depends on plants. This implies a fundamental link between man and climate, since the latter effectively governs the distribution of vegetation. In 2013, the human population of the world amounted to 7.2 billion inhabitants. A map showing its distribution over the surface of the Earth can be found at the website http://sedac.ciesin.columbia.edu/data/collection/gpw-v3/maps/gallery/search?facets=region:global. This map shows that the zones of high population density are concentrated mainly between 20°N and 60°N, a band of latitude that includes a large proportion of the continental surface and that enjoys a relatively moderate climate.

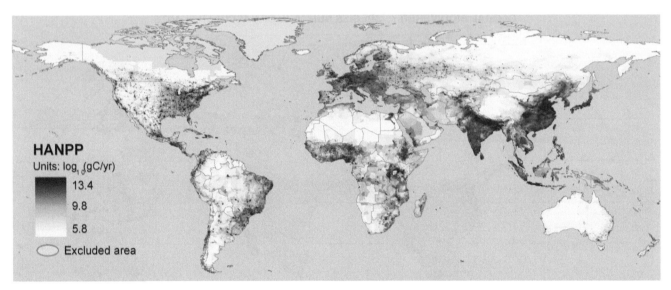

Fig. 2.7 The amount of vegetation consumed annually by humans, commonly denoted the Human Appropriation of Net Primary Production (HANPP). Not unexpectedly, this distribution is closely similar to that of the human population (which can be accessed at http://sedac.ciesin .columbia.edu/data/collection/gpw-v3/maps/gallery/search?facets=region:global). Darker areas on the map indicate greater appropriation of NPP by humans.
Source: NASA.

Food resources in continental areas come from vegetation production, mainly agricultural. The latter can be evaluated from the net primary product collected by humans (Human Appropriation of Net Primary Production, HANPP), which is expressed in terms of the amount of matter (measured by its carbon content) per unit area per year. The map in Fig. 2.7 shows how unevenly the HANPP is distributed, the lowest values being split between arid zones and regions that are partly or completely uncleared. It is no surprise that this distribution is almost the same as that of the world population: people settle where food is available.

Questions

The above considerations open our eyes to the importance of the current debate on climate change and what is at stake. In the event of rapid climate change in the near future:

- *How will the pattern of vegetation, biomes, biodiversity, agricultural production and forests change, as well as that of population?* All of these are intimately related to the distribution of the various climates.
- *More precisely, in the event of warming, would the zonal nature of the climate distributions shift the climate regions closer to the poles, or towards the Equator? . . . Would certain zones, such as deserts, extend or contract? . . . Would the climatic character of these zones (rainfall, temperature) be reinforced or attenuated? What would happen to zones of high human population density?*

Some of these questions, which pertain to climate changes in the 21st century that are caused by human activity, will be addressed in Part VI.

Chapter 3

From a patchwork of climates to an average climate

Starting from the total energy needed to sustain the various climates on Earth, we come to the concept of *average climate*. *By definition, when integrated over the Earth's surface, this average climate requires the same amount of energy.* We therefore focus on the three key factors that define climate: two of these, temperature and precipitation (rain or snow), have already been mentioned. The third is the motion of air masses (wind). For each of these physical quantities, which vary strongly with the seasons and from one region of the planet to another, we establish how they are distributed over the Earth's surface and thus define the annual average for each region. This then enables us to calculate an average climate over the whole of the Earth. Our estimates are based on the energy budget published in IPCC 2007 (Chapter 1, FAQ1.1, Fig. 1). We then examine how much energy is needed to maintain each of these three annual average values. First, however, some basic aspects of the concepts of temperature and equilibrium must be clarified.

3.1 Temperature and thermal equilibrium

Temperature is a quantity that is understood intuitively by all human beings. It remains, however, to be defined. We consider a 'system' that is at a certain equilibrium temperature, and also its surroundings. The system can be, for example, a bowl of soup, an air mass, our skin, a metal sphere, a cork, the surface of the Earth and so on. It need not, however,

Climate Change: Past, Present and Future, First Edition. Marie-Antoinette Mélières and Chloé Maréchal.
© 2015 John Wiley & Sons, Ltd. Published 2015 by John Wiley & Sons, Ltd.
Companion website: www.wiley.com\go\melieres\climatechange

consist of matter – it can also be energy, such as electromagnetic radiation, of which light is an example.

3.1.1 TEMPERATURE, A MEASURE OF AVERAGE ENERGY

The temperature of a system is a way of expressing the average amount of energy available within it; that is, available to the different constituents of the system. For matter, this energy can be found in mechanical form (the rate of displacement of the air molecules, the rotation or vibration of the molecules in solids, liquids and gases etc.). At 27°C, for example, the energy of nitrogen molecules is defined by this temperature and they move at an average speed of 476 m/s. At 0°C, they move more slowly and the average speed falls to 454 m/s. This means that less energy is available in the reservoir 'kinetic energy of translation'. Electromagnetic radiation is also present. This is thermal radiation, the spectral characteristics of which will be discussed in detail in Part II. The flux of radiated thermal energy is directly related to the temperature of the system: the hotter the body, the more intense is the radiated flux. The radiation, which is part of the system and is defined by its average energy, is constantly emitted and propagates outwards. In this book, we deal with the radiation flux emitted by the Sun (*solar radiation*) and that emitted by the Earth (*terrestrial radiation*). The flux of the Earth's radiation is directly related to its equilibrium temperature.

To define this average energy, a temperature scale is required in which the origin is at the point where the energy available in the system is zero. This is the Kelvin scale. Its relationship to the widely used Celsius scale is defined in Box 3.1.

BOX 3.1 THE TEMPERATURE SCALE

The notion of temperature is very intuitive, and to measure it a variety of temperature scales have been used. The widely employed Celsius scale is based on two equilibrium points of water, defined under normal atmospheric pressure (1013.65 hPa), to which values of temperature are arbitrarily assigned. A mixture of water and ice is associated with the value 0°C (the melting point of ice), and the value 100°C is attributed to a mixture of liquid water and water vapour (the boiling point). Between these two states, each of which corresponds to a different average energy, the scale is divided into 100 equal steps of one degree Celsius. It is measured by means of a material property that varies linearly with temperature (thermal expansion of liquids – mercury, alcohol – expansion of gases, electrical resistance of metals etc.).

We have just seen that the temperature is a means of expressing the average available energy. The scale should therefore: (i) start from zero (absolute zero), describing a state possessing zero available energy; and (ii) increase linearly. The scale defined in this way is the Kelvin scale. It is graduated in degrees (Kelvin). This is defined in such a way that one degree Kelvin (K) is equal to one degree Celsius (°C); that is, the interval between the physical state corresponding to 0°C and that of 100°C contains 100 Kelvin degrees. This defines the value of absolute zero (0 K) on the Celsius scale, which is then –273.15°C. Conversely, 0°C corresponds to 273.15 K on the Kelvin scale.

To transform one degree (either Kelvin or Celsius) into units of energy (joules, J), we must multiply by Boltzmann's constant, k (1.38 × 10^{-23} J/K).

3.1.2 THERMAL EQUILIBRIUM – ENERGY EQUILIBRIUM

These two notions describe the same equilibrium. While thermal equilibrium concentrates on the temperature of the system, energy equilibrium focuses on the mechanisms that exchange energy between the system and its surroundings. Exchange can occur through three different processes:

- *Radiation*. The exchange of thermal radiation involves, on the one hand, energy emitted by the system, which loses energy, and, on the other, energy received from the outside and absorbed by the system, which gains energy.
- *Conduction*. Heat transfer occurs by conduction through contact between two media (the system and the surroundings). This gives rise to *convection* if the medium is fluid. This is transfer of *sensible energy*.
- *A change of state* of one of the constituents when it leaves the system. When water evaporates from the system, the change of state from liquid to gas requires energy, which is removed from the liquid by breaking the bonds between the molecules in the liquid state. Evaporation involves a loss of energy from the reservoir of liquid water and hence a drop in its temperature. This is transfer of *latent energy*.

The two latter processes require the presence of matter, because it is the matter that transports the energy. By contrast, radiation energy is transported by electromagnetic waves (photons), not by matter. A planet can thus exchange energy with the rest of the universe only by radiation, because exchange of matter between the planet and the universe is extremely weak. These three types of energy enter into the energy balance of the Earth's surface.

A system is in stationary equilibrium if the total loss of energy per unit time (through radiation, conduction and change of state) is compensated by the arrival of an equal energy flux. The system is then in *thermal equilibrium*, and we may speak of its equilibrium temperature. We are now in a position to describe the basis of climate equilibrium and the mechanisms that underlie its evolution.

Equilibrium is thus based on the following relationship:

Energy flux lost by the system = Energy flux absorbed by the system

3.2 The average temperature of the Earth's surface

The temperature varies widely according to place and to season, and, of course, over the 24 hours of the day. The temperatures on the Earth's surface span a broad range, from about –90°C (the lowest temperature ever recorded, –93°C, was measured by satellite in the Antarctic plateau on 10 August 2010) to about +60°C, for example, in the Sahara (in this account, we omit volcanic zones). To grasp the concept of average temperature at the surface of the Earth (land masses and oceans), we proceed in successive steps. First, we look at how the annual average of this temperature is distributed (Fig. 3.1). The average temperature decreases from the Equator to the poles, with the Equator receiving far more solar energy over the year than the poles. To illustrate this difference between the Equator and the poles, the average temperature at different

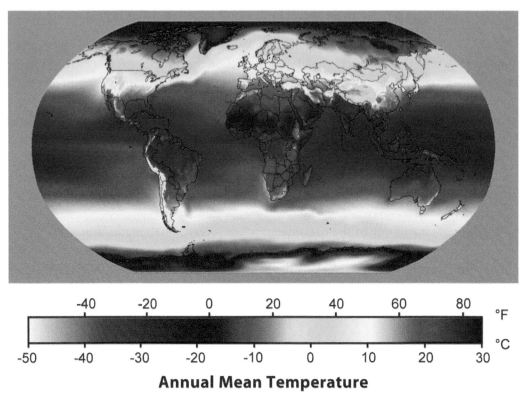

Annual Mean Temperature

Fig. 3.1 The mean annual temperature at the surface of the Earth (average from 1961 to 1990). This map was produced by combining the land surface temperature data set (Mark et al. 2000) with the sea surface temperature data set (Reynolds et al. 2002) and the 2.5°NCEP/NCAR Reanalysis version1 data set (Kalnay et al. 1996).
Source: Image created by Robert A. Rohde/Global Warming Art/CC-BY-SA-3.0/GFDL.

latitudes (called the *zonal average temperature*) is shown in Fig. 3.2. The zonal average temperature, which is related to the distribution of the continents and oceans (Part I, Note 2), reveals the dissymmetry between the two hemispheres. The final average of the annual temperature over the whole of the Earth's surface yields a value close to +15°C (current estimate +14.5°C) (Fig. 3.3). Our argumentation will be based on this mean value.

Energy cost

To maintain the Earth's surface at its average temperature throughout the year requires an uninterrupted supply of energy. The amount of energy involved is expressed in terms of the quantity of energy per unit time, or joules per second; that is, watts (W). Moreover, since it is an average value, rather than counting the whole area of the Earth, it is convenient to reduce it to one square metre of surface. The energy flux required to maintain the average temperature is evaluated in watts per square metre (W/m^2). This quantity is frequently confused with 'energy', whereas in fact it is an 'energy flux'. *So, how much energy is lost every second by radiation from a surface at +15°C?* According to the laws of thermal radiation (Chapter 6), $1\,m^2$ of the surface emits 390 W. To maintain the surface temperature of the Earth at +15°C, therefore, a permanent energy supply of $390\,W/m^2$ must be supplied to offset this loss by radiation (IPCC 2007).

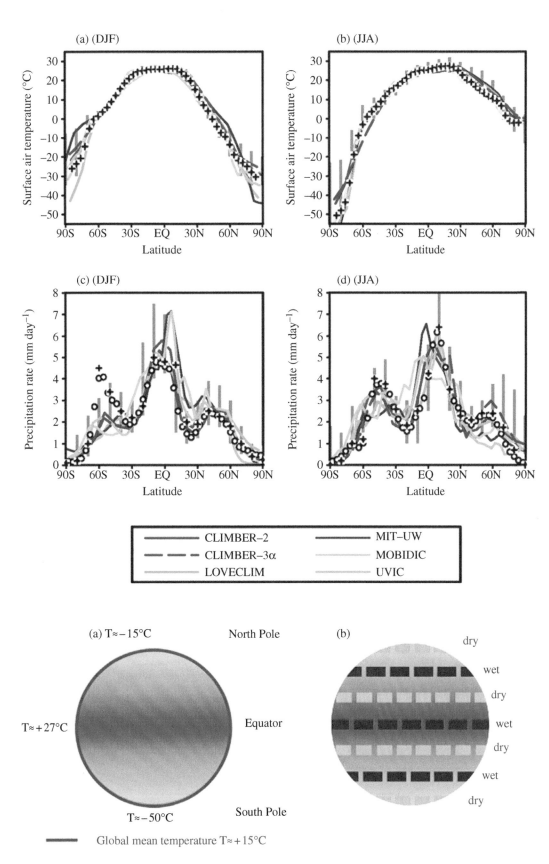

Fig. 3.2 Dependence on latitude of the zonally averaged surface air temperature (a, b) and precipitation rate (c, d) for boreal winter (a, c) and boreal summer (b, d). Black crosses and circles: measured data. Curves and grey bars are different modelling results discussed in IPCC 2007.

Source: IPCC 2007. *Climate Change 2007: The Physical Science Basis*. Working Group I Contribution to the Fourth Assessment Report of the Intergovernmental Panel on Climate Change, Figure 8.17. Cambridge University Press.

Fig. 3.3 (a): the global mean annual temperature at the surface of the planet (represented by the red circle) is about +15°C. Also shown are the mean annual temperatures at the Equator and at the poles. (b): a schematic diagram of the alternating wet and arid belts as a function of latitude.

3.3 Precipitation

Here again, the extremes on Earth are very pronounced and the amount of water delivered by precipitation (rain, snow or hail) can range from complete absence throughout a whole year, or even several years (e.g. in the Atacama Desert of the tropical Andes, one of driest in the world), to nearly 12 m per year (e.g. 10–15 m on Mount Cameroon in Cape Debuncha, Cameroon, Africa). Figure 3.4 shows the map of mean annual precipitation. Precipitation is governed by atmospheric circulation (Chapter 5). Owing to the zonal nature of this circulation (relatively uniform at a given latitude but strongly varying at different latitudes), precipitation also exhibits zonal behaviour. The result is an alternating pattern of three wet and four dry belts:

- the equatorial belt, by far the wettest, receives an annual average rainfall of almost 6 mm/day, or about 2 m of water per year;
- two dry subtropical belts, one on either side of the equatorial belt, giving rise in each hemisphere to deserts on the continents and zones of good weather over the oceans (anticyclone zones), with an annual average of about 2 mm/day (0.7 m/yr);
- at mid-latitudes in each hemisphere, a wet belt with an annual average of nearly 3 mm/day (1 m/yr) – most industrialized areas are located in this band; and
- lastly, the polar desert regions (about 2 cm/yr in the centre of the Antarctic continent).

The alternation of wet and dry belts is shown schematically in Fig. 3.3. This highly simplified diagram highlights the various questions that were raised in Section 2.3 in relation to climate warming. *If the water cycle is reinforced, will the trend in each belt be enhanced* (i.e. will wet belts become wetter and the arid regions drier)? *Or will it bring about an overall increase in rainfall? Will the belts shift,* and extend towards the poles or, on the contrary, shrink towards the Equator? What would happen, for example, in the Mediterranean basin or in Mexico, at

Fig. 3.4 Observed mean annual precipitation (cm) according to the Climate Prediction Centre Merged Analysis of Precipitation. This average is based on the climate in the period 1980–1999. Grey regions: information unavailable.
Source: IPCC 2007. *Climate Change 2007: The Physical Science Basis.* Working Group I Contribution to the Fourth Assessment Report of the Intergovernmental Panel on Climate Change, Figure 8.5. Cambridge University Press.

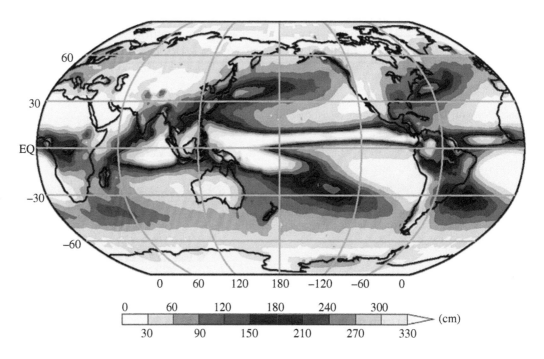

the junction between the arid tropical belt and the rain belt at mid-latitudes? These questions will be addressed at the end of our journey, in Part VI.

Energy cost

The mean annual precipitation at the Earth's surface is approximately 1 m/yr. In the water cycle, the same amount evaporates each year from the surface of the oceans and the continents in the form of water vapour, which is invisible to the eye. In the atmosphere, it condenses into droplets of liquid water or ice crystals, giving rise to clouds. Finally, it is deposited again on the Earth in the form of precipitation (rain, snow or hail). To maintain the water cycle thus requires a continuous supply of energy to the surface in order to evaporate this layer of water. The energy flux at the surface required to ensure the annual evaporation of water is $78\,W/m^2$. This evaporates about $1\,m^3$ of water (1 m of rain for each square metre of the surface). That is therefore the energy flux needed by the surface to maintain the water cycle on Earth. In the water cycle, this energy flux returns to the atmosphere in the form of *latent heat*, which is released when the water vapour condenses into clouds. In the total surface–atmosphere energy balance, this energy flux counts as a loss of energy to the Earth's surface and a gain of energy to the atmosphere. For the moment, however, we are concerned only with the Earth's surface.

3.4 Wind

The wind speed also varies greatly at the Earth's surface, from zero to 408 km/h at Mount Washington in New Hampshire (USA), one of the highest values ever recorded. The temperature of the atmosphere at ground level is on average cooler than that of the surface. Winds start when masses of air are warmed by contact with the Earth's surface when it is heated by the Sun, with the heat being transferred from the surface to the air mass by conduction. At the surface/air interface, the heat lost warms the lower layers of the atmosphere, making them less dense and causing them to rise, generating convection. In this way, energy is transferred from the surface of the Earth to the atmosphere. Exchange by conduction (and convection) depends strongly on the local wind and surface characteristics. The energy exchanged in these movements is described in Box 3.2. The energy transferred from the atmosphere through friction to the surface is negligible.

Energy cost

When averaged over the whole of the Earth's surface and over a year, the transfer of energy by conduction from the surface to the atmosphere amounts to a heat loss from the surface of $24\,W/m^2$.

BOX 3.2 ENERGY ASSOCIATED WITH MOTION OF AIR MASSES

In the energy balance discussed in this chapter, what is the contribution of the energy of the motion of the air that is generated by atmospheric circulation?

In Chapter 4, we shall see in detail that atmospheric heating takes place at different altitudes: (i) at the surface of the Earth ($24\,W/m^2$ energy flux supplied continuously by the Earth itself, through direct heating of the air masses by conduction); (ii) at medium altitudes by the formation of clouds ($78\,W/m^2$, corresponding to the release of latent heat supplied by the Earth's surface); and (iii) more diffusely, by the absorption of terrestrial radiation from the Earth's surface ($350\,W/m^2$) and by absorption of solar radiation ($67\,W/m^2$).

When it is warmed, an air mass expands and becomes less dense. According to Archimedes' Principle, it rises. In the troposphere – that is, in the first ten kilometres above the Earth's surface, where the motion of the air is turbulent – this heating permanently drags the air masses along, giving rise to atmospheric circulation. The convective ascending and descending movements mix with horizontal winds. If there were no friction – that is, if no energy loss were incurred in the movement of the air masses – these movements would have no impact on the energy balance at the surface of the Earth. The air masses would fluctuate perpetually at different altitudes with no need of an energy supply.

In fact, however, there are weak frictional forces. These amount to a loss of energy of a few watts per square metre. They are principally due to mutual friction of the air particles (molecules, aerosols etc.). This 'air-to-air' friction causes a loss of kinetic energy and heats the air. But since the energy lost is recovered in the atmosphere in the form of heat, the energy balance with respect to the Earth's surface is unaffected. Lastly, another type of friction should not be overlooked: that between the air and the surface of the Earth (the mechanical effect of wind on the ground and the sea, for example through wave formation etc.). Here again, energy degrades into the form of heat, which warms the Earth's surface. *This air–surface friction effect, however, is sufficiently small that it can be neglected in the energy balance that we outline.*

3.5 Three major items in energy consumption

We have now seen that to maintain a temperature of +15°C at the Earth's surface requires $390\,W/m^2$, that to maintain the water cycle, which in 1 year evaporates a sheet of water 1 m thick over the surface of the Earth, requires $78\,W/m^2$, and that to keep the air mass rising requires an average of $24\,W/m^2$. The Earth's climate engine thus relies on a permanent supply of $492\,W/m^2$ at the surface of the planet (Fig. 3.5).

We assume a situation in which the climate is in stationary equilibrium (the ways in which it can deviate from this equilibrium will be discussed later). We are now going to investigate the sources of energy and the mechanisms that the surface of the Earth can draw upon constantly for this energy flux.

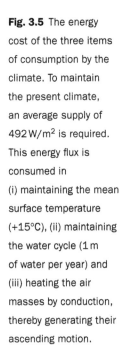

Fig. 3.5 The energy cost of the three items of consumption by the climate. To maintain the present climate, an average supply of $492\,W/m^2$ is required. This energy flux is consumed in (i) maintaining the mean surface temperature (+15°C), (ii) maintaining the water cycle (1 m of water per year) and (iii) heating the air masses by conduction, thereby generating their ascending motion.

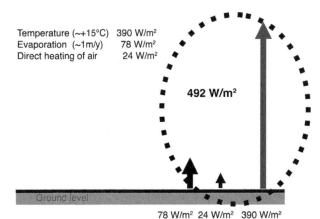

Temperature (~+15°C)   390 W/m²
Evaporation (~1m/y)   78 W/m²
Direct heating of air   24 W/m²

492 W/m²

Ground level

78 W/m² 24 W/m² 390 W/m²

Chapter 4
The global mean climate

The Earth's surface must be fed continuously with energy at an average flux of 492 W/m². *Where does this energy come from?* Many sources supply energy to the surface: solar energy radiated by our star, energy carried by cosmic rays from deep in the universe, magnetic energy radiated by the planet, *geothermal energy* from the radioactivity and the gradual cooling of the interior of planet, and so on. But one of these, the energy from the Sun, is so much greater than all the rest that the latter are completely negligible.

Solar flux arriving at Earth per square metre perpendicular to Sun

1367 W/m²

Mean solar flux per square metre of Earth's surface

342 W/m²

Fig. 4.1 The solar flux arriving at the outside of the Earth on a surface placed perpendicular to it (apparent area) is 1367 W/m². On being captured by the Earth, this energy is spread over the whole of its spherical surface. The available flux is thus four times smaller; that is, 342 W/m² over the total surface area (Part I, Note 3).

4.1 The Sun, source of energy

The solar flux striking the surface of the Earth on an area placed perpendicular to the solar rays (the *apparent surface*), when averaged over the 11-year cycle of the Sun's activity, is usually taken to be 1367 W/m². (Note that this flux has recently been re-evaluated as 1361 W/m².) The average solar flux per square metre of the total surface of the Earth, however, is obtained by dividing this number by 4 (Part I, Note 3). The incident flux averaged over the Earth's surface is thus only 342 W/m² (Fig. 4.1).

Two main sources of energy feed the Earth's surface: solar radiation and geothermal energy. Although the latter is particularly strong in volcanic regions, when the energy fluxes are averaged over the Earth's surface, it turns out that our star provides almost 10,000 times

Climate Change: Past, Present and Future, First Edition. Marie-Antoinette Mélières and Chloé Maréchal.
© 2015 John Wiley & Sons, Ltd. Published 2015 by John Wiley & Sons, Ltd.
Companion website: www.wiley.com\go\melieres\climatechange

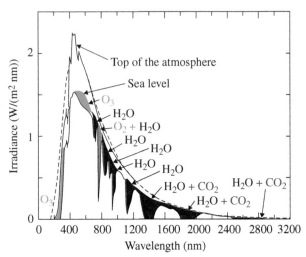

Fig. 4.2 The solar radiation flux incident at the top of the atmosphere, and at sea level, expressed in $W/(m^2\,nm)$ (black lines). The difference between the two curves is due to reflection of the solar radiation by the atmosphere (mainly by clouds and scattering by the air), and to absorption by certain molecules. The atmosphere transmits most of this radiation, which is centred at $0.5\,\mu m$. Only water vapour (blue) and ozone (green) absorb appreciably. The dashed line is the flux radiated by a black body at $5900\,K$, the approximate temperature of the Sun.

more energy at the surface of the Earth than comes from geothermal energy. Beneath the oceans, the latter is about $0.05\,W/m^2$, and $0.01\,W/m^2$ under the continents. As for the energy that enters from the rest of the universe (cosmic rays, starlight etc.), it is of the order of one millionth of the solar energy. *The main features of our climate are thus determined by the solar flux.*

What does the solar energy consist of? Visible light, from blue to red – that is, light with wavelengths between 0.4 and 0.7 micrometres (μm) – amounts to 41% of the solar energy. But there is also radiation that the human eye cannot detect: 8% in the ultraviolet range (0.2–$0.4\,\mu m$) and 51% in the near infrared range (0.7–$3.0\,\mu m$) (Fig. 4.2).

If the flux entering from outside the atmosphere is $342\,W/m^2$, how is it physically possible to collect $492\,W/m^2$ at the surface? This is a complex process involving several mechanisms, one of which, the greenhouse effect, dominates the others. In Part I, we list only the different qualitative stages in the argument and explain the nature of the mechanisms involved. A more detailed study of the question is undertaken in Part II.

4.2 The energy equilibrium at the Earth's surface

We begin by evaluating what the average temperature of a hypothetical Earth would be in three successive stages, starting from the simplest conditions. We then impose additional conditions, arriving finally at the real situation. The fundamental premise is that this Earth has no significant internal source of energy to heat its surface, and that its surface is heated only by the solar energy it receives. We take as an example the parameters that apply to the real situation of the Earth: our planet receives, on average, $342\,W/m^2$ of solar energy.

4.2.1 CASE 1: A TOTALLY ABSORBING EARTH WITHOUT AN ATMOSPHERE

We make two simplifying assumptions, which, albeit major, provide a starting point for the energy balance: (1) the planet has no atmosphere; and (2) its surface absorbs all of the $342\,W/m^2$ of incident solar radiation. This means that no rays from the Sun are reflected. The sunlight that reaches the planet is completely absorbed at the surface and is thus entirely employed in keeping the planet warm (Fig. 4.3).

Because the planet is warm, it radiates energy into space, by emission of infrared thermal radiation, which we call *terrestrial radiation*. The flux of terrestrial radiation is determined by the surface temperature: the higher the temperature, the greater is the flux of energy emitted. When the flux of energy radiated by the surface is equal to the absorbed flux, a state of stationary equilibrium is reached and the temperature stabilizes. The average temperature of the surface therefore rises until the emitted terrestrial radiation reaches 342 W/m². The laws of thermal radiation (Chapter 6) state that this surface temperature is 279 K (+6°C) and that the terrestrial radiation at this temperature lies in the infrared range. In this first case, therefore, the equilibrium temperature of the surface is governed directly by the incident solar flux.

4.2.2 CASE 2: A PARTLY REFLECTING EARTH WITHOUT AN ATMOSPHERE

We now complicate the situation slightly: the planet still possesses no atmosphere, but its surface absorbs only a fraction of the incident solar radiation, the rest being reflected by the surface of the planet (land masses and oceans) (Fig. 4.4). Only a fraction of the solar radiation is therefore used to heat the planet. We take as an example values that correspond to the Earth's real situation: 107 W/m² (31% of the solar radiation) is reflected, with only 235 W/m² being absorbed by the planet. The equilibrium temperature is thus reached when the surface radiates the same energy flux: 235 W/m². The laws of thermal radiation yield for this temperature 254 K (–19°C), and the radiation emitted into space at this temperature is situated in the infrared range.

4.2.3 CASE 3: EARTH WITH AN ATMOSPHERE

Now we assume that the planet possesses an atmosphere and that it absorbs only part of the solar radiation (235 W/m²). The remainder, 107 W/m², is reflected back into space by the atmosphere, by the continents and by the oceans. Some of the solar radiation is absorbed in its passage through the atmosphere, and some at the surface of the Earth. The planet is thus heated with just 235 W/m² (Fig. 4.5). It reaches the same equilibrium as in case 2 and radiates 235 W/m², corresponding to an equilibrium temperature of –19°C. This temperature is known as the *external temperature*, because it is associated with the outer shell of the planet. It corresponds to the average flux of energy radiated (in the infrared range) by the Earth that a satellite outside the Earth's atmosphere would observe. *So what happens at the surface of the planet, beneath the blanket of the atmosphere?*

Fig. 4.3 The energy equilibrium at the Earth's surface. Case 1: the Earth possesses no atmosphere and absorbs all the incident solar flux. At equilibrium, the energy flux radiated by the surface is equal to that absorbed (342 W/m²) and the temperature is +6°C. Solar radiation (0.2–3.0 μm) is displayed in yellow and terrestrial radiation (3–70 μm) in red.

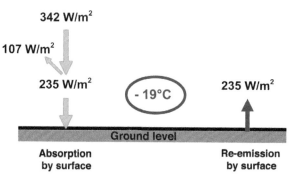

Fig. 4.4 The energy equilibrium at the Earth's surface. Case 2: the Earth possesses no atmosphere and absorbs only part of the incident solar flux (235 W/m²), the rest being reflected. The equilibrium temperature is –19°C. Solar radiation (0.2–3.0 μm) is displayed in yellow and terrestrial radiation (3–70 μm) in red.

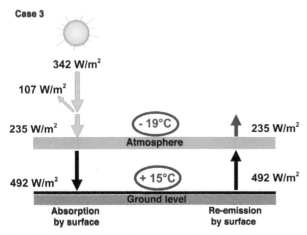

Fig. 4.5 The energy equilibrium at the Earth's surface. Case 3: the Earth possesses an atmosphere and absorbs only part of the incident solar flux, namely 235 W/m². As in Case 2, the Earth reaches equilibrium by radiating 235 W/m²; that is, the equilibrium temperature is –19°C. The different types of energy transfer (radiation, latent and sensible) through the atmosphere create an equilibrium in which the surface of the Earth absorbs a total of 492 W/m² and loses an equal amount through the various heat transfer mechanisms. Of these, the radiation flux, which is defined by the equilibrium temperature of the surface, is 390 W/m², corresponding to the temperature +15°C. Solar radiation (0.2–3.0 μm) is displayed in yellow and terrestrial radiation (3–70 μm) in red.

The passage through the atmosphere of the incident radiation from the Sun and of the terrestrial radiation from the land and oceans redistributes the transfer of energy via three different mechanisms:

- *radiation* transfer, involving both solar and terrestrial radiation;
- *latent energy* transfer, involving changes of state of water in its travels between the ground and the atmosphere (transfer of *latent heat*); and
- *sensible energy* transfer, also involving transport of matter, with air masses being warmed by contact with the surface.

The net result of this redistribution is that the surface of the Earth at ground level permanently receives an energy flux of 492 W/m². This is the quantity that defines the energy that is constantly available to maintain the average climate on Earth. It is the sum of the terrestrial radiation flux of 390 W/m² from the Earth's surface (corresponding to an equilibrium surface temperature of +15°C), the latent energy flux of 78 W/m² and the sensible energy flux, 24 W/m².

Case 3, which includes all the mechanisms involved in the climate balance on Earth, corresponds to the real situation. It is illustrated through a simple analogy, that of a bowl of soup, as detailed in Box 4.1.

BOX 4.1 THERMAL EQUILIBRIUM: THE BOWL OF SOUP ANALOGY

To illustrate the different mechanisms by which the surface of the Earth exchanges energy with the atmosphere and with space, we use the analogy of a bowl of soup. The bowl of soup is hot (+80°C). Its temperature is well above that of the room (+25°C). Since it has more energy than its surroundings (it is hotter), it gradually communicates its surplus energy to its environment – and therefore it cools. It does this through three different mechanisms, as follows:

1. **Conduction and convection** The soup exports heat by warming its surroundings by conduction: air in contact with the surface rises in the room and is replaced by cooler air. Convection thus takes place in the air above the hot soup, which transmits some of its energy to the air in the form of *sensible heat*. The expression 'sensible heat' corresponds to the fact that the energy heats the air in such a way that it can be detected by our senses (or by a thermometer).

Analogy with the Earth The same mechanism governs the transfer of heat by conduction/convection between the Earth's surface and the air masses. The Earth's surface, which is generally warmer than the atmosphere, loses heat by conduction, heating the lower layers of the atmosphere. The heat is then distributed into the air *mass by convection. This mechanism cools the surface of the land and the oceans by transferring some of their energy into the atmosphere. Averaged over the year and over the whole of the Earth's surface, the flux of energy supplied to the atmosphere is $24\,W/m^2$; that is, a loss of $24\,W/m^2$ from the surface of the Earth. In Case 3 of Section 4.2, this is called 'sensible energy transfer'.*

2. **Evaporation** Water evaporates from the soup in the bowl, changing from the liquid to the vapour state. Since energy is needed to break the bonds that hold the molecules together in the dense liquid phase, evaporation cools the liquid. Above the soup, the evaporation is made visible by the whitish plume of condensed vapour rising, and is a warning sign that you may scald yourself by drinking the soup. But if the water molecules escape in the form of a gas (this phenomenon is invisible), why then do we see the whitish plume rising? It is simply because the temperature of the air is lower than that of the soup. At this lower temperature, the air becomes saturated, so that the water vapour quickly liquefies again by condensing into very fine droplets, just as in a forming cloud. The plume is of the same nature as clouds or fog.

The heat transferred into the gas phase through the evaporation is called *latent heat*, because the energy gained by the gas phase is not perceptible by our senses. It is hidden in the gas, revealing itself only when the vapour condenses. The loss of energy when gas molecules condense is transferred into the atmosphere, thus warming it.

Analogy with the Earth The same mechanism operates at the surface of the Earth, releasing a constant flow of vaporized water molecules. The source of this evaporation is liquid water in the soil, in the vegetation and in the oceans, and is the first stage in the water cycle. Averaged over a year and over the surface of the Earth, this evaporation is equivalent to a sheet of water about $1\,m$ thick. It cools the Earth's surface by carrying away $78\,W/m^2$ of energy into the atmosphere. The air, loaded with the invisible water vapour, rises and gradually reaches cooler zones in the atmosphere. There, the water vapour condenses and gives birth to clouds. In condensing, energy is transmitted to the atmosphere, which becomes warmer. Why? The molecules, on changing back to the liquid state, release an amount of energy equal to that supplied to break the bonds in the liquid. In Case 3 of Section 4.2, this process is called 'latent energy transfer'.

3. **Radiation** Finally, the bowl of soup radiates heat, just like an electric radiator or the coals in a fire. At $+80°C$, the wavelength range of the radiation is centred in the infrared region, at about $10\,\mu m$. However, the temperature determines not only the wavelength but also the flux of energy: radiation that escapes continuously from the soup bowl carries with it an energy flux: at $+80°C$, this amounts to $880\,W/m^2$. Of course, the bowl also receives thermal radiation from its surroundings (the table, air and walls, the clothes of the person drinking the soup etc.). But as these are cooler than the bowl, the flux of energy received is smaller than that radiated by the bowl. The bowl loses energy by radiation and is therefore cooled by this mechanism as well.

Analogy with the Earth With its average temperature of $+15°C$ ($288\,K$), the surface of the planet cools in a similar way by emitting infrared radiation with a wavelength of about $10\,\mu m$ into the sky. At this temperature, the energy flux lost is $390\,W/m^2$. In Case 3 of Section 4.2, this is called 'radiation transfer'. It is the predominant mechanism in climate equilibrium.

4. **Thermal equilibrium** *How can the soup bowl remain at constant temperature?* It does so by receiving as much energy as it loses. In a kitchen, this is not generally the case, and so in everyday life the soup bowl slowly cools, until it reaches room temperature.

Analogy with the Earth The same applies to the surface of the Earth. The average surface temperature of the oceans and continents reaches equilibrium when the total energy leaving the surface in all three forms (sensible heat, latent energy and radiation) is equal to the energy it absorbs.

We have just seen how important the role of the atmosphere is in the global energy budget. It stores energy at the surface of the planet (Case 3), and increases the average temperature by several tens of degrees. The principal mechanism in the storage is the greenhouse effect. The importance of this effect, especially in view of the influence of human activity, compels us to investigate this mechanism more thoroughly in Part II, together with the energy balance at the surface of the planet. First, however, we conclude this general overview by briefly describing the atmosphere and the oceans. These play a central role in the distribution of rainfall on Earth and in the transport of heat from the Equator to the poles.

Chapter 5
Atmosphere and ocean: key factors in climate equilibrium

We have seen that by absorbing and reflecting part of the solar radiation, as well as by the changes of state of water, the atmosphere plays an important role in the energy balance of the planet. These processes depend on the composition of the atmosphere. In concert with the ocean, the atmosphere also affects energy equilibrium via the circulation of air masses. It operates through:

- the water cycle – that is, the distribution of rainfall over the Earth due to the water cycle, which is governed by the circulation of the atmosphere; and
- transport of heat from the Equator to the polar regions by fluid flow in the atmosphere and in the ocean.

A further crucial factor in the energy balance of the Earth is the large heat capacity of the ocean. It acts as a thermal reservoir, smoothing climate variations over long timescales.

Below, we describe the major features of the circulation and how they affect the climate. More detailed descriptions of the various mechanisms involved lie outside the scope of this book.

Climate Change: Past, Present and Future, First Edition. Marie-Antoinette Mélières and Chloé Maréchal.
© 2015 John Wiley & Sons, Ltd. Published 2015 by John Wiley & Sons, Ltd.
Companion website: www.wiley.com\go\melieres\climatechange

5.1 Driving forces

Two types of movement are found in the atmosphere and in the ocean:

- *Vertical movements*, caused by differences in density within the fluid. They take place within no more than several kilometres in height (or depth). In certain parts of the ocean, they can reach the bottom. These movements are driven by buoyancy forces, which cause bodies to rise or fall according to their relative density. Since the density of air decreases as the temperature increases, air will rise if the lower layers of the atmosphere are heated by the surface of the planet. In the ocean, the density depends not only on temperature but also on salinity, increasing with salinity and decreasing with temperature. In the ocean, therefore, not only one, but two parameters govern vertical movements.
- *Horizontal movements*: their overall effect over the planet is to transfer heat from the warmest regions (the equatorial zone) to the coldest (the poles). This heat transfer is a consequence of the Second Law of Thermodynamics, according to which heat travels from the warmer to the colder body. Air and water are both involved in this process. The horizontal motion builds up over thousands of kilometres. Over such large distances, the motion is strongly affected by the *Coriolis force*, which shapes large-scale atmospheric and ocean surface circulation patterns. The Coriolis force is a pseudo-force caused by the Earth's rotation about its axis. In the Northern Hemisphere (NH), it deviates bodies to the right of their direction of motion, and in the Southern Hemisphere (SH), to the left. Its effect on horizontal motion is greatest at the poles and zero at the Equator. Its influence on climate equilibrium is fundamental, profoundly modifying movements that are generated by simple temperature differences.

5.2 The atmosphere

5.2.1 COMPOSITION, PRESSURE AND TEMPERATURE

Composition and atmospheric pressure

The Earth's atmosphere is composed principally of three gases: molecular nitrogen, N_2 (78.08%), molecular oxygen, O_2 (20.95%) and argon, Ar (0.93%). These percentages are the ratio of the number of molecules of the gas to the total number of air molecules or, which is equivalent, the ratio of the volume of the gas to the total volume of dry air. The next most common component is water vapour (0.4% on average, but it is distributed very unevenly in the atmosphere), followed by carbon dioxide (CO_2) (0.039%). Low concentrations are expressed in terms of *parts per million* (ppm) – that is, the number of molecules of the substance in 1 million molecules of air – or in *parts per billion* (ppb) or in *parts per trillion* (ppt). In 2014, the amount of CO_2 in the atmosphere was close to 400 ppm. Of the remaining gases, three in particular are involved in the energy balance: CH_4, N_2O and O_3. The last of these also plays a role in the structure of the atmosphere.

N_2, O_2 and Ar make up more than 99.9% of the mass of the atmosphere. Chemically, these gases are stable and they do not exhibit changes of phase in the operating range of the atmospheric temperature and pressure. They are thoroughly mixed and, in the first 80 km of the atmosphere, their relative concentration is distributed homogeneously (Part I, Note 4). The same also applies to CO_2, CH_4 and N_2O. Although the concentration of the last three gases is affected by the seasonal cycle of the hemisphere (due largely to photosynthesis – especially in the NH, where plant biomass on land is abundant), its annual average is essentially homogeneous at each latitude. By contrast, this homogeneity does not hold either for water vapour or for ozone. The distribution of water vapour in the atmosphere varies strongly, since it participates both in evaporation and in condensation. The higher the temperature, the more water vapour can be contained in the air before it reaches saturation and the vapour condenses into cloud. The water vapour content can thus vary from a few milligrams to 35 grams per kilogram of dry air, as is the case for warm saturated air masses. Ozone, owing to its high reactivity, is located in the atmosphere mainly at altitudes between 20 and 30 km.

The atmosphere contains not only these gases, but also clouds and aerosols. Although these contribute very little to the mass of the atmosphere, their role in the energy balance of the Earth is large. Clouds, which are suspensions of water droplets or ice particles, cover about half the surface of the Earth and lie principally in a range between the surface of the planet and an altitude of about 18 km (at the Equator). Liquid or solid chemical compounds other than water form aerosols. Their distribution depends greatly on geography.

As a first approximation, atmospheric pressure decreases exponentially with altitude, at a rate of 12 hPa per 100 m in the lower layers of the atmosphere. Fifty per cent of the mass of the atmosphere is found within the first 5500 m, and 90% is below 16 km. The mass of the atmosphere and the gravity on the planet determine the pressure at its surface. The average mass of air contained in a column of cross-section $1 \, cm^2$ that rises from sea level to infinity is approximately 1 kg, and the pressure at sea level is 1013.65 hPa (1 bar). The amount of each component of the atmosphere can be expressed as the mass of that component in the column. This quantity will be used in Part II to evaluate the absorption of radiation by the different compounds in the atmosphere.

Temperature: the troposphere and stratosphere

We focus on the first few tens of kilometres, where practically the whole mass of the atmosphere is concentrated. The way in which the temperature depends on altitude defines two distinct regions of the atmosphere in which the movement of the air masses is very different (Fig. 5.1). The first region is the *troposphere*, where the temperature decreases with altitude. It stretches from the surface to an altitude of about 10 km ($\approx$ 18 km at the Equator, $\approx$ 8 km at the poles). It contains up to 80% of the mass of the atmosphere and as much as 99% of the water vapour. This region is dominated by convection, which causes continual vertical mixing of the air layers. Convection is caused by the Sun heating the surface of the Earth, which then transmits its heat to the lower layers of the atmosphere. As already noted, air is largely *transparent* to solar radiation, which means that a large part of this radiation passes through unabsorbed until it reaches the surface and warms it. The air masses, heated from below, then become less dense and rise, and convection takes place. As the air rises, the pressure decreases and it cools. If the atmosphere was dry (no water vapour), the resulting temperature gradient would be 9.8°C/km. If water vapour is present, however, it liberates heat when it condenses,

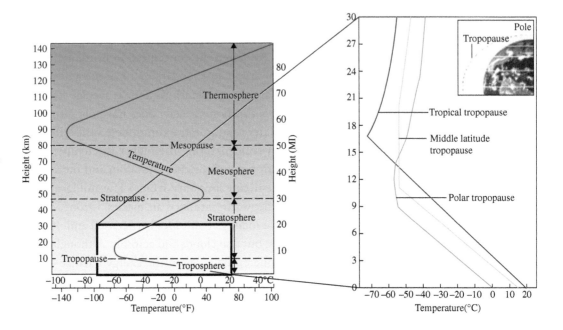

thus warming the air and reducing the thermal gradient. When averaged over the Earth, the gradient is 6.5°C/km. Convection is a characteristic property of the troposphere, and it plays an essential role in the water cycle.

The second region, the *stratosphere*, lies above the troposphere, at an altitude between 10 and 50 km. Here, the temperature starts to increase with increasing height. *Why this sudden change?* At higher altitude, the absorption of solar radiation by ozone constitutes a heat source. This feature is a peculiarity of Earth's atmosphere that is found neither on Venus nor on Mars. Since their atmospheres contain insignificant amounts of molecular oxygen, they possess no ozone: on those planets, therefore, the temperature decreases continuously with altitude. On Earth, because the temperature increases with altitude, the air in the stratosphere is stable and stratified. *Why?* If, due to any fluctuation, a pocket of air were to rise, it would expand as the pressure decreased, thus causing it to cool. Being now cooler, and hence denser than the air immediately surrounding it, it would immediately return to its original altitude, thus halting the initial rising motion. The layer is therefore stable.

The tropopause barrier

The tropopause separates the troposphere from the stratosphere. These two regions are so different that they form two distinct reservoirs in the atmosphere: convective and unstable in the troposphere, and stratified in the stratosphere. The tropopause acts as a barrier that halts upward movement of the air masses from the troposphere. Gases can pass from the troposphere into the stratosphere by the slow process of diffusion. Air masses penetrate into the stratosphere only if they possess considerable kinetic energy, such as can occur in major volcanic explosions, or thermonuclear explosions in the atmosphere. The volcanic aerosols that are injected so fast reside there for several years, leaving time for stratospheric winds to spread them all around the globe. The water cycle, however, being strongly coupled to convection, is confined mainly to the troposphere. It ensures that over a period of about 10 days (the average residence time of water molecules in the atmosphere), the atmosphere receives a regular shower, washing out dust particles and soluble components. *This permanent maintenance service constantly cleans the air we breathe.*

5.2.2 THE ZONAL PATTERN OF ATMOSPHERIC CIRCULATION

The troposphere is where convection, the water cycle and about 80% of the air mass reside. On large distance scales, the movement of air masses is the combined effect of the thermal gradient between the Equator and the poles, and of the 24-hour east–west rotation of the Earth. This yields a pattern that is essentially zonal in character; that is, common to each band of latitude (Fig. 5.2). In the complex path that the air masses take around the Earth, there is of course no beginning and no end. Here, our account starts in the region of the globe where the concentration of energy is highest, the equatorial zone.

The Intertropical Convergence Zone (ITCZ)

Heating the atmosphere from the surface of the Earth drives convection, with ascending movements of air masses. The movement is strongest where the surface receives the most solar flux or *irradiance*, namely the equatorial zone. The air masses rise to the tropopause barrier. Some air comes back down, but the rest remains at that altitude and heads towards the subtropics on either side of the Equator. In this equatorial belt, the rising movement of warm wet air and the loss of air in the subtropics generate a *low-pressure zone* at ground level. The air masses cool as they rise, and water vapour condenses progressively, releasing heat into the middle and high atmosphere and giving birth to convective clouds. This mechanism creates heavy cloud cover and intense rainfall. This is the *tropical rain belt*.

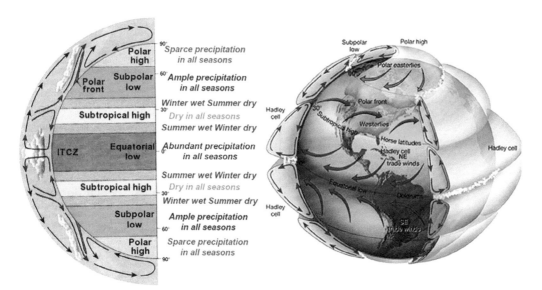

Fig. 5.2 The meridian atmospheric circulation (idealized). The three cells export part of the heat received in the equatorial zone to the polar regions. (a) The idealized precipitation patterns show the alternating bands of rain and drought as a function of latitude (adapted from NASA). (b) The major features of the global circulation. **Source:** Image courtesy of Nicholas M. Short, https://www.fas.org/irp/imint/docs/rst/Sect14/Sect14_1c.html.

In the lower troposphere, air masses from the subtropical regions are drawn towards the tropical depression zone created by the ascendant air systems, and give birth to steady winds on the surface, known as *trade winds*. In each hemisphere, these winds tend to head towards the Equator; then, deviated by the Coriolis force, they blow from north-east to south-west in the NH (and from south-east to north-west in the SH). The two systems of trade winds converge into a common zone, called the *Intertropical Convergence Zone* (ITCZ). The ITCZ coincides with the *tropical rain belt*, the position of which oscillates from one side to the other of the geographical Equator according to the season.

The large-scale cellular pattern from the Equator to the poles

On a large scale, the air masses arrange themselves into a pattern of horizontal winds that transfer some of the heat from the equatorial region towards the high latitudes. *What would the transfer pattern look like if the planet did not turn on its axis?* In each hemisphere, upon reaching the top of the troposphere in the equatorial region, the air masses would head towards the coldest region of the planet, generating a high-pressure zone in the polar region, and then plunge back to the surface. A single cell would stretch from Equator to poles, with the warm air moving at high altitude towards the polar regions and the cold air in the lower atmosphere flowing from the poles to the Equator. The polar regions would then be the site of a permanent anticyclone, and a low-pressure zone would inhabit the equatorial region. This picture is that of the atmosphere of Venus, where the rotation of the planet about its axis is extremely slow (one rotation every 243 Earth days) and, accordingly, the Coriolis force is practically absent.

On Earth, with its faster rotation, the transfer mechanism is different: the Coriolis force divides the airflow between the low and high latitudes into a succession of cells. A simplified diagram shows how the air circulation is divided into three cells (Figs. 5.2 and 5.3). We emphasize that this simplified picture overlooks certain realities, such as the presence of continents and so on.

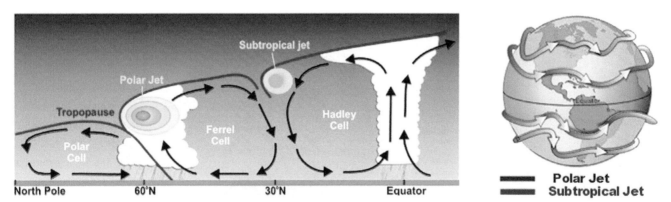

Fig. 5.3 The subtropical and polar jet streams are located in the high troposphere at latitudes 30°N/S and 50–60°N/S, respectively (left). They form at the boundaries of adjacent air masses with a significant temperature difference, such as the polar region and the warmer air to the south. Jet streams meander around the globe, dipping and rising in altitude/latitude (right).

Source: www.srh.noaa.gov/srh/jetstream/global/jet.htm

The resulting general circulation consists of bands of latitude with alternating low- and high-pressure zones, with their associated cyclonic and anticyclonic circulation patterns (Fig. 5.4).

- The *Hadley cell.* The Hadley cell is the largest cell: taken together, the Hadley cells of the NH and the SH cover half of the Earth's surface. Driven by the strong irradiance at the Equator, this cell transfers heat from the equatorial to the subtropical zone. Depending on the season, its effects on the convection are felt in the equatorial zone or in the tropics. The path of the air masses begins in the equatorial low-pressure zone, where they rise (the ascending branch of the cell) until they reach the tropopause barrier, losing their humidity on the way. Then, in each hemisphere, they begin their journey through the high troposphere towards the subtropics. As it travels, the air is increasingly deviated by the Coriolis effect. The deviation becomes more pronounced as the air moves further from the Equator, because the Coriolis force increases with latitude. The trajectory bends progressively eastward, becoming west–east at a latitude of around 30°. In the high atmosphere between 10 and 16 km, this forms a relatively stable but fierce west-to-east air current, the *tropical jet stream* (Fig. 5.3). It arises from the convergence of air masses at this latitude, and is accompanied by a belt of overpressure at ground level. On its route in the high troposphere, the dry air cools and then plunges back into the lower troposphere at about 30° latitude (the descending branch of the cell). This air in turn partly feeds the *trade winds,* closing the loop of the *Hadley cell.* The subsidence of the air generates almost permanent anticyclones in the subtropics. These are particularly firmly anchored over oceans, where, coupled with the surface currents, they generate *ocean gyres.* On continents, the scarcity of rain in this high-pressure zone creates deserts (the Sahara Desert in North Africa, the Arabian Peninsula, the Kalahari Desert in Southern Africa, the Great Victoria Desert in Australia and so on).
- The *polar cell.* The polar regions are the coldest parts of the planet. Another type of circulation takes place there, which is due to the very low temperatures in these ice-covered regions and to the very cold and dense air masses. The latter are confined to the lower layers, forming a high-pressure zone (thermal anticyclone) at very high latitudes. The cold air

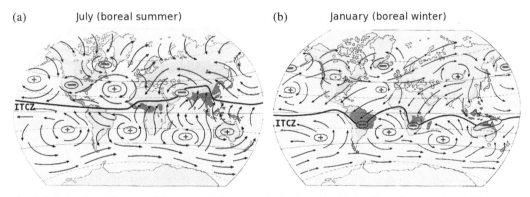

Fig. 5.4 A simplified diagram of the general atmospheric circulation in (a) July (boreal summer) and (b) January (boreal winter). The high pressure centres (blue) are indicated by '+' and low pressure (red) by '–'. Annual precipitation on continents: yellow, <50 mm; light blue, 50–100 mm; dark blue, >150 mm.

Source: Adapted from © eduscol.education.fr-MEN-droits réservés with permission. http://eduscol.education.fr/obter/appliped/circula/theme/atmo332.htm and http://eduscol.education.fr/obter/appliped/circula/theme/atmo331.htm

flows out through the low troposphere towards a low-pressure zone that regularly resides at lower latitudes. There, it encounters warmer air, giving rise to *storms* along the *polar front* at latitudes as far as 50–60°. At this point, the loop partly closes and the air rising from the low-pressure regions feeds the high-pressure polar zone. Under the effect of the Coriolis force, the wind systems in the lower atmosphere, which drive the air from the high-pressure towards the low-pressure area, blow from north-east to south-west in the NH.

The above account of air circulation in the polar regions is greatly simplified. In these regions, the air circulation is particularly complex. Although the low troposphere is a seat of high pressure, the high troposphere and low stratosphere form a pressure minimum. This zone is known as the *polar vortex*.

- *A third cell?* The existence of a third cell, the *Ferrel cell*, is often cited. It appears as a logical continuity between the high-pressure zone of the tropics and the low-pressure zone at latitudes 50–60°. In the mid-latitudes, the circulation is driven both by the southern limit of the predominant cyclonic low-pressure system around 60° (west winds) and the northern limit of the predominant anticyclones at about 30° (also west winds). It follows that the prevailing winds, the *westerlies*, blow from west to east.

In the upper troposphere between 7 and 12 km at 50–60° latitude, extremely powerful winds, grouped under the name *polar jet stream*, blow from the west (Fig. 5.3). This polar stream is stronger than the subtropical jet stream. It originates from the intense thermal gradients between the cold polar air and the warm air coming from the subtropics, and interacts with *storm* formation. Its flow is unstable, and its course around the planet is marked by north–south oscillations. In winter, it is very strong, trapping the cold polar air at high latitudes, and acting as an efficient barrier. When the *jet stream* weakens, incursions of polar air towards temperate latitudes escape more easily, bringing waves of cold air to the large continental regions in the NH and giving rise to very cold winters (North America, Europe and Asia), while warmer air penetrates into the high latitudes.

Rain belts and wind belts

From the Equator to the poles, the circulation of the atmosphere forms a pattern of successive latitude bands (Fig. 5.2). This pattern is the reason why rainfall and wind are distributed according to zones (latitude belts). Rainfall is heavier in low-pressure regions where the atmosphere is unstable; that is, in the equatorial zone (*tropical rain belt*) and in latitudes around 50–60°, which are exposed to successions of storms. By contrast, rainfall is lighter, or even scarce, in the high-pressure areas of the subtropics around 30° and in the polar regions.

In the low troposphere, the winds blow from east to west between the Equator and the subtropics (the *trade wind belt*), then from west to east at mid-latitudes (the *westerlies wind belt*), and finally from east to west at high latitudes. In the high troposphere around 30°, the *tropical jet stream* takes over and, in the two hemispheres, marks the northern and southern limits of the warm tropical air masses. In the high troposphere at about 60°, the *polar jet stream* circles the planet from east to west, preventing the cold arctic and antarctic air masses from spreading to lower latitudes.

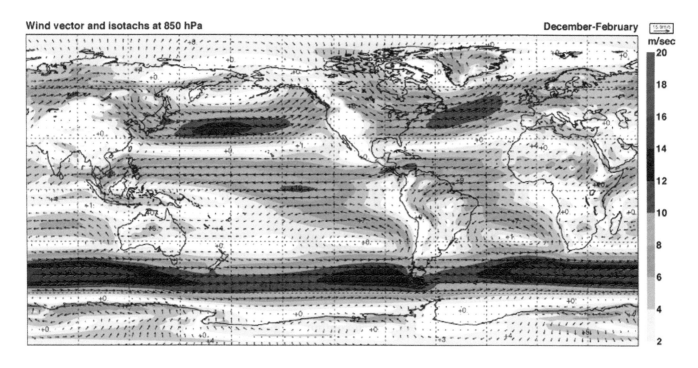

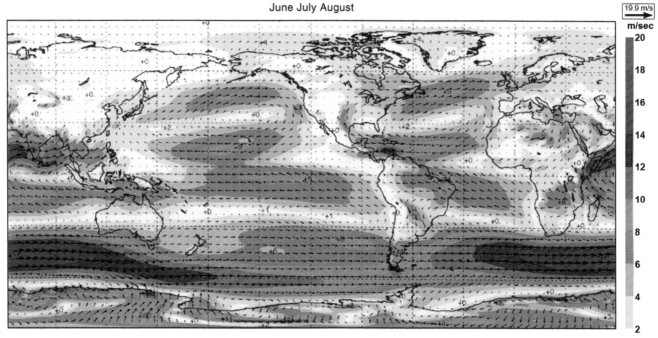

Fig. 5.5 The mean wind distribution at 1500 m above sea level (850 hPa), from the European Centre for Medium-Range Weather Forecasts reconstruction: (a) averaged over December, January and February; (b) averaged over June, July and August.

Source: Reproduced by permission of ECMWF.

The *trade winds* are steady winds that slacken on approaching the Equator. This region, called the *doldrums*, has no prevailing wind and is the site of storms and unpredictable winds. Navigators of sailing ships, who risked being becalmed, were fearful of it. At mid-latitudes the winds come characteristically from the west, the *westerlies*. The maps in Fig. 5.5 show

Tracks and Intensity of All Tropical Storms

Saffir-Simpson Hurricane Intensity Scale

Fig. 5.6 Tropical cyclone tracks across the globe recorded over a period of almost 150 years (National Hurricane Center and Joint Typhoon Warning Center). Tropical cyclones do not occur near or at the Equator, where the Coriolis force vanishes. Storms are therefore unable to generate the large-scale rotation that starts them on their path to become a hurricane. Their absence in the south-west Pacific and south-west Atlantic Oceans is related to the cold currents that run through these regions. **Source:** Image created by Robert A. Rohde/Global Warming Art

the average wind speed and direction at an altitude of 1500 m. They reveal the belt of violent winds that blow all year long from west to east between 40° and 60° in the SH (the well-known *Roaring Forties, Furious Fifties* and *Screaming Sixties*), while in the NH the circulation of air from the west is affected by the continents.

The discussion of the low latitudes cannot be closed without mentioning *tropical cyclones*. These very deep depressions develop above warm waters, and are driven by the *trade winds* from east to west. As they move over the ocean, cyclones pick up energy from evaporation and condensation, and often continue to grow until they reach a continent. There, for lack of energy supply (i.e. evaporation from the warm ocean), they decay and die. The Pacific Ocean, which offers the longest path over water and where the ocean is the warmest on the planet, is where these cyclones are the most frequent (Fig. 5.6).

Monsoon

The term *monsoon* refers to the seasonal reversal of surface winds and the accompanying rainfall patterns in certain regions of the intertropical belt. As the position of the thermal Equator swings with the season, one branch of the *trade winds* crosses the geographical Equator and enters the other hemisphere. There, it is subjected to a Coriolis force of the opposite sign. The branch of the *trade winds* that blew from south-east to north-west in the SH therefore starts gradually to blow from south-west to north-east once the geographical Equator is crossed. In some regions of the Earth, the distribution of the continents is such that the trade winds converge on warm continental regions, where they discharge the humidity that they accumulated over the ocean. This is the case for the South Asian monsoon, which affects the *Indian subcontinent* and surrounding regions, and also for West Africa, with the summer monsoon of western Sub-Saharan Africa (Fig. 5.4).

5.3 The oceans

In the climate balance, the ocean performs many functions. Notably, it:

- exchanges heat and water vapour with the atmosphere in the water cycle;
- transports heat from the Equator to the poles;
- stores energy, mainly in the form of heat (its heat capacity is about a thousand times greater than that of the atmosphere); and

- acts as a sink for CO_2 released by human agency into the atmosphere, a function that reduces greenhouse warming.

These different roles depend strongly on ocean circulation.

5.3.1 OCEAN CIRCULATION

The structure of the ocean is very different from that of the atmosphere. We just saw that the atmosphere is heated principally from underneath. This implies that temperature decreases with altitude and that the air masses mix by convection. The ocean, by contrast, is heated from above: its temperature decreases with increasing depth. On reaching the surface of the ocean after passing through the atmosphere, most of the Sun's radiation is absorbed within the first hundred metres of water (assuming clear water), thus heating the surface layer. At greater depths, absolute darkness reigns, which is lit only by a few deep-sea creatures. The temperature, which on the surface can be as high as about 30°C at the Equator, quickly falls until it approaches +3.5°C at a depth of a few hundred metres. As the ocean floor lies at an average depth of about 3750 m below the surface, this means that the great mass of the ocean is very cold. Since the warm water is less dense and lies above the cold layers, the bulk of the ocean is particularly stable and stratified. This thermal structure gives rise to two modes of circulation:

- *surface circulation*, driven by winds, which is confined to a layer a few hundred metres deep; and
- *deep circulation*, which involves the bulk of the ocean, including the intermediate and deep water.

These two types of circulation, which meet at certain points on the planet, constitute the general circulation, or the *thermohaline circulation* (THC), in which heat (*thermo*) and salt (*haline*) are the factors that change the density of the water and affect the stability of the layers. This circulation affects the entire ocean and is often referred to as *the global ocean conveyor belt*, an idea largely popularized by W. Broecker.

Salinity: a marker for the water cycle?

Does salinity vary greatly from one ocean to another? Here, we concentrate on the large masses of water – that is, the oceans – and neglect adjacent seas. Salinity is expressed in *practical salinity units* (or psu), where 1 psu corresponds to 1 gram of dissolved salt (NaCl) per kilogram of seawater. Among the world's oceans, where salinity ranges between 33 and 37 psu, the Atlantic Ocean is remarkable in its outstandingly saline surface water (Fig. 5.7). This gives it a major role in driving deep-water circulation, since high salinity implies high density. *What is the reason for its high salinity?* It is mainly the water cycle, in combination with two factors: (i) the circulation of the atmosphere, which drives warm wet air masses westwards in the tropics (with the *trade winds*), and exports part of them into the Pacific; and (ii) the shape of the Atlantic basin, which is separated from the Pacific basin by a north–south continental barrier that hems in the western border of the Atlantic Ocean. The intense evaporation E (removing fresh

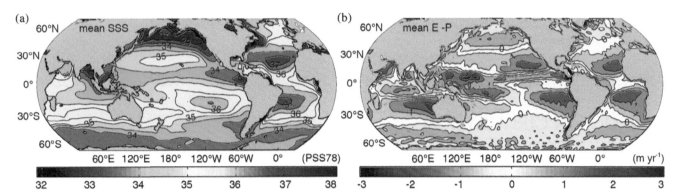

Fig. 5.7 (a) The climatological mean of sea surface salinity for 1955–2005 with seasonal and El Niño-Southern Oscillation removed. Contours every 0.5 salinity unit (psu) are plotted in black. (b) Annual mean Evaporation–Precipitation averaged over the period 1950–2000 (National Center for Environment Prediction). Contours every 0.5 m/yr are plotted in black.

Source: *IPCC 2013. Climate Change: 2013 The Physical Science Basis.* Working Group I Contribution to the Fifth Assessment Report of the Intergovernmental Panel on Climate Change, Figure 3.4ab. Cambridge University Press.

water) from the equatorial belt in the Atlantic Ocean is thus not compensated in return by precipitation P (supplying fresh water) from condensation of the same air masses, since part of this water, after passing blithely over the west Atlantic continental barrier, falls into the Pacific Ocean. The resulting evaporation surplus ($E-P>0$) makes the salinity of the Atlantic higher than either that of the Pacific or the Indian Oceans (Fig. 5.7). In the water of the *Gulf Stream*, which flows from the Caribbean Sea, the salinity between 15°N and 35°N reaches 36 psu. Furthermore, the Mediterranean Sea, which is subject to strong evaporation, ejects its salt water into the Atlantic through the Straits of Gibraltar. Finally, the Atlantic Ocean is the recipient of relatively salty water from the Agulhas Current, which starts in the Indian Ocean and flows round the Cape of Good Hope. Both of these currents add to the salinity of this ocean.

The Atlantic Ocean is an illustration of how intimately ocean salinity and the water cycle are related. Salinity can thus be used as a marker of changes in the water cycle caused by climate warming, for example. We shall return to this topic in Part V.

Surface currents

As the surface water is set in motion by winds, the resulting surface currents are an image of atmospheric circulation. These surface currents, shown diagrammatically in Fig. 5.8, involve the first few hundred metres of the depth profile of the water. A striking feature of the circulation is its pattern of large vortices, or *gyres*, between the Equator and the mid-latitudes. These correspond to permanent high-pressure areas in the subtropical zone. Like the lower atmosphere, the currents turn clockwise in the NH and anticlockwise in the SH. Water accumulates in the centre of the *gyres*, causing a bulge at the surface that can be as much as a metre high. This difference in level creates a horizontal pressure gradient force that is counterbalanced by the Coriolis effect, giving rise to what is called a *geostrophic current*. In the NH, where the continents confine the Pacific and Atlantic Oceans up to the middle and high latitudes, this phenomenon generates the *Gulf Stream* in the North Atlantic and the *Kurushio Current* in the North Pacific.

Marine surface currents transport warm water from low latitudes to the mid-latitudes, adding their contribution to the heat transfer from Equator to the poles. The surface of the

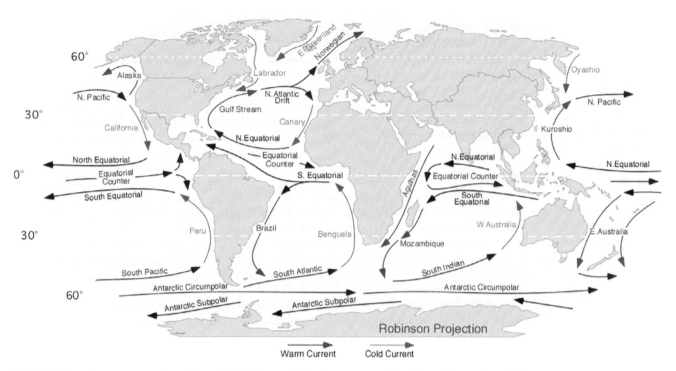

Fig. 5.8 The major ocean currents of the world. Red arrows indicate warm currents, while cold currents are in blue.

ocean, coupled with the atmosphere, also exports heat through the water cycle: when a warm ocean current enters a region where the air is cold and dry, intense evaporation takes place. The energy of evaporation (*latent heat*), extracted from the mass of liquid water in the transition to the vapour state cools the surface water. The energy is transferred into the atmosphere, and is liberated when the water vapour condenses into clouds. The condensation heats the atmosphere. The water cycle loop then closes when the rain falls back into the ocean. The latent heat is released in the mid-latitudes, where the prevailing atmospheric circulation at the Earth's surface is from west to east. Air masses heated in this way make the climate more temperate on the continents situated on the east side of the Atlantic and Pacific oceans; for example, on the west coasts of Europe and North America.

In the mid-latitudes in the SH, where continents are few and mountain ranges scarce, a belt of extremely strong west winds develops (Fig. 5.5). The effect of these winds on the mass of water gives rise to the *Antarctic Circumpolar Current*. This powerful current (also called the *West Wind Drift*) thermally insulates the Antarctic polar region by limiting the access of the tropical water masses into the middle to high latitudes. There is nothing comparable in the NH. In the Atlantic Ocean, the *Gulf Stream* separates into two branches. One turns back south and follows the west European facade. The other, the *North Atlantic Current*, is constrained by the local geometrical configuration to pursue its northward journey into the cold, high latitudes. In the Pacific Ocean, the Aleutian Island chain brings to a halt an analogous northward motion. The water of the Atlantic, which, as already noted, is highly saline, is a key factor in deep-water formation and in generating *thermohaline circulation*.

The average speed of surface currents varies greatly, ranging from centimetres per second (the *California Current* and the *North Atlantic Current*) to a metre per second or more (the

Gulf Stream). The flux of water flowing in a current is expressed in Sverdrup [Sv, where 1 Sv is 1 million cubic metres of water transported per second (10^6 m^3/s)]. As an illustrative example, the total flux of water flowing into the oceans from all the rivers in the world is of the order of 1 Sv. The water flow in surface currents can range from several Sv (the *Labrador Current* and the *North Atlantic Current*) to more than a hundred Sv (135 Sv in the Drake Passage with the *Antarctic Circumpolar Current*; 150 Sv in the northern part of the Gulf Stream). These fluxes vary strongly with time.

Upwelling

Upwelling occurs in regions where water surges upwards from a depth of a few hundred metres. *What are the mechanisms that cause this water to upwell?*

When surface water is driven by winds blowing parallel to a coast, in some well-defined areas, the Coriolis force can divert the water away from the coast. This arises in the subtropical regions where the west coasts of the continents are contiguous with the eastern parts of the subtropical ocean gyres. This makes water rise from a depth of several hundred metres, a phenomenon called *coastal upwelling*. Only a few coastal areas are the site of such surface-water enrichment: in the NH, the *California Current* in the Pacific and the *Canary Current* in the Atlantic; and, in the SH, the *Peru Current* (also called the *Humboldt Current*) in the Pacific, the *Benguela Current* in the Atlantic and the *West Australian Current* in the Indian Ocean (Fig. 5.8).

Upwelling also occurs in the equatorial region, where the west-blowing *trade winds* generate a surface current divergence. Here again, this divergence is due to the Coriolis force that deviates the surface water to the right in the NH – that is, partly towards the north (in the SH, to the left, i.e. partly southward) – drawing up to the surface an influx of deeper water. In the Pacific Ocean, the enrichment of surface water by nutrients due to this *equatorial upwelling* is clearly visible in satellite photographs, as it sometimes produces a green trail of intense growth of diatom colonies (silicate algae) at the Equator.

In the upwelling regions, the surface water is enriched in inorganic salts by water that surges upwards from a depth of a few hundred metres. This cold water, which has been exposed to a rain of organic material (*marine snow*) discharged by the biosphere in the upper ocean layers, is richer in nutrients (N, P, Si, Ca, K, S, Na, Cl, Mg, Fe, Co, Cu, Zn etc.) than the surface water. Refinements in isotopic measurement techniques have led to a better understanding of their marine cycle (Maréchal et al. 1999, 2000). It is thus a permanently replenished source of dissolved elements for photosynthetic organisms and marine organisms in general. The biomass can develop in all its forms: these waters teem with life, with the greatest abundance of fish in the world. The one exception is the Benguela Current, where, almost certainly due to overfishing, the rich marine ecosystem has become impoverished, turning into a realm of jellyfish and catfish.

Deep-water formation and deep-water circulation

Deep-water circulation is set in motion by *deep-water formation*, which occurs at just a few key locations where the surface water is dense enough to sink far into the depths. Since density increases when water becomes colder and saltier, deep-water formation tends to take place in highly saline ocean waters at high latitudes. Two regions of the planet are the site of this phenomenon: in the *Norwegian Sea* in the North Atlantic (also to a lesser extent in the *Labrador* and *Greenland Seas*), and at the edge of the Antarctic ice cap, mainly in the *Weddel Sea* and the *Ross Sea* (Fig. 5.9).

In the North Atlantic, the saline surface water is transported from the tropics by the *Gulf Stream*, and cooled by the polar air mass (with temperatures as low as −30°C). Such dense waters, at a temperature of about +2°C, sink rapidly and are replaced by warmer surface waters (above +4°C). The deep water, with its high salinity (34.9–35.0 psu), flows south at a depth between 2000 and 4000 m, while the surface waters that feed them flow northwards in the upper 1000 m thick layer of the water column (Fig. 5.10). This convective process generates a large volume of dense water. On

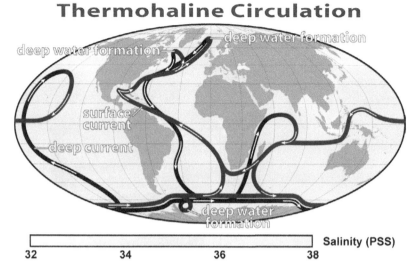

Thermohaline Circulation

Salinity (PSS)

32   34   36   38

Fig. 5.9 A schematic representation of global thermohaline circulation: blue paths, deep-water currents; red paths, surface-water currents. The oldest waters (with a transit time of about 1500 years) upwell in the North Pacific.
Source: NASA Earth Observatory, adapted from the IPCC 2001 and Rahmstorf (2002).

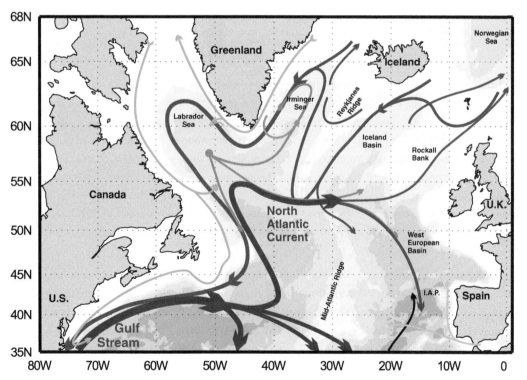

Fig. 5.10 The principal currents in the North Atlantic. At the surface, the North Atlantic Drift extends the Gulf Stream to the north (red). The deep waters (dark blue), which form mainly in the Norwegian Sea, flow southwards. The cold surface waters of the South Greenland, the East Greenland and the Labrador Currents (green), come from the north along the coasts of Greenland and Canada, and also flow southwards. Intermediate waters (purple) form at the point 'LSW' from the surface water of the Labrador Sea before joining the deep waters later. The East Coast Current, also at intermediate depths, transports the Mediterranean water (pink).
Source: Reproduced by permission of Pascale Lherminer /IFREMER Centre Bretagne France.

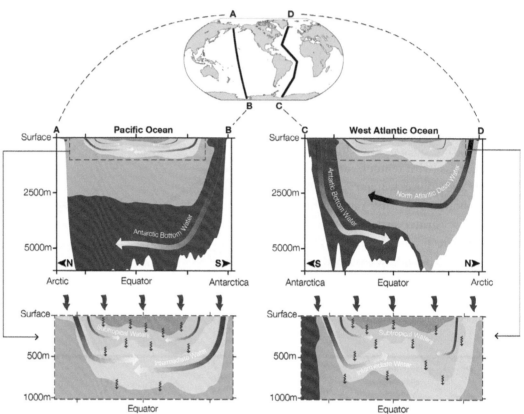

Fig. 5.11 This diagrammatic view of ocean heat uptake pathways illustrates the surface water downwelling and deep water formation in the Atlantic and Pacific Oceans, through a north–south transect (upper panels). Dark blue, cyan, light blue, green and orange denote increasing water temperature. The arrows indicate the uptake pathway of heat by the ocean in the event of global surface warming. Thick arrows fading to white indicate warming with time. In the North Atlantic, the thick red and blue arrow indicates decadal warming and cooling. Excess heat entering the ocean surface (upper curvy red arrows) also mixes slowly downward (subsurface wavy red arrows).

Source: IPCC 2013. *Climate Change 2013: The Physical Science Basis.* Working Group I Contribution to the Fifth Assessment Report of the Intergovernmental Panel on Climate Change, FAQ 3.1 Figure 1. Cambridge University Press.

the Arctic shelf (the *Kara Sea* and the *Laptev Sea*), the formation of sea ice also generates dense waters (but prevents convection, since it insulates the ocean from the air masses). Because ice crystals consist of almost pure fresh water, as sea ice forms, it expels salt into the seawater, greatly increasing its density. The deep waters of the NH constitute the *North Atlantic Deep Water* (NADW) (Fig. 5.11). The cold surface water injected into the NADW aerates the deep ocean with water that has been in contact with the atmosphere and has come to equilibrium with its partial pressure. This aerated water brings the oxygen that is essential for marine life in the intermediate and lower depths of the North Atlantic water column. Furthermore, if the composition of the atmosphere varies (injection of CFCs, CO_2 etc.), the changes are transmitted to the deep ocean from the region where the deep water forms. *So, what signs of the recent increase in atmospheric CO_2 can be seen in the ocean water?* We shall see this below, in Section 5.3.9.

At the edges of the Antarctic, convection of the water masses and the formation of sea ice both play a central role in generating dense deep water. Around the continent, catabatic winds from the interior of the land blow the ice out to sea, creating open spaces of ocean (coastal *polynias*) that are exposed to the harsh Antarctic air. In these clearings, the exchange of *sensible heat* causes intense cooling of the water, and sea ice forms. But since the wind continues to push newly formed ice out to sea, fresh ice is unable to cover the open spaces: cold, increasingly saline water continues to be produced unabated, with the salt being expelled into the water as the ice forms. The cooling (temperatures descending to $-1.7°C$) and the expulsion of salt (high salinity, from 34.6 to 34.7 psu) generate very dense water. All along the edge of the continent, this water downwells into the Southern Ocean, where it feeds the *Antarctic Bottom Water* (AABW). These water masses, the densest of the planet, descend to depths in excess of 4000 m. In the world's ocean, the flow of deep Antarctic water (AABW) is believed to be weaker (about 10 Sv) than that of the NADW (about 15 Sv). The *Antarctic Bottom Waters* move in a northerly direction, reaching as far as the equatorial Atlantic.

The present chapter cannot be concluded without mentioning the *Antarctic Intermediate Water* (AAIW). This cold water (+3 to +4°C) of low salinity (34.2–34.4 psu) is located at depths between 700 and 1200 m. It downwells from the surface in the *Antarctic Convergence Zone* (also called the *Antarctic Polar Front*), between 50°S and 60°S. In this region, particularly during the southern summer, the Antarctic surface water coming from the coast (*Antarctic Surface Water*, AASW) is cold, but of low salt content owing to mixing with meltwater from the ice sheet. This comes into contact with the north subantarctic water, which, as it comes from temperate regions, is warm and saline. The AAIW is the resulting mixture, of intermediate temperature and salinity, but its density is greater than either of its two constituents. The mixed water therefore downwells until it reaches the level of the cold, highly saline and denser NADW.

From the Equator to the high latitudes in the south, and from intermediate depths to the ocean floor, the mass of water in the Atlantic is a succession of layers of water of different density, starting with the AAIW, then the NADW and finally the AABW, with alternating flow directions (Fig. 5.11). The average velocity of the deep currents is low, of the order of millimetres per second.

The *global thermohaline circulation* is born of the junction between the surface and the intermediate/deep oceanic circulations (Figs. 5.9 and 5.11). The dense water flows into the ocean basins. While the bulk of it upwells in the Southern Ocean, the oldest waters (with a transit time of around 1500 years) upwell in the North Pacific. The *Atlantic Meridional Overturning Circulation* (AMOC) is the regional part of the thermohaline circulation in the Atlantic Ocean that connects the surface to the deep waters. As already seen, this major current involves the northward flow of warm, salty water in the upper layers (less than about 1 km deep) and the southward flow of cold, deeper waters. The AMOC is therefore a mechanism of net northward heat transport in the Atlantic. It is an important component of the Earth's climate system.

In a context of climate change, where the surface of the Earth is becoming warmer, deep-water formation is a crucial mechanism by which heat from the surface water is transferred to the deep-water reservoir, thereby attenuating over time the effects of surface warming. Figure 5.11 indicates the resulting pathway of ocean heat uptake in the Atlantic and Pacific Oceans in the situation of global surface warming. *Deep-water formation thus acts as a delaying response to climate change.*

5.3.2 CO_2 EQUILIBRIUM BETWEEN ATMOSPHERE AND OCEAN

In the ocean, carbon exists in several forms, as dissolved inorganic carbon (DIC: CO_2, H_2CO_3, HCO_3^- and CO_3^{2-}), as dissolved organic carbon (DOC: organic molecules below 0.45 µm) and as particulate organic carbon (POC: living and non-living organic materials). These different forms are in the approximate ratios DIC : DOC : POC = 2000 : 38 : 1.

The amount of gas dissolved in a liquid is defined by Henry's law: at constant temperature, the amount dissolved is proportional to the partial pressure exerted by the gas at equilibrium with the liquid. The partial pressure of CO_2 in the atmosphere thus determines the maximum amount of CO_2 dissolved in the surface water of the ocean. In the ocean, certain regions are *sources* of carbon dioxide (releasing CO_2 into the atmosphere when the partial pressure in the water is greater than in the atmosphere), and others are *sinks* (the ocean absorbs CO_2 from the atmosphere). In the surface layer of the ocean, as a result of different chemical equilibria, dissolved CO_2 accounts for only 1% of the DIC. It is in equilibrium, via acid–base reactions, with carbonic acid (H_2CO_3) (low minority form), carbonate ions (CO_3^{2-}) and bicarbonate ions (HCO_3^-) (majority form). Exchanges of CO_2 between the atmosphere and the ocean thus involve only a fraction of the carbon cycle in the ocean. The diagram in Fig. 5.12 shows this cycle before human impact, in the context of the present interglacial climate. In this cycle, carbonate chemistry allows the ocean to store a large amount of carbon.

When the cycle is at equilibrium, the annual flux exchanged between the atmosphere and the ocean is about 60 gigatonnes of carbon (GtC). The layer of the ocean in contact with the atmosphere (the first 100 m) amounts to only 2–3% of the ocean mass and its ability to dissolve the CO_2 in the atmosphere is accordingly limited: it stores about 900 GtC. It is the downwelling of the surface water into a large water reservoir that provides an efficient *sink* for the atmospheric CO_2: it feeds the intermediate and deep water with a flux of the order of 90 GtC/yr. Even though part of this flux quickly returns to the surface in places where the intermediate water rises locally, the deep-ocean reservoir is sufficiently vast to store a large amount of DIC (about 37,100 GtC). Cut off from the atmosphere, the deep water no longer exchanges its carbon (apart from the upwelling regions) and is unable to discharge its load. This dynamic mechanism is referred to as a *physical carbon pump*.

Another, more modest, mechanism also injects carbon into the deep water. A large fraction of the organic matter synthesized by the marine biosphere,

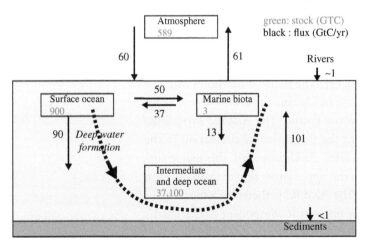

Fig. 5.12 The oceanic component of the carbon cycle in pre-industrial times. The numerical values of the different reservoir contents (gigatonnes of carbon) and flux exchanges (gigatonnes of carbon per year) are taken from the carbon cycle in the IPCC 2013 report (see Fig. 29.1).

about 10 GtC/yr, leaves the surface layer as a rain of particles (*marine snow*, aggregates of dead plankton, faecal matter etc.) and is exported to the deep water, carrying with it carbon removed from the surface by biological activity (POC). Such material is based on photosynthesis, which uses the dissolved inorganic carbon (DIC), inorganic salts and the energy from the Sun's photons to create the phytoplankton on which the trophic networks are based. This mechanism whereby the marine biosphere exports carbon is known as the *biological carbon pump*. In the course of its descent through the water column, organic matter is partly recycled by oxidation, and its particulate carbon transforms into a dissolved form that is mostly inorganic. It adds significantly to the DIC store, which increases with depth. Some of the DIC returns to the surface after a fairly short time, where it again feeds biological activity. In this particular case, the circulation loop between the ocean surface and the deep ocean is relatively short.

Carbon accumulated in the ocean returns to the atmosphere when deep water returns to the surface. We have just seen that this happens only after several centuries, or even more than a thousand years, the time for the *thermohaline circulation* to complete its cycle. In the circulation of dissolved ocean carbon, this is a slow cycle.

Atmosphere and ocean: very different carbon reserves

Let us make a last remark before closing this brief inventory. Comparisons between the carbon reserves in the atmosphere, essentially in the form of CO_2, and that of the intermediate and deep-ocean water demonstrate that in the 18th century the ocean contained about 60 times more carbon than did the atmosphere. In a climate transition, such as from an interglacial to a glacial period, it is therefore easy to appreciate how *a change in the ocean circulation and in the chemical equilibrium between ocean and atmosphere could have a strong influence on the CO_2 content of the atmosphere.*

Anthropogenic emission and the ocean sink

How does the carbon cycle in the ocean change when the concentration of CO_2 in the atmosphere changes; for example, due to anthropogenic emission? This point will be developed in Part V. Here, we present just a short overview. If we start from the beginning of the cycle, the CO_2 dissolved in the surface water increases. It has been estimated that during the recent decades the ocean sink absorbs roughly one quarter of anthropogenic emissions each year. This increases the rate of carbon export to the deep water. The CO_2-rich water thus gradually penetrates into the regions of deep-water formation (Fig. 27.6). The front of this enriched water has already started on its journey into the deep water!

5.4 Heat transport from the Equator to the poles

Atmospheric and ocean circulation, in combination with the water cycle, redistribute part of the heat received from the Sun, transporting it from the warm equatorial regions to the cold polar regions (Fig. 5.13). This transport of heat towards the poles is a maximum at latitudes of about 35°. It is observed that, except in the equatorial region, transport by the atmosphere is

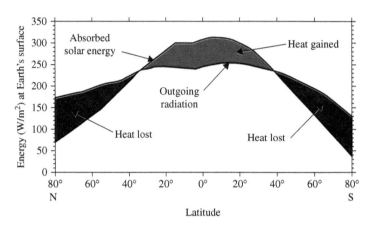

Fig. 5.13 The incoming solar radiation absorbed by the planet and the outgoing terrestrial radiation. Part of the energy absorbed at low latitudes is transferred by convection to the colder high latitudes. The redistribution causes the low latitudes to radiate less energy than they absorb (heat is gained, red) and the high latitudes to radiate more energy than they absorb (heat is lost, blue).

Source: Federation of American Scientists RST (Remote Sensing Technology) NASA https://www.fas.org/irp/imint/docs/rst/Sect14/heat_loss_heat_gain.jpg

much greater than by the ocean (Fasulo & Trenberth 2008). The result is that at low latitudes the flux of energy radiated into space by the planet is less than the solar radiation absorbed by the Earth at the top of the atmosphere. This difference in energy is transferred to middle and high latitudes, which radiate more energy than they absorb.

The dynamics of the atmosphere and of the ocean conspire with the water cycle to transfer part of the Sun's heat from the warm tropical regions to the cold regions at high latitudes, thereby reducing the temperature difference between the Equator and the poles. Their role as climate moderator is of foremost importance.

PART I SUMMARY

From the Equator to the poles, the climates on Earth are very different. They differ in available heat and in the amount of water they receive during the seasons, varying from drought to abundance of rainfall and displaying a wide range of temperatures.

Living organisms depend essentially on these two quantities, heat and water. The strategies that they must develop to adapt to available resources are such that each type of climate corresponds to a different set of ecosystems: *the pattern in which life is distributed over the Earth is intimately related to the climate zones.*

Mechanisms of selection and adaptation to climate changes that have occurred unceasingly over millions of years have been determinant in building up the diversity of species that we know today. *The biodiversity of the various climatic zones is partly the outcome of past climate history, and contains the fraction of the heritage that was able to be transmitted.*

The climate is therefore what moulds life on Earth, and any change in the climate affects the tissue of life. Humans are also dependent on the climate: their population distribution largely reflects that of plant resources.

What mechanisms determine the distribution of the different climates? How are heat and rainfall shared over the Earth? As the foremost source of heat is the Sun, the first factor that decides how the heat is distributed over the Earth is the almost spherical shape of our planet, which causes the poles to receive less heat than the Equator. Atmospheric and ocean circulation and the water cycle then conspire to transfer some of this heat towards the poles. As for rainfall, its distribution is determined by the circulation of the atmosphere and the water cycle, which, in practice, gives rise to alternating belts. Equatorial and mid-latitudes are well watered, while belts of drought and desert emerge in the subtropics and at high latitudes.

To sustain the existing climate system, a permanent flow of energy at the surface of the planet is required that maintains the temperature, the water cycle and the ascending air currents. This energy flux consists of: (i) the amount needed to keep the average temperature at 15°C ($390 \, W/m^2$); (ii) that to continue evaporating a layer of water 1 m thick ($78 \, W/m^2$); and (iii) that needed to compensate the heat loss by conduction from the surface to the air ($24 \, W/m^2$). To maintain the existing climate, a supply of $492 \, W/m^2$ must therefore be permanently available. The uncertainty in these estimates is a few watts per square metre. *Where does this energy come from?*

The primary source of this energy flux is the Sun. Other sources of energy are negligible. Only two thirds of the solar energy received by the planet is absorbed (i.e. $235 \, W/m^2$), the rest being reflected. In fact, $495 \, W/m^2$, more than twice as much as this flux, is available at the surface of the planet. This is a consequence of the greenhouse effect. This effect increases the store of energy available at the surface of the planet: *the larger the store, the greater is the amount of energy available, while the solar flux remains constant.*

The multitude of other factors are minor compared to the following three fundamental features: (i) the Sun, whose activity and distance from the Earth determines the amount of radiation that the planet receives; (ii) the reflection by the Earth, which defines the amount of radiation that the planet absorbs; and (iii) the greenhouse effect, which increases the average temperature at the surface of the Earth by about 30°C. It is these three factors that control the range of temperature on our planet. They explain the differences in climate among the various other planets of the solar system, in which planet Earth stands out as unique. With the benefit of its oceans of liquid water, this planet has been a setting in which life could evolve in immense abundance and variety of forms.

A corollary of these considerations is that *any change that modifies the flow of energy at the Earth's surface leads to a change in climate, and a change in the conditions of survival of living organisms.*

PART I NOTES

1. For example, twice as much energy is delivered at the Equator (at zenith at the equinox) when the Sun's rays fall perpendicularly on the surface than at latitudes 60°N or 60°S, when they are inclined at 60° to the vertical.

2. The South Pole seems to be much colder than the North Pole. This is true, but to compare them a correction must be applied for the fact that the temperatures must refer to the same altitude. The South Pole is located at the centre of a high continent, Antarctica, in which the central plateau rises to an altitude of 3000 m, while the North Pole, which is located in the Arctic Ocean, is by definition at sea level. To compare the two temperatures, the two sites should be reduced to the same altitude, for example, by bringing the Antarctic air mass from 3000 m to sea level. In these dry cold regions where the atmosphere contains little water vapour, the vertical temperature gradient (the lapse rate) is about −10°C/km. Bringing an air mass from 3000 m to 0 m is thus equivalent to raising its temperature by 30°C. Even after this correction, however, the South Pole is still colder than the North Pole, owing to the difference in temperature between oceans and continents and to their geographical distribution.

3. The solar flux received on Earth is collected by a disc of area πR^2, where R is the radius of the Earth. This flux is distributed over the whole of the surface of the Earth, $4\pi R^2$; that is, four times greater than the disc. The solar flux arriving from outside the Earth on a perpendicular surface, namely 1367 W/m², must therefore be divided by four in order to find the average amount received per square metre of the Earth's surface. The average energy flux per square metre of the Earth's surface is thus 342 W/m².

4. Above this height (80 km), diffusion has a stronger effect than turbulence. At such extremely low pressures, the different constituents segregate with altitude as a function of their mass.

PART I FURTHER READING

Fletcher, C. (2013) *Climate Change – What the Science Tells Us.* John Wiley & Sons, Inc., Hoboken, NJ; see Chapter 1.

Ruddiman, W.F. (2008) *Earth's Climate – Past and Future*, 2nd edn. W.H. Freeman, New York; see Part I.

PART II

MORE ON THE ENERGY BALANCE OF THE PLANET

In Part I, we gave a simple description of the processes that control the global mean temperature on the surface of the Earth. In these processes, the atmosphere plays a leading role. *How does it interact with solar and terrestrial radiation? What mechanisms are responsible for the difference in energy flux between the surface of the Earth and the outer envelope of the planet? Which are the components of the atmosphere that are principally involved?* In Part II, we outline a rudimentary approach to the energy balance of the planet that allows us to place these different questions in context. As it is greatly simplified, this approach describes merely the main traits. First, however, the properties of the thermal radiation introduced earlier must be defined, and then those of solar and terrestrial radiation.

Chapter 6
Thermal radiation, solar and terrestrial radiation

6.1 Thermal radiation from a *black body*

Earlier, we saw that when a system or a body is at temperature T, electromagnetic radiation is present in the body, sharing the available energy with the matter. This is thermal radiation, and it is constantly emitted by the body. The amount of energy radiated is determined by the average energy of the body; in other words, by its temperature T. The higher the temperature, the greater is the radiated energy flux.

The laws that govern thermal radiation are introduced below for an idealized case, the *black body*. By definition, the surface of such a body completely absorbs all the radiation that strikes it, of whatever wavelength. The name refers to the fact that when visible light (radiation of wavelength in the range between $0.4\,\mu m$, blue, and $0.7\,\mu m$, red) shines on the body, its surface appears black.

The laws that govern radiation from real bodies are derived from those of a black body (Part II, Note 1). They show that the surfaces of the planets and of the Sun radiate almost like a black body. This is the reason why we may apply the laws of black-body radiation to the Earth and to the Sun. However, since the atmosphere absorbs only at certain wavelengths, its emission is governed by the laws of real bodies.

The radiation of a black body at temperature T is defined by the flux that it emits, $E(\lambda, T)$, which is a function of the wavelength λ. The available energy excites the atoms and

Climate Change: Past, Present and Future, First Edition. Marie-Antoinette Mélières and Chloé Maréchal.
© 2015 John Wiley & Sons, Ltd. Published 2015 by John Wiley & Sons, Ltd.
Companion website: www.wiley.com\go\melieres\climatechange

electrons, causing them to emit electromagnetic radiation at a frequency ν ($\nu = c/\lambda$, where c is the velocity of light). The more energy is available, the faster is the vibration of the radiation and the shorter its wavelength. At zero degrees absolute (0 K), there is no available energy, and we consider that the wavelength is infinite.

6.2 The laws of black-body radiation

Three laws govern the radiation emitted by a black body at temperature T:

- *Planck's law* defines the distribution of the radiated flux $E(\lambda, T)$ as a function of the wavelength λ. This energy flux is expressed with respect to wavelength units (often in W/(m² μm)). When λ tends either to zero or to infinity, $E(\lambda, T)$ tends to zero, and it displays a maximum of radiated energy at a certain wavelength λ_M. Its mathematical formulation is given in Part II, Note 2. As T increases, the energy flux $E(\lambda, T)$ increases and the emission curves form a family that nest inside each other. An example of this family is shown in Fig. 6.1 for T between 213 K (–60°C) and 333 K (+60°C); in other words, roughly the range of temperatures that prevail on the surface of the Earth.

  As the temperature rises, the maximum of emission moves towards shorter wavelengths. A body at 300 K (+27°C) radiates in the infrared in a range that is not detectable by the human eye. However, when the temperature rises, the emission gradually increases and moves into the wavelength range of visible light. The body becomes luminous, emitting red light at first, and then the other visible colours are gradually added. As the temperature increases it successively emits yellow and then white light or, in other words, light that includes the visible range. Metal heated to 1000°C, for instance, is known as 'white hot' because it radiates white light.

- *Wien's law* relates the temperature to the wavelength λ_M at which the emitted energy is at a maximum. This wavelength is inversely proportional to T: the higher the temperature, the shorter is λ_M (Fig. 6.1). Wien's law is as follows:

$$\lambda_M = 2897.8/T \ (\mu m)$$

where T is in Kelvin.

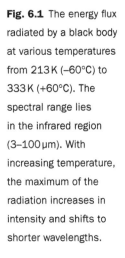

Fig. 6.1 The energy flux radiated by a black body at various temperatures from 213 K (–60°C) to 333 K (+60°C). The spectral range lies in the infrared region (3–100 μm). With increasing temperature, the maximum of the radiation increases in intensity and shifts to shorter wavelengths.

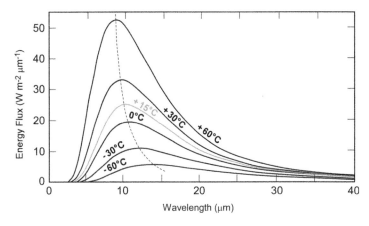

that is, roughly

$$\lambda_M \approx 3000/T \, (\mu m)$$

For the temperature range prevailing on Earth (about 300 K), the maximum of the radiated intensity is situated in a range around $\lambda_M \approx 10 \, \mu m$. By contrast, for the flux radiated from the surface of the Sun, with a temperature of about 6000 K, the maximum intensity occurs at a wavelength 20 times smaller (i.e. 20 times more energetic), namely $\lambda_M \approx 0.5 \, \mu m$. This wavelength is situated in the centre of the spectrum that is visible to the eye, which, as already noted, is between 0.4 and 0.7 μm. Note, however, that the radiated energy is often expressed as a function of the variable $1/\lambda$. With this variable, the maximum of the function $E(1/\lambda, T)$ in $W/(m^2 \, cm^{-1})$ appears at a different value of λ.

- *Stefan's law* relates the total flux radiated by the surface of the black body to the temperature T. This flux contains all the wavelengths emitted and is represented by the area situated under the $E(\lambda, T)$ curve. The total radiated energy flux is therefore expressed in watts per square metre (W/m^2). Stefan's law is as follows:

$$E(T) = \sigma T^4$$

where $\sigma = 5.67 \times 10^{-8} \, (W/m^2)/K^4$.

Let us look at some examples. If a surface behaving like a black body radiates $342 \, W/m^2$, its temperature is 279 K, or +6°C. This is the example discussed in Fig. 4.3. Similarly in the example of Fig. 4.4, the radiation emitted by the surface, $235 \, W/m^2$, implies a temperature of 254 K, or −19°C. Finally, we cite an example that will be of use later: according to Stefan's law, a surface at temperature +15°C (288 K) radiates $390 \, W/m^2$.

6.3 Solar and terrestrial radiation

The surfaces of the Sun and of the Earth, just like those of the planets, emit radiation that adheres closely to the laws of black-body radiation.

Solar radiation

The radiation emitted by the surface of the Sun, the temperature of which is almost 6000 K, is taken to be that of a black body. At this temperature, the thermal radiation is centred in the visible light range, at 0.5 μm. This range of wavelength between 0.2 μm and 3 μm accounts for more than 99% of the radiated energy (Fig. 4.2). In addition, there is a variable, but much weaker, flux of radiation in the X-ray and gamma ray range. The wavelength distribution of the solar radiation received on Earth is similar to that emitted by the Sun, but the intensity received is attenuated by the distance from the Sun to the Earth. The total flux of energy falling on a square metre perpendicular to the Sun's radiation outside the atmosphere is taken to be $1367 \, W/m^2$. If the surface temperature of the Sun, its radius and the Earth–Sun distance are known, this quantity is simple to calculate (Part II, Note 3).

Terrestrial radiation

The different surfaces of the Earth, excluding volcanic zones, lie in a temperature range between about 200 K and 350 K. The annual temperature averaged over the Earth's surface (global mean temperature) is +15°C (288 K). The spectrum of radiation emitted at this temperature, assuming that the surface is a black body, is shown in Fig. 6.1. The radiated flux is 390 W/m². It is situated in the near infrared, between 3 μm and 50–100 μm, and its maximum is centred at about 10 μm.

The range of wavelengths radiated by the surface of the Sun and that of the Earth do not overlap. The solar flux extends from 0.2 μm to 3 μm. This is the short wavelength region, often referred to as 'SW'. The radiation flux emitted by the Earth lies in the longwave (LW) near-infrared region, starting from 3 μm. The absence of overlap simplifies the interpretation of spectra seen by satellites. Observation of the Earth's surface in the SW measures the reflected solar flux. The LW region measures terrestrial radiation emitted by the surface or by the atmosphere.

Chapter 7
The impact of the atmosphere on radiation

To understand how the Sun supplies energy to the atmosphere and to the surface of the planet, and how the flux radiated by the Earth's surface escapes from the planet, we first describe how radiation interacts with matter. Whether it comes from the Sun (solar) or the Earth (terrestrial), radiation can interact with matter in two possible ways: it is either scattered or absorbed.

7.1 Scattering and reflection

First of all, the interaction of incident radiation with a particle can cause the radiation to deviate from its path without change of energy: it is scattered. Radiation is *scattered* when the particles are small or of a size comparable to the wavelength (this is the case for molecules in the air, or certain aerosols). Scattering of radiation of wavelength λ by a particle sends the radiation in all directions. In the atmosphere, the scattering particles are mainly air molecules, aerosols and fine droplets. If the size of the particles is much smaller than the wavelength of the incident radiation, the *Rayleigh scattering* law holds: the probability for radiation of wavelength λ to be scattered varies as λ^{-4}. A light ray of wavelength $\lambda/2$ thus has a 16 times greater probability of being scattered than one of wavelength λ. Air molecules, the size of which is less than a nanometre (i.e. 1000 times smaller than the wavelength of solar radiation), scatter sunlight according to the Rayleigh law: in its passage through the atmosphere, blue light is scattered approximately 16 times more strongly than red light. This is why scattered sunlight

Climate Change: Past, Present and Future, First Edition. Marie-Antoinette Mélières and Chloé Maréchal.
© 2015 John Wiley & Sons, Ltd. Published 2015 by John Wiley & Sons, Ltd.
Companion website: www.wiley.com\go\melieres\climatechange

is predominantly blue, giving the blue colour to the sky. Conversely, when the last rays of the sun strike the sky at sunset, the light rays are predominantly red, since the blue (and green) light has been scattered out more efficiently during the passage through the atmosphere. The clouds in the sky then appear to glow with crimson light. If the particles are larger, and their size is of the order of or greater than solar wavelengths (a micrometre or greater), the scattering law changes and no longer depends on the wavelength of the incident light (*Mie scattering*). In the atmosphere, the sunlight that is scattered by particles of size 1 μm or larger appears whitish: this happens in misty air, in fog and in clouds.

Scattering by molecules in the atmosphere is thus significant for solar radiation, as its wavelength is centred around 0.5 μm. By contrast, it is practically absent (because it is far too weak) for terrestrial radiation, which is centred at about 10 μm. It is only in the unusual case of very large particles, such as ice crystals or desert aerosols, that some scattering of terrestrial radiation occurs.

Radiation is scattered in all directions. In particular, the radiation that is sent back from the planet, and which consequently leaves it, is lost to the planet. In common parlance, the totality of this amount of energy is described as being 'reflected', even though the causative mechanism is scattering.

7.2 Absorption by a gas – the *cut-off approximation*

Radiation can also be absorbed by matter. Here, we briefly recall the mechanism of absorption. Radiation consists of photons. When matter absorbs radiation, the photons enter the atoms or molecules and, in transferring their energy, the photons disappear. This energy is then redistributed among the other molecules and ultimately contributes to the reservoir of 'average energy' that defines the temperature. The absorption of the incident radiation by the molecules in the atmosphere is governed by the Beer–Lambert law.

7.2.1 THE BEER–LAMBERT ABSORPTION LAW

When radiation of wavelength λ passes through a gaseous medium, its absorption depends on two factors:

- The ability of the gas molecule to absorb the radiation; that is, to absorb photons of the energy associated with the wavelength λ, namely hc/λ (where h is Planck's constant and c is the velocity of light). For this, the molecule must possess another energy level that is situated at an energy difference hc/λ above its initial state. This condition is characterized by the absorption coefficient $K(\lambda)$ of the gas at wavelength λ. The curve $K(\lambda)$ is composed of the different absorption lines of the gas. We take the example where $K(\lambda)$ is a single absorption line centred at the wavelength λ_0 (Fig. 7.1).
- The number of molecules encountered in its path. This number is proportional to the quantity ρz, where z is the length of the path and ρ is the density of the gas. In the present case, where the path of the radiation passes through the whole of the atmosphere, ρz is equal to the mass of gas contained in a column of surface area 1 cm^2 in the atmosphere, and is expressed in grams per square centimetre (g/cm^2). Here, and in subsequent chapters, we refer to ρz as the *column mass*. In Part II, Note 4, ρz is evaluated for the example of CO_2.

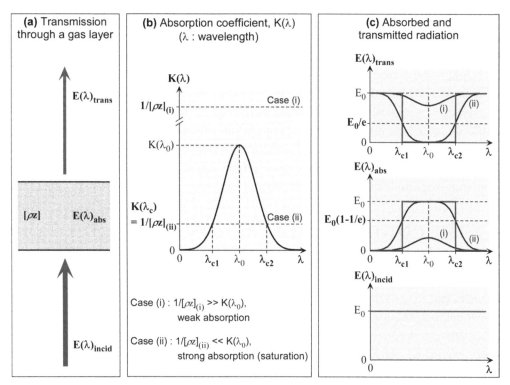

Fig. 7.1 (a) The radiation flux $E(\lambda)$ transmitted through a gas layer, where the column mass ρz defines the mass of absorber in the path length z of the radiation (see text). Part of the incident flux E_{incid} is absorbed (E_{abs}), and part is transmitted (E_{trans}). (b) In this example, the absorption coefficient $K(\lambda)$ of the gas has a single absorption line centred at λ_0. Case (i): $\rho z \ll 1/K(\lambda_0)$ (weak absorption). Case (ii): $\rho z \gg 1/K(\lambda_0)$ (strong absorption) causes total absorption around λ_0. In the *cut-off* approximation, λ_{c1} and λ_{c2} are the cut-off wavelengths, where $(\lambda_c) = 1/\rho z$. (c) The incident radiation is assumed to be independent of wavelength. $E(\lambda)_{abs}$ and $E(\lambda)_{trans}$ are shown for Cases (i) and (ii). In the *cut-off approximation*, absorption is taken to be total in the wavelength region where $K(\lambda) \geq K(\lambda_c)$, and zero outside. Absorption is therefore total between λ_{c1} and λ_{c2} and zero elsewhere. The absorption curve (black) is approximated by the red curve, of closely similar area.

The energy transmitted at the wavelength λ, $E(\lambda)_{trans}$, is then expressed as a function of the incident flux, $E(\lambda)_{incid}$, by the Beer–Lambert law:

$$E(\lambda)_{trans} = E(\lambda)_{incid} \exp[-K(\lambda)\rho z] \qquad (II.1)$$

The flux of energy absorbed by the gas is then

$$E(\lambda)_{abs} = E(\lambda)_{incid} - E(\lambda)_{trans} = E(\lambda)_{incid}\{1 - \exp[-K(\lambda)\rho z]\}$$

The higher the value of the argument $K(\lambda)\rho z$, the weaker is the flux of transmitted energy and the greater is the absorbed flux. For a given value of ρz, which is determined by the composition of the atmosphere, the absorption thus depends on the wavelength λ of the incident radiation, and is maximum for the value λ_0 at which $K(\lambda)$ is a maximum.

Figure 7.1 illustrates the absorption and transmission in the simple case where the incident flux is independent of the wavelength, for two values of *column mass* ρz. In Case (i), ρz is much smaller than $1/K(\lambda_0)$; that is, absorption at λ_0 is not total. In Case (ii), ρz is much larger than $1/K(\lambda_0)$, and accordingly absorption at λ_0 is total.

The atmosphere contains many different gas molecules. Some of these are able to absorb part of the solar and terrestrial radiation that passes through it. To get an idea of the importance of the absorption of these different gases, we concentrate only on the case where the absorption is total. We use the *cut-off* or *rectangular lineshape approximation*. This provides a quick estimate of the wavelength regions in which the gases totally absorb the incident radiation, and gives the order of magnitude of the energy absorbed.

7.2.2 THE CUT-OFF APPROXIMATION

We return to Case (ii) discussed above (Fig. 7.1). The regions of total absorption are those in which $K(\lambda_0)$ is much greater than $1/\rho z$. We call λ_c the *cut-off* wavelength, at which the absorption coefficient $K(\lambda) = 1/\rho z$. At this wavelength, according to eqn. II.1, the transmitted energy flux is equal to $E(\lambda)_{incid}/e$, where $e = 2.72$.

Two values of λ correspond to this situation, λ_{c1} and λ_{c2}. Absorption is then considered to be total between λ_{c1} and λ_{c2}, and zero elsewhere. This means that the absorption curve (in black in the figure) is approximated by the red curve, of a closely similar area. This area represents the total flux (at all incident wavelengths) of energy absorbed in the passage of the radiation through the gas, E_{abs}. In the simple case considered here, where the incident flux E_{incid} is constant for all wavelengths, the area is equal to $E_{incid} \times \Delta\lambda$, where $\Delta\lambda$ is the difference $\lambda_{c2} - \lambda_{c1}$. Thus

$$E_{abs} \sim E_{incid} \times (\lambda_{c2} - \lambda_{c1})$$

To evaluate the wavelength regions in which an atmospheric component totally absorbs the incident radiation, we only need to know:

- the distribution of the incident radiation $E(\lambda)_{incid}$ (in W/(m^2 μm));
- the absorption coefficient $K(\lambda)$ of the molecule (in cm^2/g); and
- the column mass ρz of the molecule (in g/cm^2).

With this approximation, we may now evaluate the wavelength ranges in which solar and terrestrial radiation are totally absorbed.

7.3 Absorption of solar and terrestrial radiation by atmospheric gases

7.3.1 ESTIMATES

In which ranges of wavelength is the solar radiation totally absorbed in its trajectory from the upper atmosphere to the Earth's surface? We also ask the same question for the terrestrial radiation from the Earth's surface to the top of the atmosphere.

TABLE 7.1 THE COLUMN MASS ρz AND ABSORPTION COEFFICIENT $K(\lambda_c)$ AT CUT-OFF FOR THE PRINCIPAL GASES DISCUSSED IN THIS CHAPTER, WHERE ρz IS THE MASS OF THE STATED MOLECULE IN A 1 CM^2 COLUMN OF ATMOSPHERE ABOVE THE EARTH'S SURFACE.

Molecule	ρz (g/cm^2)	$K(\lambda_c) = 1/\rho z$ (cm^2/g)
O_2	231	4.3×10^{-3}
N_2	755	1.3×10^{-3}
H_2O (vapour)	~2.5	~0.4
CO_2	~0.6	~1.7
O_3	~0.0006	~1.7×10^3

The gases that make up 99% of the mass of the atmosphere are listed in Table 7.1, as well as the most important minor compounds that influence radiation transfer. The table also lists the corresponding values of the column density ρz, together with the quantity $1/\rho z$ that determines $K(\lambda_c)$, as well as the cut-off wavelength λ_c defining the total absorption threshold of the compound in the cut-off approximation.

The absorption coefficient $K(\lambda)$ of each of these gas molecules at atmospheric pressure is given in Fig. 7.2 for λ between 0.2 μm and 100 μm. The range of the incident solar radiation, from 0.2 μm to 3 μm, is highlighted in yellow, while that emitted by the surface of the Earth, from 3 μm to 100 μm, is highlighted in pink. The latter is equivalent to that of a black body at

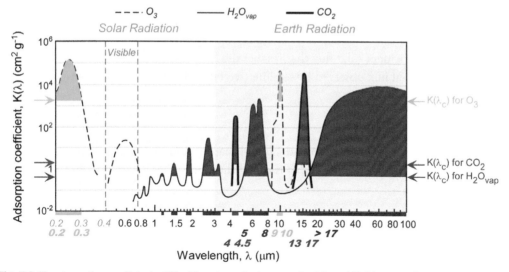

Fig. 7.2 The absorption coefficients $K(\lambda)$ of the atmospheric gases O_3, CO_2 and H_2O between 0.2 and 100 μm. The solar radiation range (0.2–3.0 μm) is highlighted in yellow and terrestrial radiation (3–100 μm) in pink. Note that O_2 has a narrow absorption line around 0.7 μm, (not shown here), and that the absorption of the two majority gases, N_2 and O_2, lies outside this range at shorter wavelengths (<0.2 μm). Of the minority gases, only H_2O, CO_2 and O_3 can absorb totally at certain wavelengths. The values of $1/\rho z$ giving a *cut-off wavelength* λ_c (see text) are listed for each of these gases in Table 7.1. The corresponding ranges of total absorption in the *cut-off* approximation are indicated (O_3, green; H_2O, blue; CO_2, red). Note that the wavelength scale is linear from 0.2 to 0.4 μm and logarithmic thereafter.
Source: Absorption coefficients modified from Queney (1974).

15°C (Fig. 6.1), and spans from 3 μm to 50–100 μm. In the cut-off approximation, we determine the pairs of values of λ_c that delimit the zones of total absorption for each of the gases; that is, the values of λ for which $K(\lambda)$ is greater than $1/\rho z$. The different values of $K(\lambda_c)$ for three of the compounds in Table 7.1 are plotted in Fig. 7.2, together with the corresponding zones of total absorption.

The two majority components, N_2 and O_2, absorb at shorter wavelengths ($< 0.2\,\mu m$) and therefore are not shown here, with the exception of a very sharp O_2 absorption line at 0.7 μm (omitted here) in the solar radiation spectrum. Among the minority gases, only H_2O, CO_2 and O_3 can totally absorb certain wavelengths.

7.3.2 ABSORPTION OF SOLAR AND TERRESTRIAL RADIATION

In the wavelength range of solar radiation, ozone absorbs totally between 0.2 μm and 0.3 μm (and partially between 0.5 μm and 0.7 μm). As for water vapour, it appears in a few absorption bands in the near infrared (IR) (with two total absorption bands centred at 1.4 μm and 1.9 μm, respectively), in a region where the solar radiation is already less intense.

In the wavelength range of terrestrial radiation, H_2O and CO_2 take the lion's share. Water vapour absorbs totally in two regions, from 5 μm to 8 μm and beyond 17 μm. CO_2 absorbs totally between 13 μm and 17 μm. It also possesses a narrow total absorption line at 4.5 μm, but in this region, where terrestrial radiation is very weak, its impact on absorption is small. Ozone absorbs totally in a narrow line centred at 9 μm (absorption exists on either side of this line, but it is only partial).

The different regions where absorption by the atmosphere is total are indicated in the spectra of the incident solar and terrestrial radiation, respectively (Fig. 7.3). These regions are identifiable in the incident solar flux that reaches the Earth's surface, shown in Fig. 4.2, and in the emitted terrestrial flux observed by satellite (see Fig. 8.4).

The atmospheric window

The wavelength range between 8 μm and 12 μm is hardly absorbed by the atmosphere, except by the small absorption line of O_3. This is called the 'atmospheric window'. In this window,

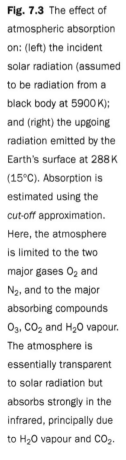

Fig. 7.3 The effect of atmospheric absorption on: (left) the incident solar radiation (assumed to be radiation from a black body at 5900 K); and (right) the upgoing radiation emitted by the Earth's surface at 288 K (15°C). Absorption is estimated using the *cut-off* approximation. Here, the atmosphere is limited to the two major gases O_2 and N_2, and to the major absorbing compounds O_3, CO_2 and H_2O vapour. The atmosphere is essentially transparent to solar radiation but absorbs strongly in the infrared, principally due to H_2O vapour and CO_2.

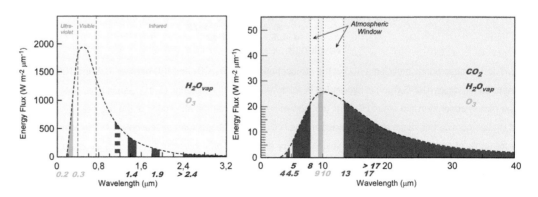

satellites outside the atmosphere can observe the radiation emitted by the surface of the Earth. Since this radiation can be approximated by that of a black body, satellites have direct access to the surface temperature. Conversely, this is the spectral window through which astronomers can observe the universe without hindrance from the atmosphere (in good weather of course, i.e. in the absence of clouds).

The cut-off approximation can be used to assess roughly what fraction of the radiation emitted from the Earth's surface is absorbed by the atmosphere. In Box 7.1, the absorption by CO_2 is evaluated for a situation in which the atmosphere is reduced to a layer of CO_2 of column mass $0.6\,g/cm^2$. That value corresponds to the present level of CO_2 in the atmosphere, about 400 ppm (395 ppm in 2013) (Part II, Note 4).

BOX 7.1: THE ABSORPTION OF TERRESTRIAL RADIATION BY ATMOSPHERIC CO_2

To evaluate the proportion of the flux radiated by the Earth's surface (terrestrial radiation) that is absorbed by the layer of CO_2 in its passage through the atmosphere, we adopt the following simplifying assumptions:

- the average temperature at the Earth's surface is 288 K (15°C);
- the surface radiates like a black body;
- the atmosphere is assumed to be a single layer of CO_2, of column mass $\rho z = 0.6\,g/cm^2$; and
- absorption is calculated using the *cut-off approximation*.

The absorbed fraction of the incident radiation is represented in Fig. 7.3 by the red area between $13\,\mu m$ and $17\,\mu m$. It is given approximately by

$$E_{abs} \sim E(\lambda) \times \Delta\lambda$$

where $E(\lambda)$ is the energy radiated at $15\,\mu m$ (about $20\,W/m^2\,\mu m$) and $\Delta\lambda = 4\,\mu m$. The energy flux absorbed by the layer of CO_2 is therefore roughly $80\,W/m^2$.

The flux radiated by the Earth's surface is defined by Stefan's law (see Chapters 6 and 8):

$$E_{emit} = \sigma T^4 = 390\ W/m^2$$

With the above assumptions, the fraction of the Earth's radiation that is absorbed by the layer of CO_2 is approximately $(80\,W/m^2)/(390\,W/m^2)$, or about 20%.

How does this fraction increase when the CO_2 layer becomes thicker?
When the column mass ρz increases, the cut-off coefficient $K(\lambda_c) = 1/\rho z$ decreases. The range from λ_{c1} to λ_{c2} over which the cut-off applies therefore broadens and the amount of energy absorbed increases accordingly. But the increase operates only at the edges of the red area, since no extra absorption can occur between λ_{c1} and λ_{c2}, where absorption is already total. Thus, if ρz doubles (doubling the atmospheric concentration of CO_2), the absorption does not double, but increases instead only by roughly 10%.

Mapping the atmospheric CO_2 concentration
Satellite measurements of IR absorption are now commonly used to determine the CO_2 concentration in the atmosphere, as illustrated in Fig. B7.1?

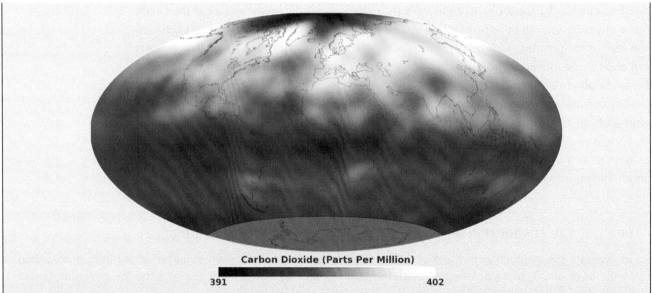

Fig. B7.1 The global carbon dioxide distribution in the atmosphere. This image was generated from data collected by the Atmospheric Infrared Sounder (AIRS) in May 2013, when carbon dioxide levels reached their highest point in at least 800,000 years. The highest concentrations, shown in yellow, are in the Northern Hemisphere. Concentrations are lower in the Southern Hemisphere. In May, the NH growing season was just beginning, so plants were removing little carbon dioxide from the atmosphere.
Source: NASA Earth Observatory.

7.4 Direct transfer by the atmosphere

Figure 7.4 illustrates the effect of scattering and absorption by atmospheric gases on the incident solar and terrestrial radiation. In addition to the gases already mentioned, the minor role played by CH_4 and N_2O, which absorb only very weakly in the terrestrial radiation spectrum, is also included. Absorption by clouds, however, is not included. The figure shows the downgoing solar radiation and the upgoing terrestrial radiation after their journey through the atmosphere. The solar radiation that strikes the surface of the Earth has already been partly absorbed by the O_3 (in the ultraviolet) and by the water vapour (in the near IR). It has been partly 'reflected' after being Rayleigh scattered by the molecules of the air, which attenuate the transmission of this radiation between about $0.3\,\mu m$ and about $0.5\,\mu m$.

The directly transmitted terrestrial radiation is limited mainly to the region of the spectrum situated between $8\,\mu m$ and $12\,\mu m$, after absorption by H_2O and CO_2. *But beware! This is not all the terrestrial radiation that leaves the atmosphere.* The total terrestrial radiation that is observed by satellites includes, in addition, the radiation emitted by the atmosphere.

Although the above approach is highly simplified (homogeneous atmosphere, cut-off approximation), it captures the essential effects of the gas molecules in the atmosphere on solar and terrestrial radiation. The model shows why solar radiation is only weakly perturbed in its journey through the atmosphere, while the terrestrial radiation is strongly absorbed. This is the basis of the greenhouse effect, which substantially increases the surface temperature of the planet. Moreover, with this simplified approach we can estimate the solar and terrestrial

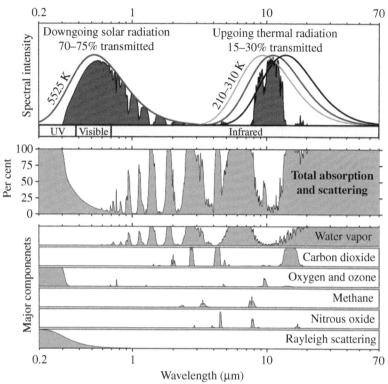

Fig. 7.4 The absorption band pattern in the atmosphere and its effect on downgoing solar radiation and on upgoing terrestrial radiation. Top: part of the downgoing solar radiation and upgoing terrestrial radiation directly transmitted through the atmosphere. The continuous lines are the incident solar radiation (red) and the terrestrial radiation (blue) emitted by the surface at three different surface temperatures (from 210 K to 310 K). The black lines are the incident radiation transmitted after scattering and absorption in the atmosphere. Note that the right-hand curve is not the total flux radiated into space by the atmosphere: to this must be added the infrared emission by the atmosphere. Middle: the percentage of flux absorbed and scattered by the atmosphere at different wavelengths. Bottom: the contribution of the major components in the atmosphere. Scattering occurs mainly for solar radiation at short wavelengths. In the terrestrial infrared, absorption is mainly by H_2O vapour and CO_2. The data used for these figures are based primarily on the work of Gorley et al. (1994), Rothman et al. (2004) and Peixoto and Oort (1992). **Source:** By courtesy of R. Rhode, Rhode Global Warm Art http://www.globalwarmingart.com/wiki/File:Atmospheric_Transmission_png

absorption due to the different constituent gases in the atmosphere. Below, we give a brief review of the characteristics of the principal gases involved.

7.5 Major atmospheric constituents involved in radiative transfer

We have just seen that oxygen and nitrogen molecules, which make up more that 98% of the mass of the atmosphere, have the important property of being practically transparent both in the wavelength range of the Sun's radiation and in that of the Earth's IR radiation. Sunlight, which supplies the Earth with energy, and terrestrial radiation, which allows the Earth to cool, both pass unimpeded though these gases. This property is related to the fact that such gases are diatomic molecules (containing only two atoms). Owing to their structure, they absorb in the shorter-wavelength range; that is, at higher energy. The third component, argon, is a monatomic gas that absorbs neither solar nor terrestrial radiation. These three gases constitute 99.9% of the mass of the atmosphere.

Numerous other compounds are present, but, in much smaller amounts. The principal species are molecules with more than two atoms, generally three. In spite of their low concentrations, as we have just seen, these interact strongly with the Earth's radiation by absorbing it, sometimes even absorbing solar radiation in the near IR. They are as follows.

Water (H_2O)

Water is found in the atmosphere mainly in the form of a gas, water vapour. Its distribution over the Earth is not uniform, because the maximum amount of water vapour that the air can hold before it condenses into droplets is determined by the air temperature. For example, at an atmospheric pressure of 1013.24 mbar (1 atm) at +25°C, air can contain as much as 20 g of vapour per kilogram of dry air. If additional water vapour is injected into this mass of air, it condenses. At 0°C, the maximum amount falls to 3.8 g per kilogram of dry air. Given that the temperature varies strongly between the Equator and the poles, the water vapour content also varies greatly, decreasing towards the poles. For the same reason, the content also varies with altitude, since, as already noted, air cools at a rate of 6.5°C/km. The result is that the water vapour content in the atmosphere ranges between 0 and 4% by volume.

Since the availability of liquid water varies widely among the different areas of the Earth, from oceans to deserts, air masses at the same temperature in different regions do not carry the same load of water vapour. It is nonetheless possible to average the total amount of water vapour in the atmosphere and to reduce it to the mass in a column of air 1 cm^2 in cross-section. The *column mass* of this water vapour amounts to about 2.5 g/cm. If all the water vapour were to condense, it would be enough to cover the whole of the Earth with a sheet of liquid water of thickness 2.5 cm. We recall that this component absorbs only a small part of the solar radiation, but the major part of the terrestrial radiation. Between 5 µm and 8 µm, and beyond 17 µm, absorption by water vapour is total.

When water condenses as the air masses cool, it forms clouds. Here too, cloud cover varies over the surface of the Earth, and as a function of time. Its distribution is related to the circulation of the atmosphere (Fig. 5.2). Its average value is about 60%. The total amount of water present in the atmosphere in liquid (droplets in the clouds) or solid form is far smaller than that of the water vapour, but its role is no less crucial to the transfer of radiation in the atmosphere. Clouds absorb all the terrestrial radiation, but they also interact with the solar radiation by scattering it, thereby contributing to the albedo of the planet. They also absorb some of it.

Carbon dioxide (CO_2)

The concentration of CO_2, when integrated over the height of the atmosphere, is even four times smaller than that of water vapour. Its column mass is 0.6 g/cm^2. Its distribution varies with the seasons, but the yearly average of this almost inert gas is evenly spread over the surface of the planet. Its concentration increases regularly, and at present (2014) stands at around 400 ppm (Part II, Note 5). As the mixing time of the atmosphere between the two hemispheres is of the order of a year, any increase in this gas in one hemisphere is carried across to the other hemisphere after 1 year. Absorption of the terrestrial radiation by CO_2 is strong, and between 13 μm and 17 μm it is total.

Methane (CH_4) and nitrous oxide (N_2O)

The concentrations of these molecules are far lower. The present values are 1.8 ppm and 0.3 ppm, respectively. Throughout the year, these two gases are well mixed in the atmosphere. Both absorb weakly in the terrestrial IR spectrum (Fig. 7.4) through unsaturated absorption lines. These gases therefore contribute only moderately to the greenhouse effect.

Ozone (O_3)

Ozone is far less concentrated, with an average column mass of $0.0006 \, g/cm^2$. This gas, which is chemically very reactive, is not uniformly distributed either in altitude or over the surface of the Earth. It is concentrated in the upper atmosphere (the stratosphere), in a layer that stretches on average between 10 km and 40 km in height, in which the ozone concentration is greatest between 20 km and 35 km. In the polar regions, where the concentration is on average twice that at the Equator, the altitude of the layer falls to 10–20 km. If this gas layer were brought to ground level – that is, atmospheric pressure – its average thickness would be about 3 mm. O_3 is also present in the lower atmosphere at ground level, but this tropospheric layer contains only about 10% of the total ozone. Here again, the distribution of tropospheric ozone at the surface of the Earth is not uniform, with the Northern Hemisphere being richer than the Southern Hemisphere. Between 0.2 μm and 0.3 μm, and also between 9 μm and 10 μm, absorption of both solar and terrestrial radiation by O_3 is total.

Aerosols

Finally, aerosols must be taken into consideration. This category includes solid particles (dust) raised from the soil by wind, particles of salt from the waves in the sea and particles produced by condensation of various gases released into the atmosphere, as well as other components such as pollen, spores and so on. Particularly notable are aerosols produced by condensation of gases such as oxides of sulphur that give rise to sulphated aerosols. These gases (and the resulting aerosols) are emitted naturally on a large scale in certain volcanic eruptions. They are also generated continuously by human agency through the burning of sulphur-rich fossil fuels. Other aerosols are also involved, such as *black carbon* aerosols from combustion of organic matter (urban pollution, fires etc.), or from photochemical reactions with volatile organic compounds in the atmosphere. Their interaction with radiation from the Sun and from the Earth is complex, and depends on the type of aerosol.

All the above-mentioned compounds are present naturally in the atmosphere, but the concentration of many of them can be strongly modified by human activity. Since the beginning of the 20th century, new synthetic compounds that interact with the Earth's radiation and reinforce the greenhouse effect, such as chlorofluorocarbons (CFCs), have been injected into the atmosphere.

Chapter 8
Radiative transfer through the atmosphere

8.1 Three radiative mechanisms that heat or cool the Earth's surface

In this chapter, we take into account neither the transport of energy from the surface to the atmosphere, via the water cycle, nor the transfer of energy due to heating of air masses in contact with the surface of the Earth. We concentrate only on radiation heat transfer by solar and terrestrial radiation within the system of the atmosphere and the Earth's surface. In this transfer, reflection and absorption contribute in three ways: (i) reflection of solar radiation, (ii) its absorption and (iii) absorption of terrestrial radiation. Each of these produces either warming or cooling of the Earth's surface. The parameters used in the following estimates are based on the energy budget presented in the IPCC 2007 report.

8.1.1 REFLECTION OF SOLAR RADIATION

The case considered below is similar to Case 2 that has already been discussed in Section 4.2.2. There is no atmospheric absorption and only part of the incident solar energy, E_s, is reflected by the atmosphere and the Earth's surface. Figure 8.1 illustrates the effect of reflection on the energy balance. The reflected solar radiation is AE_s, where A is the *albedo; that is, the solar radiation reflection coefficient of the planet.* (Note that this coefficient refers to the energy of all wavelengths of the solar spectrum; that is, from $0.2\,\mu m$ (UV) to $3\,\mu m$ (near IR). The

Climate Change: Past, Present and Future, First Edition. Marie-Antoinette Mélières and Chloé Maréchal.
© 2015 John Wiley & Sons, Ltd. Published 2015 by John Wiley & Sons, Ltd.
Companion website: www.wiley.com\go\melieres\climatechange

definition is by no means restricted to the visible portion of the solar spectrum.) The remainder, $(1-A)E_s$, is totally absorbed by the Earth's surface, and heats it. The mean temperature of the surface, T, comes to equilibrium in such a way that the radiation flux emitted by the surface, E_t, is equal to the absorbed solar flux $(1-A)E_s$. Thus

$$E_t = (1-A) \times E_s$$

The flux E_t is determined by the temperature of the surface, which radiates like a black body, and is equal to σT^4. (For simplicity, we adopt the approximation that the mean of the fourth power of the temperature is equal to T^4.) The equilibrium temperature is then defined by the relationship

$$\sigma T^4 = (1-A) \times E_s$$

In the absence of reflection (i.e. $A = 0$), the equilibrium temperature would be T_0 and would satisfy the equation $\sigma T_0^4 = E_s$. When reflection is present, the equilibrium temperature T of the surface can be expressed in terms of T_0 as follows:

$$T = T_0 \times (1-A)^{1/4} \tag{II.2}$$

T is smaller than T_0. As reflection reduces the amount of absorbed solar energy, it causes the surface to cool.

How much is this cooling?

On Earth, roughly one third of the incident solar energy is reflected back into space without heating the planet. According to IPCC 2007, the solar flux E_s arriving at the Earth is $342\,\text{W/m}^2$. Without solar reflection ($A = 0$), the temperature at the Earth's surface would be $T_0 = 279\,\text{K}$ (eqn. II.2). However, as the albedo is $A = 0.31$, only $235\,\text{W/m}^2$ is absorbed at the Earth's surface. The energy available at the surface is therefore no longer E_s, but only $0.69E_s$. The new equilibrium temperature T is $254\,\text{K}$, which means that reflection of solar radiation cools the surface of the Earth by $25°\text{C}$.

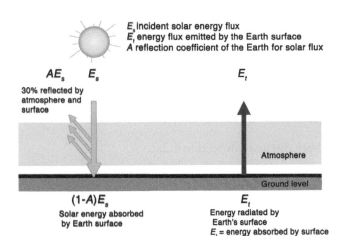

E_s Incident solar energy flux
E_t energy flux emitted by the Earth surface
A reflection coefficient of the Earth for solar flux

AE_s E_s E_t

30% reflected by atmosphere and surface

Atmosphere

Ground level

$(1-A)E_s$ E_t
Solar energy absorbed   Energy radiated by
by Earth surface   Earth's surface
                                                   E_t = energy absorbed by surface

Fig. 8.1 Part of the incident solar flux, AE_s, is reflected by the atmosphere. In this model with albedo $A = 30\%$, the flux heating the surface is not E_s, as it would be without reflection of the solar radiation, but $0.7E_s$. Reflection cools the Earth's surface.

8.1.2 ABSORPTION OF SOLAR RADIATION IN THE ATMOSPHERE: THE *ANTI-GREENHOUSE EFFECT*

We now refer to Case 3 of Section 4.2.3, which involves the atmosphere. Nevertheless, we simplify this case by considering only the effect of absorption of the solar radiation (i.e. $A = 0$). The atmosphere then absorbs part of the solar radiation, but allows the terrestrial radiation emitted by the Earth's surface to pass through.

Part of the incident solar radiation E_s is absorbed by the atmosphere; this is $a_s E_s$, where a_s is the absorption coefficient of the solar radiation in the atmosphere. The remainder $(1 - a_s)E_s$ is totally absorbed by the Earth's surface, and heats it. *What becomes of the energy absorbed by the atmosphere? What is its effect on the equilibrium temperature of the Earth's surface? Is it warmed or cooled?*

It is useful to think in terms of a highly simplified model in which the atmosphere is a layer surrounding the Earth that is sufficiently thin for its two sides to be at the same temperature. The thermal radiation emitted, fixed by its temperature, will then be the same on each side. This model belongs to the *single-layer model* category, and the mechanism is illustrated in Fig. 8.2. The layer is heated only by the absorbed solar energy. At its equilibrium temperature, it constantly re-emits radiation in the

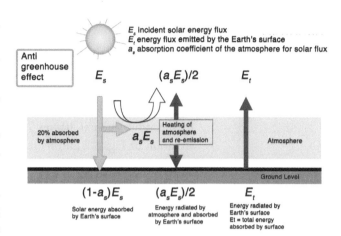

Fig. 8.2 Part of the incident solar flux, $a_s E_s$, is absorbed by the atmosphere. In this model, the surface absorbs $(1 - 0.5 a_s E_s)$ (see text). With $a_s = 20\%$, the flux heating the surface is not E_s, as it would be without absorption of the solar flux, but $0.9 E_s$. Absorption of solar flux by the atmosphere cools the Earth's surface.

infrared range (centred at about $10\,\mu m$); that is, the wavelength range for temperatures of a few hundred degrees K. This energy is re-emitted partly into space and partly towards the Earth's surface. In this simple scheme where the temperature in the atmosphere is uniform, the atmospheric layer re-emits it in equal parts, one half ($a_s E_s/2$) going into space and the other half ($a_s E_s/2$) to the Earth's surface, where it is totally absorbed. The Earth's surface thus receives and absorbs, on the one hand, the direct solar flux, $(1 - a_s)E_s$, transmitted through the atmosphere and, on the other, the infrared flux, $a_s E_s/2$, re-emitted by the atmosphere. This makes a total of $(1 - a_s/2)E_s$. The temperature T of the surface comes to equilibrium when the radiation flux emitted by the surface, E_t, is equal to the absorbed solar flux $(1 - a_s/2)E_s$; that is, when

$$E_t = (1 - a_s/2)E_s$$

The flux E_t is determined by the temperature of the surface. The latter radiates as a black body, emitting a flux equal to σT^4. The equilibrium temperature is thus governed by the relation

$$\sigma T^4 = (1 - a_s/2)E_s$$

Without absorption of solar radiation by the atmosphere (i.e. $a_s = 0$), this equilibrium temperature would be T_0, and would satisfy the equation $\sigma T_0^4 = E_s$. When the atmosphere absorbs solar radiation, the equilibrium temperature of the surface, expressed as a function of T_0, becomes

$$T = T_0 \times (1 - a_s/2)^{1/4} \tag{II.3}$$

T is smaller than T_0. As absorption of solar radiation by the atmosphere reduces the solar energy absorbed by the surface, it causes the surface to cool.

How much is this cooling?

We must now estimate a_s and also define T_0 in the present context. First, to calculate the effect of absorption of solar radiation, we must take into account the albedo. We already saw that owing to albedo, the solar flux absorbed by the Earth is $235\,W/m^2$ and the temperature of the Earth's surface is $254\,K$. According to IPCC 2007, of these $235\,W/m^2$, $67\,W/m^2$ are absorbed by the atmosphere (Chapter 9) and $168\,W/m^2$ by the Earth's surface. We are therefore in the situation of Fig. 8.2 and its related results; that is, an incident solar flux $E_s = 235\,W/m^2$, and an initial temperature $T_0 = 254\,K$, when absorption of the solar flux by the planet is total, and the absorption coefficient of the atmosphere is equal to 0.29 (67/235). The energy available at the surface is now no longer E_s, but $(1 - a_s/2)E_s = 0.86E_s$. The new equilibrium temperature T (eqn. II.3) becomes $244\,K$. This means that the surface of the Earth cools by $10°C$.

We call this effect the *anti-greenhouse effect*, because the mechanism is analogous to that of the *greenhouse effect*. It involves *absorption* of radiation by the atmosphere, and *at equilibrium, re-emission* of this flux by the atmosphere *in the infrared wavelength region*. In the *anti-greenhouse effect*, the absorption of solar radiation causes the surface to cool; as we shall see in the greenhouse effect, absorption of the thermal radiation from the Earth's surface makes it warmer.

8.1.3 ABSORPTION OF TERRESTRIAL RADIATION BY THE ATMOSPHERE: THE GREENHOUSE EFFECT

We now refer again to Case 3 of Section 4.2.3, where the atmosphere is involved, but we consider only the effect of absorption of radiation emitted by the Earth's surface. Again, we assume thermal equilibrium with the same simplified *single-layer model* of the atmosphere as before. The atmosphere allows the incident solar radiation E_s to pass through without hindrance, which is thus absorbed by the surface. It absorbs part of the terrestrial radiation, E_t, emitted by the surface; that is, the quantity $a_t E_t$, where a_t is the absorption coefficient of the terrestrial radiation in the atmosphere. The remainder, $(1 - a_t)E_t$, is transmitted directly into space and leaves the Earth. *How is the absorbed energy shared between the atmosphere and the surface? Will the Earth's surface be warmed or cooled?*

The mechanism is illustrated in Fig. 8.3. The absorbed energy flux, $a_t E_t$, constantly warms the atmospheric layer. At equilibrium, its temperature is stable and the layer continuously re-emits an energy flux equal to that which is absorbed. As before, this energy is re-emitted in the infrared range. Half of it, $a_t/2 \times E_t$, is radiated into space, while the other half, $a_t/2 \times E_t$, is radiated towards the Earth's surface. It follows that the Earth's surface receives and absorbs the solar flux that is transmitted through the atmosphere, E_s, and also the

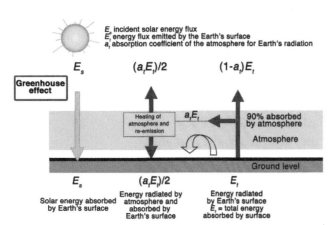

Fig. 8.3 The incident solar flux E_s is fully transmitted by the atmosphere and absorbed by the surface. At equilibrium, the Earth's surface radiates E_t. Part of this flux, $a_t E_t$, is absorbed by the atmosphere. In this model, the surface absorbs $E_s/(1 - 0.5a_t)$ (see text). Since a_t is ~90%, the flux that heats the surface is not E_s, as it would be without absorption of terrestrial radiation by the atmosphere, but about $1.8E_s$, which is much greater. Absorption of the Earth's radiation by the atmosphere warms the Earth's surface.

infrared flux re-emitted by the atmosphere, $a_t/2 \times E_t$. This makes a total of $E_s + a_t E_t/2$. The temperature T of the surface comes to equilibrium when the flux radiated by the surface, E_t, is equal to the absorbed flux:

$$E_t = E_s + a_t/2 \times E_t$$

and hence

$$E_t = E_s/(1 - a_t/2)$$

The flux E_t is determined by the temperature of the surface, which radiates like a black body, and is equal to σT^4. The equilibrium temperature is thus governed by the relation

$$\sigma T^4 = E_s/(1 - a_t/2)$$

Without the absorption of terrestrial radiation by the atmosphere (i.e. $a_t = 0$), this equilibrium temperature would be T_0 and would satisfy the equation $\sigma T_0^4 = E_s$. When absorption of the Earth's radiation by the atmosphere is present, T can be expressed as a function of T_0 as follows:

$$T = T_0 / (1 - a_t/2)^{1/4} \tag{II.4}$$

Thus, T is greater than T_0. The surface is warmer because it receives not only the incident solar flux but also part of the terrestrial radiation, $a_t E_t/2$. The net result of this mechanism is to heat the surface of the Earth.

This effect is called the *greenhouse effect* by analogy with garden greenhouses, where the transparent windows allow solar radiation to pass through, while absorbing part of the infrared radiation emitted by the ground. It should, however, be noted that in the latter case the heating of a greenhouse is largely due to the fact that the warm air cannot escape. For this reason, the greenhouse analogy is imperfect. Another, more intuitive, analogy can be drawn, however – that of filling a bathtub. This 'bathtub analogy' is described below.

How much is this warming?

In the present context, we must estimate a_t and define T_0. On the basis of the previous argument, we consider the case where the atmosphere absorbs $235\,\text{W/m}^2$ and the temperature is $T_0 = 254\,\text{K}$. IPCC 2007 reports that the radiation flux emitted by the Earth's surface is $390\,\text{W/m}^2$, of which $350\,\text{W/m}^2$ are absorbed in its passage through the atmosphere. This yields for the absorption coefficient $a_t = 0.9$. Such a large absorption coefficient produces a very large heating effect, since the heating flux at the surface is no longer E_s but $E_s/(1 - a_t/2)$, or roughly $1.8 E_s$. From eqn. II.4, the new equilibrium temperature of the Earth's surface due to the greenhouse effect is $T = 295\,\text{K}$, which corresponds to a warming of $41\,°\text{C}$.

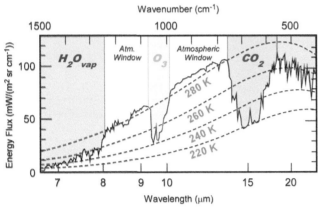

Fig. 8.4 The satellite observation of thermal emission from the Earth's atmosphere above the Mediterranean Sea. The regions where H₂O (vapour), CO₂ and O₃ contribute to radiation are shown. The contributions of CH₄ and N₂O, around 8 μm, are smaller and not visible here.

Source: Hanel et al. (1971). Reproduced with permission of The Optical Society.

Terrestrial radiation emitted by the planet

Satellite observations give a direct reading of the radiation emitted by the Earth's surface and of its absorption by the atmosphere. These illustrate the dominant role of greenhouse gases. Figure 8.4 shows a satellite observation from a region of the Mediterranean Sea, displaying the absorption by various atmospheric gases (Hanel et al. 1971). As the radiation from the surface is approximately that of a black body and it is totally transmitted in the atmospheric window between ~8 μm and ~12 μm, the spectrum in this region provides an estimate of the surface temperature: in this case, it is about 15°C. Absorption is total for H_2O, O_3 and CO_2 in the wavelength ranges indicated by the coloured areas in Fig. 7.3. The part of the spectrum seen by satellites in each of these ranges corresponds to the energy flux re-emitted into space by the atmosphere, more precisely by the external layers of the different gases. This energy flux depends on the temperature of the outer layer. In the satellite measurements, the radiating layer for CO_2 is at 220 K (–53°C), while that of H_2O vapour, located at a lower altitude, is much warmer, about 260 K (–13°C).

8.1.4 COMMENTS

The above estimates were made under highly simplified assumptions in which the atmosphere is reduced to a homogeneous layer of uniform temperature. In this model, half of the radiation from the layer is emitted downwards and half is emitted upwards. In fact, however, the atmosphere cannot be regarded as a thin layer of uniform temperature, because its density and its temperature both decrease with altitude. The flux that is re-emitted outside the planet is therefore smaller (due to the lower temperature) than that re-emitted towards the surface. In the final energy balance, treated in Chapter 9, re-emission is much greater towards the Earth's surface (about two thirds) than into space (about one third).

In spite of its simplicity, this *single-layer model* captures the order of magnitude of the different mechanisms in the transfer of solar and terrestrial radiation. In warming the Earth's surface by about 40°C, the greenhouse effect more than compensates for the other two effects, which cool the surface by about 25°C (reflection of solar radiation) and by about 10°C (absorption of solar radiation by the atmosphere), respectively.

Finally, we emphasize that these partial balances include only the solar and terrestrial flux. The total energy balance must take into account the water cycle and convection of air warmed by the Earth's surface. These involve transfer of latent and sensible heat. This topic will be discussed in Chapter 9.

8.2 The greenhouse effect

The greenhouse effect is not a recent subject of concern. As early as the 19th century, scientists were already becoming aware of the effects of atmospheric CO_2 concentration, notably through the pioneering work of Joseph Fourier (1824), John Tyndall (1861) and Svante Arrhenius (1896).

The large difference between the radiative flux emitted by the Earth's surface ($390\,\text{W/m}^2$) and the solar flux that the planet actually absorbs ($235\,\text{W/m}^2$) is not intuitively obvious. Indeed, the greenhouse effect, which is responsible for this increase, looks at first sight like the miracle of the five loaves and the two fishes. In reality, however, it is only a question at the school certificate level of our grandparents, where problems involving water taps could amply illustrate a large part of the complexity of the world. This is the approach that we use here.

8.2.1 THE GREENHOUSE EFFECT: THE BATHTUB ANALOGY

Taking a bath can sometimes clarify one's ideas. It is said that a certain Archimedes in ancient times once leapt out of his bathtub, shouting 'Eureka!' It is not our purpose here, however, to linger over his famous Principle. Instead, we use the bath as a means of illustrating the notion of equilibrium temperature and of understanding in practical terms the fabulous, albeit complex, *greenhouse effect*. We draw a parallel between the amount of energy available at the surface of the Earth and the amount of water in the bath, and we examine how it changes. The analogy is more straightforward if we consider a bathtub with vertical walls. In this case, since the surface area remains constant when the volume of water in the bath changes, the volume is determined by the height of the water. Our illustration is thus based on an analogy between temperature, which measures the average amount of energy on Earth, and the level in the bath, which measures the amount of water in it.

First of all, it is essential not to confuse the flux of energy (solar flux, terrestrial flux) with the energy itself (the available energy, i.e. the temperature according to the Stefan–Boltzmann relation). This confusion stems from our everyday use of language: we speak of solar energy received when in fact it is the *energy flux* received. This confusion ends up by masking the difference between the rate of supply (the arrival of a quantity – in this case, joules received per unit of time by the Earth's surface, in W/m^2) and the content (the energy available at the Earth's surface, as measured by its absolute temperature, in kelvin). This shift in language makes it more difficult to grasp how it is that, by keeping the energy input constant (the rate of supply) as well as its output (the radiation flux), the available energy can be increased.

The bathtub

Let us imagine the following situation. A bathtub is being filled with water from a tap, and at the same time water leaves the bath through the plughole. This outlet, which is partly open, is located at the bottom surface of the bath. At the start of the observation the system is at equilibrium (i.e. in a stationary condition): the volume of water arriving per unit time is equal to that draining out through the plughole, and the level of water in the bath is therefore constant. The volume of water in the bath (volume = surface area × height) is directly proportional to the height of the water. The water is continually renewed, but equilibrium has been reached and the volume of water in the bath remains constant (Fig. 8.5a).

This initial situation is defined by the incoming flow rate D_0 (in litres per second), by the outgoing flow rate (which is identical to D_0) and by the height h_0 of the water in the bath. The rate of outflow is proportional to the cross-sectional area S_0 of the plughole, and to the velocity

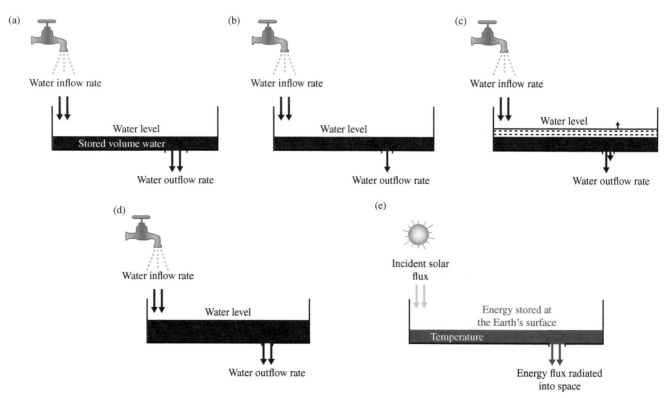

Fig. 8.5 The analogy between the volume of water in a bathtub and the store of energy at the surface of the Earth in the greenhouse effect. (a) Equilibrium condition: the volume of water remains constant because the water outflow rate is equal to the inflow rate. (b) The start of the experiment: the cross-sectional area of the outlet is divided by two, reducing the outflow rate by a factor of two. (c) The water level rises, increasing the outflow rate. (d) The level becomes stationary when the outflow rate is again equal to the inflow rate: the volume of water in the bath is then doubled. (e) The analogy with the radiative equilibrium at the Earth's surface (see text and Table 8.1).

of the water V_0 going through it. It is equal to the product $V_0 \times S_0$. Moreover, the pressure of the water at the bottom of the bath is directly proportional to height of the water level. If, for example, the height of the water were doubled, the pressure at the bottom, the velocity V and the rate of flow would also be doubled. The following question now arises. *Can we change the amount of water in the bath without changing the input rate? In other words, for the same input, can we have more water available in the bath?* But of course! The immediate common-sense solution is to close the outlet completely. If the plughole is closed, all the water collects in the bath. This particular case, however, is not the one of interest to us here.

The experiment

If now we reduce the surface area of the plughole by a factor of two, it becomes $S_0/2$, and the outflow will also be reduced by a factor of two, since the new rate of flow out of the bath becomes $V_0 \times S_0/2$ (Fig. 8.5b). In each second, half as much water will leave the bath as enters it: the volume of water in the bath must therefore increase, and the level rises. The rise in the height of the water increases the pressure at the bottom of the bath. As the pressure rises, the velocity of the water increases and, accordingly, the flow out the bath also increases (Fig. 8.5c).

The outgoing flow rate, which was divided by two at the beginning of the experiment, starts to increase continuously in response to the increase in height of the water level. When the height doubles, the pressure at the bottom is also multiplied by two, which makes the velocity at the outlet rise to $2V_0$. The output flow rate then becomes $2V_0 \times S_0/2$: in other words, it recovers its initial value $V_0 \times S_0$ (Fig. 8.5d). The incoming and outgoing flow rates are again in balance: the volume of water in the bath stops increasing and remains stationary. A new equilibrium is reached. Thus, with no change in the incoming flow rate, the amount of water in the bath has doubled, simply by altering the cross-section of the outlet. The mechanism by which this new equilibrium comes about stems from the fact that the increase in the volume of water produces an increase in the flow rate at the outlet (in this model, the relationship is linear).

Relevance to the greenhouse effect

The mechanism of the greenhouse effect can be compared to the process we have just described, with, however, the following difference: it is not water, but energy. Consequently, it is not the same physical laws that apply (Fig. 8.5e). Here, we are comparing the amount of water in the bath (expressed as a volume) with the average amount of energy available at the surface of the Earth (expressed in terms of temperature). The role of the inlet tap is played by the Sun. The inlet flow is the analogue of the flux of solar energy (E_{s0}) that supplies the reservoir of available energy at the Earth's surface (the absorbed solar flux). The rate of evacuation of water draining through the outlet is the analogue of the flux of thermal energy emitted by the Earth (E_{t0}), and which is radiated into space. The equilibrium situation where the temperature at the surface is stationary is reached when the absorbed solar flux is equal to the flux radiated by the Earth ($E_{s0} = E_{t0}$). In this analogy, the question we asked at the start of the bathtub experiment now becomes: *Can we change the amount of energy available at the surface of the Earth (i.e. the temperature) without changing the energy supply from the Sun and, at the same time, become warmer?*

The stages described in the case of the bathtub can be taken point by point. Reducing the size of the outlet pipe is now analogous to the presence (or increase in the amount) of greenhouse gases in the atmosphere. This presence (or increase) at first reduces the energy flux E_t radiated by the planet. The response then follows. The water level, which in the case of the bathtub, increased and raised the pressure at the bottom, and with it the outflow rate, is replaced here by the increase in available energy at the surface (the surface temperature).

The rise in temperature increases the flux radiated by the surface, just as the increase in water level raised the output flow. This rise in temperature in turn produces an increase in the total flux radiated by the planet. In summary, the temperature of the surface of the planet gradually increases, increasing the flux E_t radiated by it, which finally returns to its initial value E_{t0}. Equilibrium is once again attained ($E_t = E_{s0}$), but the amount of stored energy is now greater. Meanwhile, the Sun continues to supply energy at the same rate E_{s0} to heat the planet. The different stages in the analogy are listed in Table 8.1.

The major difference between the two experiments, of course, is that the first describes the changes occurring in a reservoir that contains matter (water), while in the second the reservoir contains energy. But there is a further difference: in the first case the effect is linear (the flow rate is proportional to the height of the water in the bath), while in the second case the effect varies with the fourth power (according to Stefan's law of black-body radiation, the radiated flux is proportional to the fourth power of the temperature).

TABLE 8.1 THE ANALOGY BETWEEN THE INCREASE IN TEMPERATURE AT THE SURFACE OF THE EARTH DUE TO THE GREENHOUSE EFFECT AND THE INCREASE IN THE VOLUME OF WATER THAT ACCUMULATES IN THE BATHTUB.

Bathtub	Surface of the Earth
Reducing the cross-section of the outlet pipe causes a (temporary) decrease in the outflow rate	Increasing greenhouse gases causes a (temporary) decrease in the flux radiated into space by the planet
A (temporary) decrease in the outflow increases the amount of water collected in the bathtub (the water level rises)	A (temporary) decrease in emitted radiation increases the amount of available energy (the surface temperature rises)
An increase in the height of the water level increases the pressure at the bottom of the bathtub (pressure proportional to the height of the water)	An increase in surface temperature increases the flux radiated by the surface (radiated flux proportional to the fourth power of T)
The pressure increase at the bottom of the bathtub increases the rate of outflow	The increase in flux radiated by the surface increases the flux radiated into space
The rate of outflow stops increasing when it reaches its initial value, equal to the rate of inflow of the water: equilibrium is reached	The flux radiated into space stops increasing when it reaches its initial level, equal to the absorbed incident solar flux: equilibrium is reached

In conclusion, just as with the bathtub, the temperature of the Earth's surface can be increased without changing the supply of energy by modifying the parameter that controls the evacuation of the energy (the quantity of greenhouse gases in the atmosphere). It turns out that the compounds present in the atmosphere can play more than one role in the transfer of the solar and thermal energy. Some compounds, such as water, contribute both to the greenhouse effect and to mechanisms that cool the Earth's surface. By contrast, CO_2, the next most important GHG after water vapour, contributes in the energy balance only to the greenhouse effect: it can be described as a 'pure' greenhouse effect gas. Any increase of its concentration in the atmosphere can only raise the temperature of the Earth's surface.

8.2.2 VENUS AND THE GREENHOUSE EFFECT

There is one planet in our solar system on which the greenhouse effect is larger than that on Earth. Our neighbour Venus, sister to the Earth in terms of its size and its gravity, is located closer to the Sun. The atmospheric pressure at ground level is 92 times greater than on Earth, and its atmosphere consists of more than 96% carbon dioxide, the greenhouse gas *par excellence*. This means that the *column mass* ρz of CO_2 is roughly $100\,\text{kg/cm}^2$. Although Venus is closer to the Sun than is the Earth, and it therefore receives more solar flux ($2614\,\text{W/m}^2$ compared to $1367\,\text{W/m}^2$ on Earth), the absorbed solar flux is lower than on Earth because of reflection by the very high albedo of the Venusian atmosphere. This is the reason why Venus radiates into space as a black body at temperature $184\,\text{K}$, which is much lower than that of the Earth ($254\,\text{K}$). The temperature at ground level on Venus, however, is about $460\,°\text{C}$ due to the intense greenhouse effect. At that pressure and temperature, water cannot exist in the liquid state. There is therefore no water cycle and the immense oceans that harboured the beginnings of life on Earth are absent from the surface of Venus.

8.3 Radiative transfer: the roles of the different constituents

Here, we briefly resume how the different components of the atmosphere and the surface of the Earth enter into the above mechanisms.

8.3.1 REFLECTION OF SOLAR RADIATION: PRINCIPAL FACTORS

The Earth and the Moon are visible from space because of reflected solar radiation, and it is for the same reason that the planets shine brightly in the night sky. Venus, with an albedo higher than 0.9, thus appears so brilliant that it is popularly referred to as a star, the Evening Star.

For the Earth, most of its planetary albedo is due to atmospheric reflection. Surface reflection contributes relatively little. Of the $342 \, W/m^2$ of solar flux that the Earth receives, $107 \, W/m^2$, or roughly 31%, is reflected back into space. The atmosphere returns $77 \, W/m^2$, the majority of which is by reflection (~80%), mainly by clouds and, to a lesser extent, by aerosols. The remainder (~20%) is backscattering by air molecules (Rayleigh scattering). Reflection by the Earth's surface of solar radiation is attenuated in its passage through the atmosphere (through reflection, backscattering and/or absorption), so that only $30 \, W/m^2$ of it escapes into space. This reflection depends strongly on the nature of the surface covering. Oceans absorb sunlight strongly (albedo 0.05–0.10), thereby making them appear black in satellite photos, while reflection by the ground depends on its composition. Deserts, with an albedo of about 0.3, reflect most and appear bright in satellite photos. Then come soils with different types of vegetation cover, among which forests (albedo ~ 0.1) absorb more solar radiation than cultivated land and prairie (albedo ~ 0.2). Lastly, land covered with snow or ice is an excellent reflector when the snow is fresh (albedo sometimes as high as 0.9), but the albedo falls to 0.5 when the snow is old. These long-lived features can be temporarily perturbed by aerosols injected into the atmosphere by volcanoes or by humans (Part II, Note 6).

Albedo thus varies strongly according to the region of the planet, and reflection by clouds is much greater than surface reflection.

8.3.2 ABSORPTION OF SOLAR RADIATION BY THE ATMOSPHERE: THE ANTI-GREENHOUSE EFFECT

The absorption of solar radiation in its passage through the atmosphere is illustrated in Figs. 4.2, 7.3 and 7.4. It amounts to $68 \, W/m^2$. We recall that the two principal constituents of the atmosphere, N_2 and O_2, absorb hardly any solar radiation. Only O_2 exhibits very weak absorption bands in the visible range, but these have a negligible effect on the radiation balance. The first compound to intercept the incident solar radiation is ozone (O_3) in its stratospheric layer, the densest part of which is situated at an altitude of about 25 km, far above the main mass of the atmosphere. The absorption of solar radiation by this extremely thin layer (average *column mass* $0.0006 \, g/cm^2$, compared with the $1 \, kg/cm^2$ of the atmosphere) has important consequences. It completely absorbs the harmful ultraviolet solar radiation between $0.2 \, \mu m$ and $0.3 \, \mu m$ (Part II, Note 7) and creates a warm pocket that stratifies the air in the stratosphere. Deeper in the troposphere, absorption is dominated by water vapour, which removes about 17% of the incident solar flux, followed far behind by the clouds (about 3%). Ozone in the troposphere plays practically no role, since the solar radiation in its absorption range has already been removed by the stratospheric ozone layer. Finally, aerosols also absorb part of the solar flux; this varies as a function of their composition. In total, the fraction of incident solar radiation that is absorbed by the atmosphere is estimated at about 20%.

8.3.3 ABSORPTION OF TERRESTRIAL RADIATION BY THE ATMOSPHERE: THE GREENHOUSE EFFECT

Absorption of terrestrial radiation by the different gases in the atmosphere amounts to about $350 \, W/m^2$, and is illustrated in Figs. 7.4 and 8.4. As with solar radiation, the two main constituents, N_2 and O_2, do not absorb any radiation in the terrestrial wavelength range. Few atmospheric components absorb terrestrial radiation strongly. Essentially, there are three: water vapour, carbon dioxide and clouds. Other gases also contribute, but to a much lesser degree – principally ozone and, to an even lesser degree, methane, nitrous oxide, chlorofluorocarbons (CFC) and so on. *What are the respective roles of the major greenhouse constituents in the greenhouse effect?*

The individual role of the different constituents (gases, clouds and aerosols) in the greenhouse effect is not easy to evaluate, since the roles tend to overlap when different constituents absorb in the same wavelength range.

A first possibility is to consider an atmosphere that contains only a single compound, and to evaluate what fraction of the radiation from the surface of the Earth would be absorbed. With this hypothesis, and using our *single-layer absorption model*, we compare the absorption by the two most significant gases:

- *Water vapour*. If the water vapour were reduced to a uniform sheet of $\rho z = 2.5 \, g/cm^2$ surrounding the Earth, it would absorb about half of the radiation flux emitted by the surface. Although this water vapour constitutes only 2.5 per thousand of the mass of the atmosphere, its role is disproportionately large.
- *Carbon dioxide*. When reduced to a uniform sheet of $0.6 \, g/cm^2$ around the Earth, the CO_2 would absorb about one fifth of the radiation flux emitted by the surface. Here again, it plays a major role even though it constitutes only 0.6 per thousand of the mass of the atmosphere.

Water vapour and carbon dioxide do not absorb at the same wavelengths. Their absorption of the infrared radiation emitted by the Earth is complementary and therefore additive. Together, they absorb about 70% of the radiation emitted by the surface of the Earth. There is, however, a difference between these two active ingredients: their distribution over the surface of the Earth is not the same. While the yearly averaged atmospheric CO_2 content is practically the same at all latitudes, that of water vapour (and consequently, its absorption of terrestrial radiation) decreases strongly on going from the Equator to the poles, and also varies regionally according to the vegetation cover.

Finally, let us consider the *role of clouds*. Wherever clouds are present, they absorb practically 100% of the terrestrial radiation flux emitted by the surface. As this absorption spans the whole wavelength range of terrestrial radiation, it does not add to, but instead combines with, that of water vapour and CO_2.

To evaluate the absorption of the terrestrial radiation by the atmosphere, we make a bold simplification. Where clouds are absent, the combined absorption by water vapour and CO_2 in our single-layer model is roughly 70%. Where clouds are present, 100% is absorbed. With the additional assumption that clouds cover about 60% of the Earth's surface, the average absorption of surface radiation by the atmosphere then amounts to roughly 90%. This conclusion is in agreement with the observed value of the absorption coefficient, namely 0.9 (Chapter 9).

The minor gas ozone, which contributes more modestly to the greenhouse effect, absorbs in a narrow saturated band at 9–10 µm. The roles of methane and nitrous oxide are also minor. These gases are, however, remarkable in that their absorption lines are unsaturated, which implies that their absorption could increase dramatically if their concentration were to increase.

8.3.4 THE MULTIPLE ROLES OF THE COMPONENTS

In the above inventory, certain compounds in the atmosphere contribute both to heating and cooling of the Earth's surface. In the first place is ozone, which absorbs from solar radiation an amount of energy that is comparable to what it absorbs from terrestrial radiation. Ozone in the stratosphere does not play the same role as ozone in the troposphere, however, since the former captures solar radiation and the latter radiation from the surface of the Earth. Then there are aerosols, whose role is complex. Sulphated aerosols, for example, cool the Earth's surface by reflecting more solar radiation. And especially, there is water, which plays a role everywhere. In its three phases – gas, liquid and solid – it contributes strongly to cooling the surface of the planet: by reflection of a large part of the solar energy from clouds and frozen areas and, more modestly, by absorbing solar radiation in the vapour phase. At the same time water contributes to heating, via the greenhouse effect: in its vapour form it can absorb half of the terrestrial radiation emitted by the Earth's surface. Whenever clouds are present, water, in its liquid or frozen state, can totally absorb the terrestrial radiation, thus contributing strongly to warming the surface of the planet. Finally, carbon dioxide plays a role that is both unambiguous and unique: it contributes only to the greenhouse effect. Carbon dioxide absorbs the Earth's radiation in a wavelength range (13–17 μm) where no other gas absorbs. In particular, in the high atmosphere, where the concentration of water vapour is negligibly small, it is the only gas engaged in the greenhouse effect.

8.4 The radiation balance of the Earth

In summary, the above three mechanisms contribute both to cooling and to heating the surface of the Earth. *Cooling* occurs because part of the solar radiation does not reach the surface, due to:

- reflection of the solar radiation, principally by clouds, and also by the air, aerosols and the Earth's surface; and
- absorption of the solar radiation by the atmosphere, principally by water vapour, then by stratospheric ozone, clouds and aerosols.

Heating occurs because part of the energy radiated by the surface of the Earth returns to the surface after being absorbed and re-emitted by the atmosphere. This role is played primarily by water vapour, carbon dioxide and the clouds and, secondarily, by ozone, methane, nitrous oxide, CFCs, aerosols and so on.

To establish the equilibrium temperature of the planet, we must determine the total energy budget at its surface. Up to now, we have limited energy exchange between the surface and the atmosphere to radiative processes. In reality, however, two other processes exist: transfer of latent heat via the water cycle, and transfer of sensible heat by air convection from the surface. These must also be included.

What, then, is the equation for the equilibrium temperature of the surface of the planet?

Chapter 9
The energy balance

9.1 The energy balance at the surface of the Earth in the *single-layer model*

9.1.1 THE SINGLE-LAYER APPROXIMATION

We now examine in detail the last situation, Case 3 (Fig. 4.5), where there is an atmosphere. Compared to Case 2, the existence of an atmosphere adds to the complexity of the energy balance at the Earth's surface in three different ways:

- By its interaction with solar and terrestrial radiation, as discussed in the previous chapter. This warms the surface of the planet by the greenhouse effect, at the same time as cooling it, through reflection and the anti-greenhouse effect.
- By the water cycle. Evaporation of water at the Earth's surface cools it.
- By heating colder air in contact with the warm surface of the Earth, thereby also cooling the surface.

The last two of these energy fluxes leaving the surface, which we denote as E_{lat} (latent heat flux) and E_{sens} (sensible heat flux), are entirely absorbed by the atmosphere. The former warms the air when the water vapour condenses into clouds, and the latter heats the air directly. This heating adds to the flow of energy that warms the atmosphere, and it is totally re-emitted as thermal radiation, partly back to the Earth's surface and partly into outer space. We now return to the simplified *single-layer model*, discussed in Chapter 7,

Climate Change: Past, Present and Future, First Edition. Marie-Antoinette Mélières and Chloé Maréchal.
© 2015 John Wiley & Sons, Ltd. Published 2015 by John Wiley & Sons, Ltd.
Companion website: www.wiley.com\go\melieres\climatechange

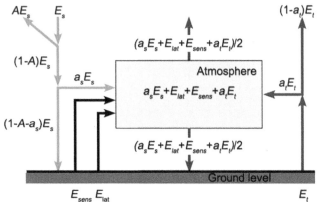

Fig. 9.1 The energy balance in the single layer atmospheric model (see text). The solar flux E_s is partly reflected (AE_s) and partly absorbed (a_sE_s) by the atmosphere, the rest being absorbed by the surface. E_{sens} and E_{lat} are, respectively, the fluxes of sensible and latent energy that leave the Earth's surface and are absorbed by the atmosphere. E_t is the terrestrial radiation emitted by the surface, part of which is absorbed by the atmosphere (a_tE_t). Half of the energy absorbed by the atmosphere is radiated towards the surface, where it is absorbed, and half is radiated into space.

in which the atmosphere is taken to be a homogeneous layer of uniform temperature. In this model, the absorbed energy is radiated in equal amounts back to the surface and into outer space. A diagram of the energy balance is shown in Fig. 9.1. When the energy budget is balanced, the solar energy absorbed by the planet taken as a whole is offset by the terrestrial energy radiated into space.

At ground level, equilibrium also requires that the energy lost by the surface be equal to that absorbed. The following relationship therefore holds:

$$E_t + (E_{lat} + E_{sens}) = (1 - A - a_s) \times E_s + (a_s \times E_s + a_t \times E_t + E_{lat} + E_{sens})/2 \tag{II.5}$$

In eqn. II.5, E_s is the incident solar flux and E_t is the flux emitted by the Earth's surface, while E_{lat} and E_{sens} are, respectively, the latent heat and the sensible heat flux emitted by the surface. A is the albedo of the planet, and a_s and a_t are the atmospheric absorption coefficients in the solar and terrestrial wavelength ranges, respectively. As already noted, E_t is determined by the surface temperature T, and is equal to σT^4.

The left-hand side of eqn. II.5 is the energy lost by the surface (the radiated flux E_t and the latent and sensible heat fluxes, $E_{lat} + E_{sens}$). The right-hand side is the energy absorbed by the surface. It comprises the solar energy that is directly transmitted through the atmosphere and absorbed at the surface, $(1 - A - a_s) \times E_s$, and the energy radiated by the atmosphere towards the surface. In our model, the latter is equal to half of the total flux absorbed by the atmosphere (i.e. part of the terrestrial flux emitted by the surface, $a_t \times E_t$, and part of the solar flux, $a_s \times E_s$, together with the latent and sensible heat fluxes, $E_{lat} + E_{sens}$).

9.1.2 SURFACE TEMPERATURE AS AN ADJUSTABLE PARAMETER

The flux of energy emitted by the surface of the Earth depends on its temperature. When it is colder, evaporation decreases, heating of the layers of air becomes less effective, and the radiation flux emitted by the surface diminishes. The surface temperature is thus the key parameter that determines the transfer of energy from the surface to the atmosphere. For a given solar energy received by the planet, after account has been taken of the different transfer mechanisms through the atmosphere, it is the surface temperature that adjusts until the energy flux that it emits becomes equal to the energy flux that it absorbs.

In this simplified single-layer approximation, if the energy fluxes E_s, E_{lat} and E_{sens} are measured and the absorption and reflection coefficients are known, then, since $E_t = \sigma T^4$, eqn. II.5 can be used to calculate the equilibrium temperature of the Earth's surface. Thus

$$T = [\sigma \times (1 - 0.5a_t)]^{-1/4} \times [(1 - A - 0.5a_s) \times E_s - 0.5(E_{lat} + E_{sens})]^{1/4} \tag{II.6}$$

What do the measurements tell us?

9.2 The Earth's energy balance at equilibrium

9.2.1 THE DIFFERENT PIECES OF THE PUZZLE

The pieces of the puzzle are now ready. It only remains to put them together. The measured values of the different components of the energy budget are displayed in Fig. 9.2, with a zoom of the surface in Fig. 9.3. These values are based on IPCC 2007. The state of knowledge in 2013 will be seen in the next paragraph. The energy budget shown is balanced, but let us not be misled by the precision of the values of the flux, which are accurate to only a few watts per square metre. The three fluxes on which this energy balance is based were measured by satellites. They are the incident solar flux A (342 W/m²), the reflected solar flux B (107 W/m²) and the terrestrial radiation flux emitted by the planet, which is equal to the absorbed solar radiation C (235 W/m²).

What does our single-layer model give?

The values of A (0.31), a_s (0.2) and a_t (0.9) are found from the energy balance measurements shown in Fig. 9.2, together with the three fluxes E_s (342 W/m²), E_{lat} (78 W/m²) and E_{sens} (24 W/m²). But it is important to note that here the value of $a_s = 0.2$ is the absorption with respect to the incident solar flux, $E_s = 342$ W/m². By contrast, in the discussion of the anti-greenhouse effect in Section 8.3.2, the absorption of the atmosphere was expressed with respect to the absorbed flux (235 W/m²) and, in that earlier context, was therefore equal to 0.29.

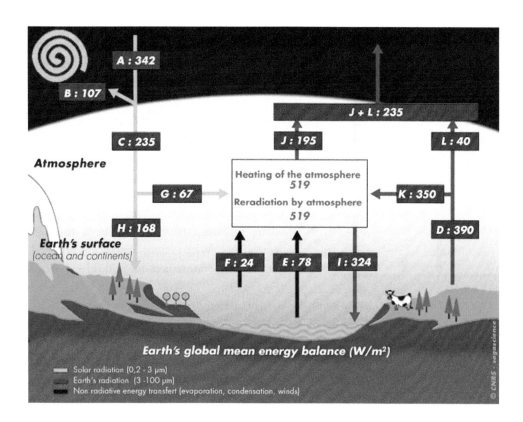

Fig. 9.2 The global mean energy budget of the Earth at equilibrium. The incoming solar flux absorbed by the planet is equal to the long-wavelength radiation flux emitted by the planet into space. Values (W/m²) are taken from IPCC 2007. Items in the budget are listed in Section 9.2. For simplicity, the reflection of solar radiation by the atmosphere and the surface are shown above the atmosphere. **Source:** From 'Bilan énergétique de la planète' © CNRS. Reproduced with permission.

Equation II.6 yields for the surface temperature $T = 264$ K. A more realistic value is obtained on noting that (Fig. 9.2) the flux radiated by the atmosphere towards the Earth's surface (324 W/m²) is not 50% of the total flux absorbed by the atmosphere (519 W/m²), but 62%. Similarly, the flux radiated into space is not 50%, but 38% (195/519). Equation II.6 must accordingly be modified. This yields

$$T = [\sigma(1 - 0.62a_t)]^{-1/4} \times [(1 - A - 0.38a_s)E_s - 0.38(E_{lat} + E_{sens})]^{1/4} \tag{II.7}$$

The surface temperature accordingly becomes 288 K. Not unexpectedly, the revised set of parameters is in agreement with observation (288 K).

Comments

In fact, E_{lat} and E_{sens} are temperature dependent and the calculation of T from eqn. II.6 (and eqn. II.7) is a crude approximation, since these two terms are parameterized. Nevertheless, the equation provides a good approximation for the temperature deviation from the present climate state (excluding feedbacks) when a perturbation is applied (solar change, increase in greenhouse gases, albedo change and so on). Feedback effects must be estimated afterwards (Section 10.3, eqn. II.8).

9.2.2 A BRIEF SUMMARY

In Fig. 9.2, the *surface of the planet* is heated by the flux of solar energy (H); in other words, the unreflected fraction of the incident solar flux that is transmitted directly through the atmosphere, and also by the thermal flux (I) radiated by the atmosphere towards the surface. It is cooled by three flux emissions: the latent heat flux (E) due to evaporation, the flux of sensible heat (F) from heating the layers of air in contact with the surface, and the thermal flux (D) radiated by the surface. The radiation flux (D) is related directly to the surface temperature by Stefan's law. The temperature can be deduced from satellite measurements of this flux in the atmospheric window (8–12 μm), where radiation passes unhindered through the atmosphere. When energy balance is attained, the surface temperature is stable and (H+I) is equal to (F+E+D).

The *atmosphere* is heated in several ways: (i) by absorbing a part (G) of the incident solar flux and a part (K) of the thermal flux emitted by the surface; (ii) by the latent heat flux (E) from condensation of water vapour; and (iii) by direct heating of the air in contact with the surface (F). It is cooled by

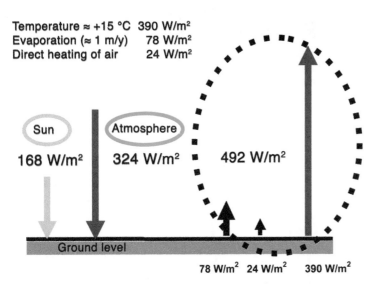

Temperature ≈ +15 °C   390 W/m²
Evaporation (≈ 1 m/y)   78 W/m²
Direct heating of air   24 W/m²

Sun
168 W/m²

Atmosphere
324 W/m²

492 W/m²

Ground level

78 W/m²   24 W/m²   390 W/m²

Fig. 9.3 To sustain the present climate, the Earth's surface requires a permanent flux of 492 W/m². This comes from two sources, the solar flux (168 W/m²) and infrared radiation from the atmosphere (324 W/m²). The size of the latter, almost twice the solar flux, illustrates the important role of the greenhouse effect.

the flux of thermal radiation towards the surface (I) and into space (J). Equilibrium requires that $(E+F+G+K)=(I+J)$. This situation is summarized in Figure 9.3.

Seen from space, the Earth receives the solar flux (A), part of which (B) is reflected. It radiates into space the sum of the flux (J) emitted by the different components of the atmosphere (clouds, gases etc.) and the surface radiation flux (L) directly transmitted through the atmosphere. The external temperature of the planet is defined by the flux $(J+L)$ that it radiates. This is the temperature of a black body that radiates the same flux. When energy balance is reached, $(A-B)$ is equal to $(J+L)$.

9.3 The impact of human activity

We have just seen that the surface temperature is one of the basic parameters of climate. It is the resultant of a balance involving many processes and the exchange of different energy fluxes (terms A–L). Any change, whether natural or anthropogenic, in any of these terms will bring about a change in climate. One of the processes (involving terms K and I) is the greenhouse effect, which heats the Earth's surface and has the greatest impact on the temperature.

Even though humans have modified their environment for a long time, the Industrial Revolution has, to an unprecedented extent, deeply modified both the composition of the atmosphere and the surface of the planet. The climate equilibrium parameters have changed accordingly. This matter will be discussed more fully in Part V. For the climate, the most critical modification is without question the increase in greenhouse gases. This alters the heat flux at the surface and generates *radiative forcing*, a concept that will be explored further in the following chapter. The modification disturbs the energy balance, which is no longer in equilibrium.

9.4 The present unbalanced global energy budget

Since IPCC 2007, understanding of the flows of radiative energy in the climate system has improved, and an updated diagram of the global mean energy balance is required. The incident solar flux arriving on a surface perpendicular to it at the outside of the Earth's atmosphere is referred as the *Total Solar Irradiance* (TSI). This flux has been measured with unprecedented accuracy. The observed TSI, $1360.8\pm0.5\,\text{W/m}^2$, is now believed to be more credible than the previous value, $1365.5\pm1.3\,\text{W/m}^2$ (Kopp & Lean 2011). The revised value corresponds to an incident solar radiation, when averaged over the Earth, of about $340\,\text{W/m}^2$ (term A). The uncertainty of the different measured fluxes in the energy budget of IPCC 2013 ranges from 1 to $17\,\text{W/m}^2$. The smallest uncertainty ($1\,\text{W/m}^2$) is in the incident solar radiation (term A). The largest uncertainty, $17\,\text{W/m}^2$, is in the solar flux absorbed by the atmosphere (term G). The uncertainties of the other terms lie between 4 and $12\,\text{W/m}^2$.

The energy budget of the Earth's surface is at present out of balance. Fig. 2.12 of the IPCC 2013 report presents a schematic diagram of the Earth's energy budget, based on the work of Wild et al. (2013). Most of the energy that arrives at the Earth's surface is absorbed by the continents and oceans. It heats them and is re-emitted. But another smaller part penetrates

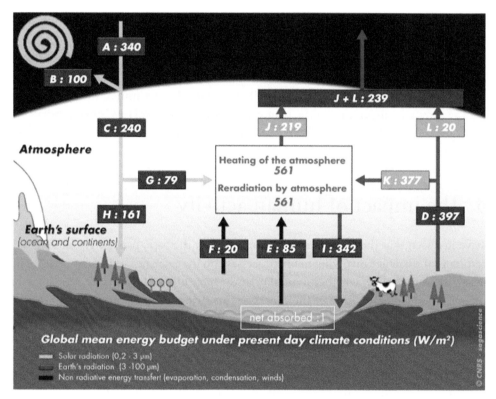

Fig. 9.4 The present global mean energy budget of the Earth. Unlike Fig. 9.2, it is out of balance, with the ocean accumulating about 1W/m^2. The figure is based on the IPCC 2013 energy budget. In conformity with recent estimates, the terrestrial flux transmitted directly through the atmosphere is taken to be 20W/m^2 (term L, pale purple) (Costa & Shine 2012). This yields the values of the terms K and J (pale purple). The uncertainty of the different measured fluxes ranges from 1 to 17W/m^2 (see text).

Source: From 'Bilan energétique de la planète' © CNRS. Reproduced with permission.

into the ocean and is transferred to the deeper waters. This is of the order of magnitude of 1W/m^2 (Hansen et al. 2011; Loeb et al. 2012); the IPCC 2013 estimate is 0.6W/m^2. The imbalance in the energy budget is as follows: a small part of the energy received at the Earth's surface does not heat it and does not contribute to the radiative flux emitted by the planet's surface, but is stored continuously, heating the intermediate and deep ocean water masses. This loss to the net energy budget at the Earth's surface affects the global energy budget of the planet. It creates a deficit of about 1W/m^2 in the energy that the planet radiates into space.

The energy budget scheme in IPCC 2013 is adjusted for consistency with these independent estimates, and it takes the imbalance into account. The solar radiation absorbed by the Earth (term C, 240W/m^2) is almost, but not totally, compensated by the thermal emission into space (term J+L, about 239W/m^2). The present energy budget is shown in Fig. 9.4. This figure is based on IPCC 2013, but in addition we have assumed that the term L is equal to 20W/m^2, in agreement with recent estimates (Costa & Shine 2012). Unlike Fig. 9.2, the budget is not balanced, and the ocean is accumulating a flux of about 1W/m^2 (Part II, Note 8).

Chapter 10
Climate forcing and feedback

10.1 Climate forcing

We have described the energy balance at the surface of the Earth (i.e. ground level). This defines the surface temperature of the planet (Fig. 9.1). Any factor that modifies a component of this energy balance will force a change in the climate, making it warmer or cooler when the new equilibrium is reached. Two categories of factor can be distinguished: internal processes and external forcing.

10.1.1 INTERNAL PROCESSES

Internal processes affect the energy balance through causes that originate within the interacting climate system. On the timescale of a year, a decade or a century, these arise from dynamic interactions between the atmosphere and the oceans. Climate fluctuations due to El Niño (or La Niñā) events are a typical outcome of such interactions. Over longer time-scales, these processes incorporate changes in the cryosphere (ice sheet, sea ice), changes in ocean circulation and so on. Over geological timescales, modifications of mountain patterns and rearrangements of the continents alter the atmospheric and oceanic dynamics. This affects energy transfer from the Equator to the poles and the distribution of energy over the Earth's surface.

Climate Change: Past, Present and Future, First Edition. Marie-Antoinette Mélières and Chloé Maréchal.
© 2015 John Wiley & Sons, Ltd. Published 2015 by John Wiley & Sons, Ltd.
Companion website: www.wiley.com\go\melieres\climatechange

10.1.2 EXTERNAL FORCING AND RADIATIVE FORCING

External forcing is so called because its cause lies outside the interacting climate system. It includes factors that affect the global energy budget, pushing it out of equilibrium by introducing *radiative forcing* (RF). The forcing is expressed as the net change in radiation at the top of the troposphere. The climate responds by restoring the radiative balance between incoming and outgoing radiation. Positive radiative forcing tends to warm the surface and negative forcing tends to cool it.

Radiative forcing can be caused by:

- Changes in the solar radiation, E_s.
- Changes in the albedo of the planet. The principal external causes are changes in the composition of the atmosphere, mainly from injection of sulphate aerosols (through explosive volcanism or human activity), which increase the atmospheric albedo.
- Changes in the composition of the atmosphere imposed from outside, such as a change in the concentration of greenhouse gas (GHG), aerosols and so on. Increasing the GHG content, for example, enhances the warming due to the greenhouse effect.

Note that the notion of radiative forcing does not refer to changes in water vapour content (a powerful GHG) or cloud coverage (which essentially affect the albedo and greenhouse effect), and so on. These changes are not imposed from outside the climate system: they are secondary changes that are the consequence of an initial external change (see Section 10.2).

What about the radiative forcing generated by human activity since the Industrial Revolution, starting in the late 18th century, when agriculture, manufacturing, transport and so on underwent fundamental changes? Since the 19th century, these activities have significantly disturbed the composition of the atmosphere, mainly by increasing its GHG content (Fig. 29.4). The CO_2 content increased from its equilibrium level of the last millennium (~280 ppm) to 395 ppm in 2013. Smaller but appreciable increases occurred in two other GHGs, CH_4 (from ~0.7 to 1.8 ppm) and N_2O (from ~0.27 to 0.32 ppm). The *radiative forcing* from these three gases is estimated to be about +3.3 W/m² in 2011 (IPCC 2013). When the various changes in atmospheric compounds with positive and negative RF are taken into account (Fig. 29.10), the total RF is estimated as +2.3 W/m² (between 1.1 and 3.3).

The CO_2 doubling scenario

We now turn to the future, where one notable scenario assumes that the CO_2 content doubles with respect to the pre-industrial period. In this $2 \times CO_2$ scenario, the CO_2 content rises from 280 ppm to 560 ppm. According to Dufresne and Bony (2008), the associated RF is about +3.7 W/m² (an average of 12 models being 3.7, with extreme values 3.5 and 4). The resulting increase in the Earth's temperature can be evaluated by holding fixed all the other components of the climate system. Under these conditions, forcing by +3.7 W/m² produces a warming of +1.2°C. In reality, further feedbacks come into play, causing even greater warming (see below).

The direct action of a forcing factor on climate can be amplified or attenuated by the presence of feedback in the climate system. What are the main feedbacks?

10.2 Feedbacks

The initial warming or cooling due to external forcing can modify some of the components in the climate system. *Feedbacks* are the consequence of these modifications. Feedback mechanisms are a response of the climate system to external forcing, in which the initial forcing is either enhanced (positive feedback) or attenuated (negative feedback). *But a feedback does not change the sign of the initial forcing: it just reduces or amplifies it!* Feedback mechanisms can operate over a wide range of timescales, from slow to fast.

10.2.1 SLOW FEEDBACKS

The following examples are illustrations of slow feedback in continents and oceans. On geological timescales, different feedback mechanisms can also operate, but these lie outside the scope of this book.

Ocean circulation

Circulation in the oceans spreads heat into the whole climate system. Changes in the circulation therefore affect the transport of heat. Such modifications could be triggered by changes in climate. At the same time, the Gulf Stream pattern would be modified, further reducing the supply of heat to Europe. On a larger scale, deep-water formation, which occurs mainly in the North Atlantic Ocean, could also be altered by the expanding sea ice, and heat transport to the north would be reduced. Positive feedback mechanisms such as these intervene at the onset of a glacial climate.

The biosphere

Continental biosphere surfaces play a role mainly through the vegetation cover. Vegetation affects not only the reflection of the Sun's radiation (vegetation-covered soil reflects less than bare soil), but also the water cycle (vegetation enhances evaporation through transpiration) and the carbon cycle (by carbon storage in organic matter), on which the greenhouse gas content in the atmosphere depends. When the climate changes, these components of the climate system are affected, because vegetation responds to climate change. In a warmer climate, for example, forested areas of the northern regions will expand further north and replace tundra, thus decreasing solar reflection.

The carbon cycle

An important feedback mechanism that also operates on a longer timescale is that of the carbon cycle. This cycle determines the CO_2 content of the atmosphere through different mechanisms with time constants that vary from a few years to millions of years. The biosphere and the oceans are both involved in this cycle. Here are two examples:

- At short timescales, the topic of warming on soils is attracting increased attention. Excess CO_2 of anthropogenic origin is at present partly absorbed by the continental biosphere (Chapter 29). Further warming, however, could decrease its absorption capacity, thereby increasing the CO_2 content in the atmosphere and causing further warming.

- On a timescale of 1000 years, a change in climate can modify ocean circulation and ocean temperature. Since gases become less soluble when temperatures rise, warming releases CO_2 from the ocean and increases the atmospheric content of this greenhouse gas. Some more complicated feedbacks are also involved, and consequently, during glacial–interglacial transitions, the atmospheric CO_2 content increases by more than 30%. The same feedback acts in the reverse direction during interglacial–glacial transitions. This mechanism reinforces climate changes that are instigated by changes in the orbital parameters (Chapter 20).

10.2.2 FAST FEEDBACKS

The principal feedbacks come from changes in the water cycle. Their impact in the models becomes clearer if we distinguish four sources of feedback. The first three are related to changes in the physical state of water: vapour in the atmosphere, liquid in the form of clouds and solid as snow or ice at the Earth's surface. The fourth source has to do with changes in the transfer of energy between surface and atmosphere through the water cycle; it is referred to as the lapse rate effect. Each of these changes either amplifies or attenuates the initial change.

Water vapour

The water vapour content of the atmosphere increases as the temperature rises. This is because at higher temperature water molecules possess more energy. As more molecules are able to overcome their binding energy in the liquid state, the concentration in the vapour phase increases. The consequence is that a warmer atmosphere is able to retain more water vapour.

This is the source of the strongest positive feedback mechanism, as water vapour is a major greenhouse gas. According to present estimates, it more than doubles the increase in surface temperature that would occur under fixed water vapour conditions.

The effect of clouds

Clouds contribute to the energy balance in a variety of opposing ways that are related principally to their high reflectivity and their greenhouse effect. These effects complicate estimates of their feedback in climate changes. Low-lying clouds, for example, tend to cool the Earth by virtue of their high albedo, but the effect of blanketing by high clouds tends to warm it. In the context of a warming climate, changes in the cloud coverage are also accompanied by modifications of the water cycle and of atmospheric condensation mechanisms.

Snow–ice albedo

Snow cover and ice surfaces (sea ice, ice sheets and glaciers) reflect solar radiation strongly, with reflectivity (albedo) ranging between 50% and 90%. Atmospheric warming reduces this coverage, by reducing either the duration of winter snow or sea-ice cover, or the extent of glaciers. The surfaces thus liberated are darker (the albedo of continents ranges from 30% to 10%, and that of oceans from 5% to 10%) and absorb more solar radiation, thus further warming the surface. This feedback is positive.

The lapse rate

The lapse rate is the rate at which the temperature of the atmosphere decreases with altitude. Warming the surface decreases the lapse rate: the feedback is negative. A simplified picture of this mechanism can be seen as follows. When the surface is warmer, evaporation from it increases, and the flux of latent heat leaving the surface therefore also increases, thus causing the surface to cool. At the same time the lapse rate decreases. *Why?* Because, as condensation in the atmosphere increases, the atmosphere is warmed by the latent heat released. This reduces the temperature difference between the surface and the atmosphere; that is, the lapse rate decreases. The principal effect of this feedback is thus to cool the surface: the feedback is negative. In reality, however, the situation is more complex.

Evaluation of these feedbacks in the $2 \times CO_2$ scenario

Here, we evaluate these feedbacks in the context mentioned earlier of doubling the CO_2 content in the atmosphere: without feedbacks, the rise in temperature from the resulting additional flux of $3.7 \, W/m^2$ would be $1.2°C$. The work of Dufresne and Bony (2008) itemizes the evaluations of 12 Ocean Atmosphere–General Circulation models. The effect on the average temperature of each of the above feedback mechanisms is as follows:

- the increase in water vapour introduces an extra warming of $+1.7°C$, more than doubling the initial warming;
- the modification of the cloud cover, in spite of the increase in cloud albedo, is accompanied by warming of $+0.9°C$;
- the decrease in the albedo of the planet due to reduced cover of snow and ice, and also to changes in vegetation, gives rise to an additional warming of $+0.3°C$; and
- finally, the reduction in the lapse rate cools the planet by $-0.8°C$.

The sum of these feedbacks is $+1.9°C$, which is to be added to the initial warming of $+1.2°C$, yielding a total increase of $+3.1°C$. This is close to the value stated in IPCC 2013.

The transient climate response

In the case of CO_2 doubling, the equilibrium global mean temperature thus increases by about $+3°C$. But this result must be treated with caution, since it refers to the final state of the planet, when the energy balance is re-established. During the transitory period before equilibrium is attained, however, some of the energy is absorbed by the ocean and warms it, as it has not yet reached its equilibrium temperature. This reduces the heat available to the surface. This corresponds to the present situation (Fig. 9.4). To estimate the transient warming, or the *transient climate response*, a scenario is assumed in which the atmospheric CO_2 content increases by 1% per year and doubling is accordingly attained after 70 years. At that time, the temperature rise is only $+2°C$ (IPCC 2013). After that date, it continues to rise as time advances, and at equilibrium it reaches about $+3°C$.

10.3 Climate sensitivity

Climate sensitivity is a measure of how the global temperature responds to *radiative forcing*. It is the change in the global mean surface temperature at equilibrium (ΔT_{equil}) in response to the radiative forcing RF, where feedbacks are involved:

$$\Delta T_{equil} = \lambda \times \text{RF}$$

The *equilibrium climate sensitivity parameter* λ depends upon climate feedbacks. For coupled Atmosphere–Ocean Global Climate Models (AOGCMs), climate sensitivity is a crucial property. It is not a model parameter: it is an emergent property that characterizes the model.

Equilibrium climate sensitivity is often expressed as the temperature change for the $2 \times CO_2$ scenario mentioned earlier. We saw that doubling the CO_2 content produces a RF of $+3.7 \text{W/m}^2$. On the basis of the results of 12 AOGCMs, Dufresne and Bony (2008) estimated for λ the mean value of $0.79°C/(W/m^2)$, in which the range spanned from 0.63 to $1.27°C/(W/m^2)$. IPCC 2013 presents different types of estimates. The conclusion is that the equilibrium temperature after doubling of CO_2 is likely to lie in the range 1.5–4.5°C, where the principal uncertainty is attributable to cloud feedbacks. This implies that the value of λ *lies between 0.4° C/(W/m²) and 1.2° C/(W/m²)*:

$$\Delta T_{equil}(°C) \approx 0.8 \times \text{RF}(\text{W m}^{-2})$$

This relationship yields an estimated warming of $3 \pm 1.5°C$ when the CO_2 content doubles. Since the warming would be 1.2°C if feedback did not exist, the result can easily be understood by considering that *feedbacks multiply the warming by a factor of 2.5, in which the range of uncertainty is between 1.25 and 3.75*:

$$\Delta T_{equil}(°C)(\text{includingfeedbacks}) \approx 2.5 \times \Delta T_{equil}(°C)(\text{withoutfeedbacks}) \quad (II.8)$$

Can the CO₂ content double in the near future?

The future scenarios discussed in Part VI anticipate that this will occur in the years around 2050 for scenario RCP8.6, and around 2070 for scenario RCP6 (IPCC 2013).

Chapter 11
Climate modelling

When we try to calculate how our climate will evolve in the future, numerical models cannot be avoided. In spite of their incomplete account of the complexity of natural environments and of their attendant uncertainties, they nevertheless provide an overall view of the transformations happening to our planet. They also reveal the basic mechanisms that govern climate equilibrium and how it varies. In simulating human activities and their consequences, they fulfil an important function in society: in order to make informed political decisions that affect our global environment, there is often no alternative to modelling. We shall briefly mention the different steps that lead from the simplest models, the zero-dimensional (0D) models, in which the unknown (the temperature) is independent of any variable, to the three-dimensional (3D) global climate models in which the unknowns depend on the three dimensions of the planet.

11.1 The Energy Balance and Radiative–Convective Models

11.1.1 ENERGY BALANCE MODELS (EBMs)

The Earth's climate is determined primarily by its radiation balance. The average temperature at the surface of the Earth varies according to the radiation input. The surface temperature is controlled by the balance between the energy absorbed and that emitted by the Earth's thermal energy. Energy Balance Models (EBMs) use a simple scheme to describe the radiation

Climate Change: Past, Present and Future, First Edition. Marie-Antoinette Mélières and Chloé Maréchal.
© 2015 John Wiley & Sons, Ltd. Published 2015 by John Wiley & Sons, Ltd.
Companion website: www.wiley.com\go\melieres\climatechange

energy balance at the surface of the Earth, in which the unknown quantity is the average temperature T. This type of model was developed in the 1960s, in particular by M.I. Budyko (1969) and W.D. Sellers (1969). The models do not incorporate atmospheric dynamics (i.e. convection is not explicit), and hence transport and mixing of air masses are omitted. In spite of such simplifications, these conceptual models highlight the principal mechanisms and processes that govern the climate. This is the kind of model that is described in Chapters 8 and 9, and by which the average temperature at the surface of the planet was evaluated. To illustrate the different steps in the construction of such a model, we return to the extremely simplified model. The same steps are found in the more complex models.

An example

The aim of the model described in Chapter 9 was to calculate the average temperature T of the surface of the Earth. It contains a single unknown quantity, T, that is independent of any variable: it belongs to the simplest category, that of zero-dimensional models (0D). To calculate T, the system was expressed in terms of an *equation* that describes the energy balance at the surface of the planet, and in which T is the unknown. This type of equation is a member of the EBM. An energy balance is established between two terms, the solar flux received by the Earth and the radiation flux emitted by its surface. The Earth system was defined in terms of the transfer of radiation from the Sun and from the Earth through the atmosphere: this is a *radiation transfer model*. Such a description of the planet is highly simplified: the atmosphere is pictured as consisting of one single layer. The state of the planet is defined by *parameters*, such as the solar reflection by its albedo, the composition of the atmosphere by the sum of the components in the column of air above it and the water cycle by the annual rainfall, and the Earth's surface is taken to be a black body. We then set the incoming solar flux (the *boundary condition*). Instead of fixing the values of the different absorption coefficients of the atmosphere for the radiation from the Sun and from the Earth, we introduced a computational module that calculates these from the properties of the molecules. This method enables us, for example, to vary the concentration of greenhouse gases and to recalculate the new absorption coefficients. In this way, we can investigate how the climate depends on the concentration of these greenhouse gases.

This model elucidates the role and the influence of the various mechanisms involved in the equilibrium temperature; for example, the principal cause of the difference between the temperature at the surface and the temperature at the exterior of the planet (the greenhouse effect). It offers a means of estimating how forcing affects the temperature (e.g. doubling the CO_2 in the atmosphere, or the impact of polar icecaps, which change the albedo); it also explains the sign of the associated feedback mechanisms.

The extreme simplification of the model can be overcome in successive steps. First, the 0D model can be supplemented by allowing the temperature $T(\theta)$ to vary with latitude θ, and by allowing energy to be transported horizontally from the Equator to the poles. A mechanism that describes heat transfer between different latitudes must therefore be devised and parameterized. The unknown $T(\theta)$ then depends on one variable (θ) and the model belongs to the class of one-dimensional (1D) models. The next crucial improvement, however, opens the way to the dynamics of the atmosphere: for this, a one-dimensional (1D) radiative–convective model is required.

11.1.2 RADIATIVE–CONVECTIVE MODELS (RCMs)

These models are still centred on the transfer of solar and terrestrial radiation through the atmosphere, but they introduce atmospheric dynamics by including vertical transport of heat by convection. The atmosphere is sliced into superimposed homogeneous layers, among which air masses exchange with each other; the unknown $T(z)$ then depends on one variable, the altitude z (1D model). With such models, the water cycle and the formation of cloud layers can be taken into consideration. In view of the role of the water cycle, of the clouds and of water vapour in the energy balance, this step is essential. Here too, further refinement can be introduced by allowing for transfer of energy between different latitudes. With the addition of another variable, the latitude, the model acquires a 2D status. Radiative–convective models were widely developed in the 1960s, in particular by S. Manabe and R.T. Wetherald (1967). These are conceptual models that to this day remain a valuable means of determining the order of magnitude of various mechanisms involved in climate equilibrium.

11.2 Three-dimensional Atmosphere Global Circulation Models

Now comes the third step. It consists in portraying as faithfully as possible all the complexities of the Earth's atmosphere, and describing the changes in the temperature, in the water cycle and in the winds over the Earth (the three parameters by which we described the climate in Chapter 3). The atmosphere is defined in three dimensions and the general movement of the air masses is expressed in terms of the laws of physics. Such are the Atmosphere Global Circulation Models (AGCMs). They originate from efforts made in the 1920s to provide numerical predictions of the weather. It was only from the 1960s, however, that the power of computers had grown sufficiently to provide a description, however slender, of the weather.

11.2.1 THE STRUCTURE OF THE MODEL

The atmosphere

The atmosphere is now no longer divided into a series of overlying homogeneous layers, but into a network of elementary volumes, with their centres located at altitude z, latitude θ and longitude Φ. In models that are coupled with the ocean, the size of these boxes is between 100 and 300 km in the horizontal plane, and from a few tens to a few hundred of metres vertically. Models designed for regional applications have a horizontal resolution of about 30 km.

Variables

In each box of the network, the model calculates the three basic variables (temperature, humidity and horizontal wind speed) from their average values in the mesh. All other variables (pressure, vertical wind speed, energy flux, precipitation, clouds etc.) are deduced from these.

Equations

In a three-dimensional global climate model, the basic variables are found by numerically solving the equations that contain the fundamental conservation laws of physics and the equation of state. These laws are as follows:

- conservation of energy;
- conservation of mass (of air and water);
- conservation of momentum; and
- the gas laws – that is, the equation of state.

Parameterization

Phenomena that occur on a scale finer than the mesh size cannot be treated explicitly and must be parameterized. This applies mainly to turbulence, cloud formation and precipitation. In this way, parameterization is used to represent the effect of each process at the length scale of the network mesh.

Physical parameters

The equations are solved for the physical parameters that correspond to the current state of the system on Earth. In the present case of the AGCM, which describes the state of the atmosphere and hence also the temperature at the surface of the planet, this includes the following:

- The composition of the atmosphere (often reduced to the content of greenhouse gases).
- The irradiance (the amount of energy entering at the top of the atmosphere).
- The surface conditions:
  - the distribution of the different types of surface (oceans, land masses, icecaps, various types of vegetation);
  - the orography (terrain relief); and
  - the ocean surface conditions (temperature and sea-ice cover).

These properties, which express the *boundary conditions*, may depend on the choice of forcing selected by the modeller according to the question that is asked. One may, for example, wish to calculate the effect on the surface temperature of the solar flux, of the albedo, of a variation in the temperature of the ocean surface, of a different greenhouse gas content and so on.

The initial state

At the start of the experiment, an initial state is selected for all the variables that are directly involved in the solution of the equations (i.e. the starting values of these variables are defined). These variables are then calculated as a function of time (the model fixes the time step). From this initial state, changes in the atmosphere are calculated over a sufficiently long period of time for its characteristics (temperature and rainfall) to become stationary under the specified boundary conditions.

11.2.2 METEOROLOGICAL MODELS VERSUS CLIMATE MODELS

The aim of meteorological models is to forecast the weather on the planet for the hours or days that follow the moment t_0 at which the model is initialized. These models are continuously reinitialized. Such regular updating is based on data received from a network of sensors all around the globe. By contrast, once they are initialized, climate models simulate the climate over a sufficiently long period, at least of the order of several weeks, for the memory of the initial moment to be lost. This procedure removes the variability of the climate system, yielding an *average state that is independent of the initial conditions.*

11.2.3 COUPLED ATMOSPHERE–OCEAN GLOBAL CIRCULATION MODELS (AOGCMs)

In general atmospheric circulation models, the surface temperature of the ocean was at first taken as a fixed parameter. Interactions between the climate system and the surface were therefore very limited. What was obtained was, in fact, the circulation of the atmosphere at equilibrium with these fixed surface conditions. We know, however, that the ocean does not remain inert during a change of climate. To take account of changes in the ocean, the next major step involves constructing a model with three-dimensional ocean circulation, the Ocean General Circulation Model (OGCM). This describes the ocean completely, and in particular the temperature at its surface. As the surface temperature varies, thus influencing the energy equilibrium as well as the atmospheric conditions, it was a natural step to couple the ocean model with the atmospheric model, thereby giving birth to the AOGCMs. These became operational in the early 2000s.

11.3 Three-dimensional models: ever-increasing refinements

Just as the state of the ocean interacts with the climate system and adjusts to its changes, other physical parameters that are defined *a priori* at the outset are modified by the influence of the climate, on timescales that range from decades to thousands of years. Such parameters include, for example, the state of the biosphere, of the cryosphere, the chemistry of the components of the atmosphere and so on. The AOGCMs were gradually refined to include these interactions, which took on greater importance as more questions were asked. Models of the continental surface, which began as a very simple hydrological system with a fixed albedo and surface roughness, grew into models with interactive vegetation, in which surface properties could be calculated as a function of changes in vegetation, either as a result of changes in the climate or by the hand of man. Finally, global climate models now increasingly incorporate the chemistry of the atmosphere and of aerosols, as these have a strong effect on radiation, and major biogeochemical cycles such as the carbon cycle. In these models, the concentration of CO_2 in the atmosphere is no longer predefined, but is calculated as a function of emissions.

Finally, a new class of models has emerged, the 'Earth System Models of Intermediate Complexity', which bridge the gap between conceptual and general circulation models. These models contain a more complete description of the climate system than the simple ocean–atmosphere 'kernel', while their computation time is shorter than that of general circulation models. Changes in the climate system can thus be investigated over long timescales or, alternatively, numerous scenarios can be explored.

11.4 Climate models – what for?

Models have been developed mainly with regard to current observations relating to the past few decades. Their value lies in their ability to simulate our present climate and recent changes to it. The following three paths beckon to be pursued, each responding to a different concern – two of them refine our understanding of the workings of the climate engine, while the third addresses matters of current concern for human society:

- *Sensitivity tests* for climate forcing, processes or feedback mechanisms. For example, various trial values are attributed to a forcing, and the response of the climate system to these different conditions is analysed. By this means the signature – that is, the influence of each forcing on the ensuing climate change – can be discriminated from among several forcing mechanisms or changes to the boundary conditions.
- *Simulation of past climates* and comparison with reconstructions, using different proxies. This tests the ability of the model to simulate climates that can be very different from our own, on different timescales. Not only does it demonstrate the operational capacity of the models, but it also lends credibility to simulations of the future climate. Experiments range from climates similar to the present one – involving, for example, the changes over the past 1000 years – to climates that are far removed from ours, such as the glaciations of the Quaternary and glacial–interglacial transitions.
- *Changes in the future climate* that stem from different natural and anthropogenic forcing mechanisms. This area is concerned principally with the climate changes during the 21st century that follow in the wake of greenhouse gas emission and other disturbances from human activity. This is the mandate of the IPCC. An account of the IPCC 2013 report simulations is given in Part VI.

PART II SUMMARY

This part of the book describes the different processes that ensure energy equilibrium at the surface of the Earth, where $490\,W/m^2$ are required to maintain the climate, but only $340\,W/m^2$ are supplied directly by the source of energy, the Sun. To this end, we adopt a highly simplified model in which the atmosphere is represented by a homogeneous layer at uniform temperature. We then analyse the role of the constituents of the atmosphere as the radiation from the Sun and from the Earth passes through it.

At the surface, the balance of energy is based on the fact that the energy flux absorbed by the surface compensates for the flux that it loses. The flux supplied to the surface ($490\,W/m^2$) comes, on the one hand, from the thermal radiation emitted by the atmosphere toward the surface and, on the other, from the incident solar radiation. The energy flux that leaves the surface ($490\,W/m^2$) can be decomposed into:

- The thermal radiation emitted by the surface.
- The flow of latent heat required by the water cycle to evaporate the liquid water.
- The flux of perceptible heat that warms the atmosphere.

Apart from solar flux, all the fluxes in the surface energy budget depend on the surface temperature. This temperature is therefore a key factor that governs the energy budget, and adjusts accordingly in such a way that equilibrium is reached.

The flux of energy from the Sun and from the atmosphere that is absorbed by the surface ($490\,W/m^2$) is determined by the combined effect of three independent processes: reflection of solar radiation, which cools the surface of the planet; absorption of the solar radiation by the atmosphere (20%), which also cools the planet by a mechanism that we call the 'anti-greenhouse effect'; and, finally, absorption by the atmosphere (90%) of the radiation emitted by the Earth's surface. The last is responsible for the greenhouse effect mechanism and it warms the surface of the planet. A rough estimate of each of these processes is made for the highly simplified case in which each of these processes acts separately: it shows how the greenhouse effect is far greater than the two other processes. Although this approach is vastly simplified, it illustrates the crucial role played by the greenhouse effect in maintaining the mild climate that prevails at the surface of the planet.

With the help of this simplified approach, the detailed description of the interaction between the atmosphere and radiation shows how little of the solar radiation the atmosphere absorbs, and how much it adsorbs of the radiation from the Earth. We also establish the list of components in the atmosphere that are involved. The greenhouse effect appears to depend principally on water vapour, on CO_2 and on the clouds. Since the clouds also play a central role in solar reflection, however, the heating that they produce through the greenhouse effect is more or less compensated by the cooling that they produce through solar reflection.

How does a climate change start? When a change occurs that modifies any of the terms in the climate equilibrium, thereby generating *radiative forcing*. The warming (or cooling) that ensues introduces feedbacks. If it is positive, the total feedback amplifies the initial warming (or cooling).

The most refined of the current climate models concludes that these feedbacks will amplify an initial change by a factor between 2 and 3. Current climate models (three dimensions) couple the simulation of the atmosphere to that of the ocean and to other mechanisms. It is with these models that the simulations of the future climate were made and the results published in the recent IPCC 2013 report.

PART II NOTES

1. The absorption coefficient, $a_{abs}(\lambda)$, of a real body at temperature T depends on the wavelength λ of the incident radiation, because, unlike a black body, it absorbs only at certain wavelengths. The thermal radiation emitted at wavelength λ is that of a *black body* at temperature T, weighted by $a_{abs}(\lambda)$. In other words, a real body emits only in the wavelength range where it is able to absorb.

2. The energy flux E radiated at wavelength λ by a *black body* at equilibrium temperature T is given by

$$E(\lambda, T) = (2hc^2\lambda^{-5})/[\exp(hc/\lambda kT) - 1]$$

where the universal constants h, k and c are, respectively, Planck's constant ($h = 6.62 \times 10^{-34}$ J s), Boltzmann's constant ($k = 1.38 \times 10^{-23}$ J/K) and the velocity of light in vacuum ($c = 3.00 \times 10^8$ m/s). T is in Kelvin. The spectrum is continuous. This expression defines the flux (W) emitted per unit surface area (m^2), per unit solid angle (steradian) and per unit wavelength (m).

3. The surface of the Sun radiates like a black body at temperature T_S. The flux of energy radiated per unit area is σT_S^4 (W/m^2) and the total flux radiated into space is $\sigma T_S^4 \times 4\pi R_S^2$ (W), where R_S is the radius of the Sun. At the Earth–Sun distance D_{ES}, this flux crosses a sphere of radius D_{ES} and of surface area $4\pi D_{ES}^2$. An area of 1 m^2 located at this distance therefore receives the total flux divided by this surface area; that is, $\sigma T_S^4 \times R_S^2/D_{ES}^2$ (W). This is the flux that arrives perpendicularly at the outside of the Earth's atmosphere. With $R_S = 0.6956 \times 10^6$ km, $D_{TS} = 149.6 \times 10^6$ km and a surface temperature of $T_S = 5780$ K, the calculated flux is 1368 W/m^2, which is close to the value stated in IPCC 2007.

4. At present, the fraction of CO_2 in the atmosphere is about 400 ppm (395 ppm in 2013), or 400 molecules of CO_2 per million molecules of air. Since the molar mass of air is 29 g/mol, and that of CO_2 is 44 g/mol, there are 44×400 g of CO_2 per 29×10^6 g; that is, a ratio of 0.6×10^{-3} between the masses of these two gases. But above a surface area of 1 cm^2 at sea level, there is about 1 kg of air (the total mass of air in the column of atmosphere), part of which is the mass of CO_2. In this column, therefore, the amount of CO_2 is 0.6 g; that is, 0.6 g/cm^2.

5. The units ppmv, ppbv and pptv stand for 'parts per million', 'parts per billion' and 'parts per trillion by volume', respectively. When combined with the ideal gas law (which states that one molar volume is the same for every gas), they indicate the number of molecules of a given gas in 1 million (10^6), 1 billion (10^9) and 1 trillion (10^{12}) molecules of air, respectively. The reference to volume is often implicit, and the simplified expressions used here become ppm, ppb and ppt.

6. The release of SO_2 into the atmosphere by human agency gives rise to aerosol sulphates in the first few kilometres of the atmosphere. They remain for a week or two before being washed out by rain, thereby acidifying the rain. Some kind of 'justice' may be at work here: owing to the zonal character of atmospheric circulation, during this couple of weeks the air masses remain confined to the same latitudes and the acid rain falls mainly in the same latitudes as emit the SO_2. During the 1970s and 1980s, these emissions were particularly copious (Chapter 29) and gave rise to the infamous 'acid rain' episode, until at last vigorous measures were taken. The impact on the ecosystem was such, for example, that in many lakes in Sweden (where the underlying crystalline rock cannot buffer the acidity of the rainwater) the fauna was deeply perturbed. Aerosol sulphates generated in this way contributed to cooling the Earth's surface for several decades.

7. The ozone layer absorbs solar radiation between about 0.20 μm and 0.31 μm. The UV radiation that reaches the ozone layer is divided into UV-A (0.400–0.315 μm), the harmful UV-B (0.315–0.280 μm) and the very harmful UV-C (0.280–0.100 μm). UV-C is entirely screened out at an altitude of around 35 km by a combination of dioxygen (O_2) (< 0.20 μm) and ozone (O_3) (greater than about 0.20 μm). UV-B causes damage at the molecular level to the fundamental building block of life, deoxyribonucleic acid (DNA). UV-B can also be harmful to the skin, causing sunburn. Excessive exposure can cause skin cancer. The ozone layer is highly effective at screening out UV-B. Ozone is transparent to most UV-A.

8. The terrestrial radiation emitted by the Earth's surface that is directly transmitted through the atmospheric window was usually estimated to be 40 W/m^2 (IPCC 2007). Atmospheric absorption by the water vapour continuum decreases this value to approximately 20 W/m^2.

PART II FURTHER READING

Archer, D. (2012) *Global Warming – Understanding the Forecast*. John Wiley & Sons, Hoboken, NJ; see Part 1.

Climate Change 2013: The Physical Science Basis. Working Group I Contribution to the Fifth Assessment Report of the Intergovernmental Panel on Climate Change (IPCC), Cambridge University Press; available at http://www.ipcc.ch/report/ar5/wg1/-; see Chapter 8.

Houghton, J. (2009) *Global Change – The Complete Briefing*, 4th edn. Cambridge University Press, Cambridge, UK; see Chapters 2, 3 and 5.

Neelin, J.D. (2010) *Climate Change and Climate Modeling*. Cambridge University Press, Cambridge, UK; see Chapters 2–6.

PART III

THE DIFFERENT CAUSES OF CLIMATE CHANGE

Chapter 12
The choice of approach

Now that all the ingredients of the system are known, it is natural to ask how and why their effects can vary in the course of time. In a later chapter, we investigate the 'how' by looking at changes that have taken place in the past. For the moment, we concentrate on the 'why', on the various causes that can bring about changes in the climate. These are numerous and their origins are very diverse. They can be introduced in several ways: according to their properties, or according to the timescale over which they act and so on. One such method, for example, is adopted in Fig. 12.1, with the causes being divided into four different categories (astronomical and geological forcing, intrinsic and anthropogenic causes), together with their associated characteristic time constants (Bard 2006). The approach that we employ here, however, is different.

The basis of the argument we use is the flux of solar energy, since it is the main source of energy at the surface of the Earth. We seek to know how changes in the flux can arise, from its starting point to its arrival at the surface of the planet. We conclude the account by examining the mechanisms involved in the distribution of energy at the surface of the planet. The successive stages are summarized in Fig. 12.2.

Our starting point is therefore the solar energy radiated by our star and its changes in the course of time: this energy can exhibit short-term fluctuations or it may gradually change in the long term as a result of the internal dynamics of the Sun. We then consider the fraction of the energy emitted by the Sun that reaches the Earth: this depends on the distance to the Earth and is a function of the characteristics of the Earth's orbit, whether in its annual average or according to the different seasons. Then comes the passage of the Sun's radiation through the atmosphere, which, through mechanisms described in the previous chapter, can bring about changes in the average energy available at the Earth's surface. Changes in the composition of

Fig. 12.1 The different causes of climate change are grouped here into four classes. Astronomical and geological forcing, as well as human causes, are 'external' forcings because, unlike intrinsic causes (internal forcing), they are independent of the climate.
Source: Bard (2006). Reproduced with permission of E. Bard.

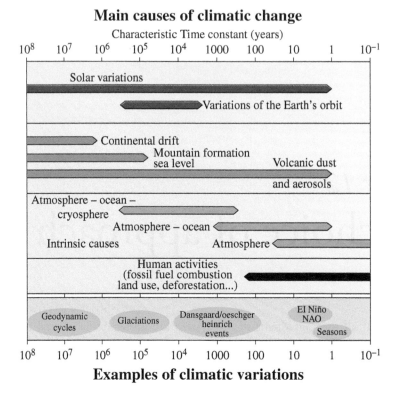

Main causes of climatic change

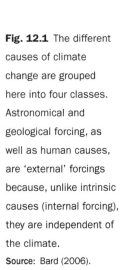

Examples of climatic variations

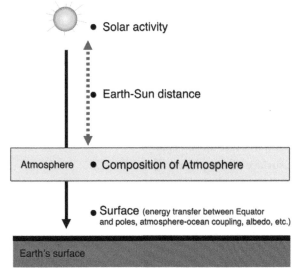

Fig. 12.2 The principal stages in the transmission of energy from the Sun to the Earth's surface: (a) solar activity determines solar emission; (b) the distance from the Sun to the planet determines the solar flux arriving on Earth; (c) the composition of the atmosphere modifies the energy transfer from the Earth's exterior to ground level; and (d) mechanisms acting at the surface of the Earth determine how the incoming energy is distributed.

the atmosphere, be it over billions, millions or even over only tens of years, have an impact on the average climate on Earth. Lastly, at the surface of the Earth, various parameters and mechanisms can influence the solar energy received, either by modifying the ability of the surface to absorb it, or by changing the long-term surface energy distribution (e.g. by altering the transfer of energy from the Equator to the poles), or again through the oscillations that couple the vast reservoirs of fluid formed by the atmosphere and the oceans.

These different stages will be recalled in turn, each with its corresponding timescale. In this part, therefore, we do not discuss the consequences on the climate of changes in the different parameters. This is treated in Part IV. Instead, here we give an overview of the natural causes of climate change.

The factors discussed in this part are not exhaustive. One other at least should be mentioned: types of land cover. This affects the absorption of the solar flux at the Earth's surface and also, through latent heat transfer (the water cycle), the energy distribution. Although solar radiation is absorbed almost completely by oceans and lakes, it is partly reflected by vegetation and more by deserts, while snow and ice-covered surfaces can reflect practically all of it. In addition, the different surface covers do not have the same ability to maintain the water cycle. Changes in the surface conditions therefore modify the energy budget at the surface and can produce a change in climate. Surface conditions can reinforce a climate change through feedback. The impact of snow-covered surfaces during a climate change, for example, neatly illustrates one feedback mechanism: a colder climate creates a larger surface area under snow, which reinforces the cooling by reflecting the solar radiation more strongly. This powerful mechanism is most active at middle to high latitudes. Likewise, changes in the albedo when deserts extend or, conversely, when vegetation develops, influence the climate: the life cycle, which responds to the climate, influences it in turn. Humankind, through its recently acquired ability to modify continental surfaces on a large scale, has been a contributing factor in recent decades.

Among the factors we have enumerated above, we omit climate changes that could arise from events such as meteorite impacts or other natural catastrophes. These lie outside the scope of this book.

Chapter 13
The Sun's emission

13.1 The impact on the climate

As the energy supplied by the Sun is almost the only source of energy available for heating the surface of the planet, a reduction in solar emission is necessarily accompanied by a drop in temperature of the Earth's surface. The energy radiated by the Sun is the result of its activity and it varies in the course of time. Whether our Sun 'flickers' or evolves gradually depends on its complex internal dynamics. This solar activity has changed slowly since the birth of the solar system 4.57 billion (thousand million) years ago, increasing by about one third during the first 2 billion years of its life. We confine ourselves here to timescales covering the last hundreds of millions of years. Over this period, solar activity was in the main stable, but it underwent cycles and fluctuations, of which we shall indicate only the most important.

13.2 How emission varies

13.2.1 OVER A DECADE

The principal cycle of the Sun is the 11-year cycle. Since the 1980s, satellites have accurately measured the flux of solar energy received at the exterior of the Earth's atmosphere (Fig. 13.1). The average value the solar flux arriving at each square metre (i.e. on a surface placed perpendicular to the solar flux) was until recently taken to be $1367\,W/m^2$. Very recently, however, more precise satellite measurements have revised this average value to about $1361\,W/m^2$

Climate Change: Past, Present and Future, First Edition. Marie-Antoinette Mélières and Chloé Maréchal.
© 2015 John Wiley & Sons, Ltd. Published 2015 by John Wiley & Sons, Ltd.
Companion website: www.wiley.com\go\melieres\climatechange

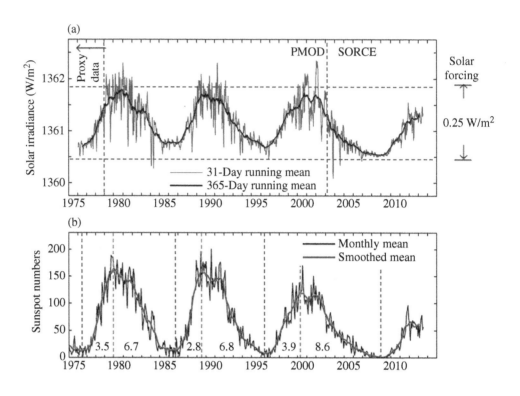

Fig. 13.1 (a) The solar energy flux arriving at the Earth per square metre perpendicular to the Sun's rays, or the Total Solar Irradiance (W/m²) from composite satellite-based time series from 1975 to 2013. The irradiance varies with an amplitude of about 1‰ in an 11-year cycle. (b) The number of sunspots at the surface of the Sun follows the same cycle. **Source:** Reproduced by permission of Makiko Sato.

(Kopp & Lean 2011). During the 11-year cycle, the amplitude of the variation in the solar flux is no greater than 1.5 W/m². This corresponds to a variation in the energy absorbed by the total surface of the planet of about 0.25 W/m² (Part III, Note 1). We saw in Chapter 10 that, due radiative feedbacks, a radiative forcing of 1 W/m² produces a global warming of +0.8°C (between 0.4 and 1.2°C) when the equilibrium temperature is reached. The temperature change induced by a radiative forcing of 0.25 W/m² per cycle would therefore be no greater than 0.2°C: the effect of the cycle on the climate is very weak.

This cycle is detectable to the naked eye (BUT NEVER LOOK DIRECTLY AT THE SUN WITHOUT A FILTER!), with darker regions sometimes appearing at the surface of the Sun. The more sunspots there are, the greater is the energy radiated by the Sun (Fig. 13.2). During the 11-year cycle, the number of sunspots falls almost to zero when solar activity is at a minimum, then increases again to more than a hundred when the activity is at a maximum.

Fig. 13.2 Sunspots at the Sun's surface on 29 April 2000. Photograph taken by the SOHO space probe. Left, natural light emission; right, UV emission. Sunspots are sites of intense activity. **Source:** NASA Earth Observatory.

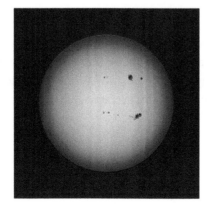

13.2.2 DECADES AND CENTURIES

The number of sunspots at the emission maximum of the 11-year cycle was the first indicator to be used to reconstitute the qualitative changes that have occurred in solar emission over the past centuries. Starting from 1609, with the invention of the Galilean telescope, observations of this sunspot number were achieved with ever-greater precision. After that date, variations in the sunspot number were found to reveal periods of weaker activity that extended over several decades, or even centuries. These reductions in solar activity spill over into the climate, thereby affecting the whole of the planet. Climate fluctuations that have occurred over the past millennium, with its periods of advancing and retreating glaciers, are most likely a direct illustration of this. In addition to visual observations of the number of sunspots, other techniques exist that track solar activity over several centuries and several millennia.

13.2.3 THE PAST MILLENNIA

Reconstructions of solar activity over earlier centuries and millennia are mainly based on the amounts of certain radionucleides created in the very high atmosphere under the action of cosmic rays (the flux of highly energetic particles emitted by stars or supernovae that are constantly received by the Earth). The two radionucleides most commonly used are ^{14}C, the half-life of which is 5730 years (Part III, Note 2) and ^{10}Be (half-life 1.5×10^6 years). *What is the relationship between solar activity and the amount of radionucleides formed?* The penetration of cosmic rays into the Earth's atmosphere is controlled by the magnetic field of the Sun, which acts as a shield, attenuating this cosmic radiation. The greater the field, the fewer cosmic rays penetrate and the fewer radioneucleides are formed. Thus periods of strong solar activity, and hence of a strong solar magnetic field, are associated with reduced formation of ^{14}C and ^{10}Be. The amounts of these radionucleides, which are deposited regularly on the Earth's surface, are recorded in different types of sediment. These enable the solar activity to be retraced and its variations reconstituted as a function of time. The reconstruction involves numerous intermediate steps in which the uncertainties add to each other.

All the different reconstructions are in agreement with the finding that during the Holocene (the warm period that started about 12,000 years ago), solar activity has fluctuated slightly, comparably to what occurred in the Little Ice Age (LIA) (the recent cold spell that marked recent centuries).

13.3 What are the consequences?

Solar fluctuations must thus be responsible for fluctuations of the climate that can last from several decades to several centuries. On the timescale of several millennia – for example, during the Holocene – these fluctuations are considered to be the cause of the advancing and retreating phases of the glaciers. The most recent example, which is also the most widely studied, is that of the Little Ice Age that took place during the last millennium. In Part IV, we discuss this fluctuation in more detail, because the episode illustrates how a natural fluctuation at some time in the future would cause planet-wide cooling.

Since the fluctuations are inherent to our star and independent of the climates that are generated on Earth, they must also have existed with comparable frequency and intensity under earlier climates, including glaciations. The interest of this topic, which is a focal point of study and debate, is to determine to what degree the fluctuations that occurred during the last great glaciation (which lasted several tens of thousands of years) are due to solar fluctuations or are the result of other mechanisms.

Chapter 14
The position of the Earth with respect to the Sun

14.1 An overview

In Part I, we saw that the Sun is the source of energy that feeds the Earth's 'climate machine'. The greater the distance from the Sun, the smaller is the flux of energy that reaches the Earth. The distance between the Sun and the Earth, D, is therefore the major factor in the climate. The flux that strikes a surface perpendicular to the Sun's radiation decreases as $1/D^2$ (Part III, Note 3). *How does this distance vary with time? In other words, how does the energy available for the climate vary?* Here, we restrict ourselves to the past few million years. During this period, the annual average of the distance between the Earth and the Sun has been practically stable. This implies that the total amount of energy received by the Earth in 1 year is almost constant. *Does this mean that this distance has no effect on the climate?* No, because over the millennia, the different seasons do not take place at the same distance from the Sun. Due to the attraction of the different bodies in the solar system, in particular the Sun, the Moon and neighbouring planets, the distance D in a given season varies cyclically, with several periods that range from thousands to hundreds of thousands of years. These oscillations in distance add to those of the tilt of the Earth's rotational axis (obliquity), which gives rise to the seasons. Over the millennia, this produces a variation in the seasonal solar energy flux received at each latitude. We now look at how the solar flux varies with the different seasons at different latitudes over the millennia.

Climate Change: Past, Present and Future, First Edition. Marie-Antoinette Mélières and Chloé Maréchal.
© 2015 John Wiley & Sons, Ltd. Published 2015 by John Wiley & Sons, Ltd.
Companion website: www.wiley.com\go\melieres\climatechange

14.2 Irradiance, determined by orbital parameters

The solar energy flux or *irradiance* received at each latitude on Earth during the year is determined mainly by the combination of three orbital parameters that define the position of the Earth with respect to the Sun. These parameters, which are discussed in the following paragraphs, are as follows:

- *Obliquity*. This is the tilt of the axis of rotation of the Earth (the axis that joins the geographical poles) with respect to the perpendicular to the ecliptic plane of the Earth's orbit around the Sun (the *z*-axis) (Box 1.1, Fig. B1.1).
- *The shape of the Earth's orbit* around the Sun. This is an ellipse, almost a circle, in which the Sun occupies one of the foci. The ellipse is characterized by its *eccentricity* ε.
- *Precession of the axis of rotation* of the Earth around the *z*-axis perpendicular to the ecliptic plane.

Parameter 1, the obliquity, is the cause of the seasons. The greater its value, the more pronounced are the seasons. At middle and high latitudes, for example, the difference in irradiance between summer and winter is thus greater. High obliquity brings hot summers. The combined effect of the other two parameters (the eccentricity of the ellipse and precession) defines the distance between the Sun and the Earth in each season (parameter 2), and how it changes over the millennia. Here, we recall that the calendar of the year is defined with respect to the seasons: the dates of the summer and winter solstices are fixed at 21 June and 21 December in the Northern Hemisphere (NH), and vice versa in the Southern Hemisphere.

The combined effect of parameters 1 and 2 – that is, the obliquity and the distance between the Earth and the Sun in each season – determines the solar flux received throughout the seasons at different latitudes of the planet. Their cyclic variations produce cyclic variations in the climate.

14.3 Changes in obliquity: the impact on the seasons

The first parameter, the *obliquity*, is the angle between the axis of rotation and the normal O*z* to the plane of the ecliptic. This angle varies over time in response to the gravitational forces of the large planets (mainly Jupiter). It varies between about 24.5° and 21.9°, with a period of almost 41,000 years. Its qualitative behaviour over the past million years is shown in Fig. 14.1, and in greater detail in Fig. 14.2. Its present value is 23°27′. About 11,000 years ago (i.e. 11 kyr, expressed in the commonly used unit kiloyears, used particularly for astronomical periods), the obliquity passed through a maximum (more than 24°); now it is decreasing towards its minimum value, 21.6°, which will be reached in about 10,000 years, one half period (about 20,000 years) after the maximum. This variation has three important consequences.

Obliquity and annual irradiance

How does annual irradiance vary with changes in the obliquity? First, we look at the solar energy received over the year by the whole planet. This annual solar energy depends only on the Earth–Sun distance, and does not change when the obliquity changes. The solar energy received by the planet, whether it is the instantaneous value or integrated over the whole year, is therefore independent of the obliquity.

What happens, then, to the annual solar flux arriving at a given latitude when the obliquity varies? To answer this question, we consider two extreme situations (in which the Earth's orbit round the Sun is taken to be circular) and compare in each case the annual energy received, for example, at the poles. In the first case, the obliquity is zero (0°): the Sun's rays arrive tangentially at the poles all year long and the energy received at the poles is zero. In the second case, the obliquity is 90°. During the 6 months of the year around the winter solstice, the pole receives no rays from the Sun. But between the spring and the fall equinoxes, the pole is exposed to sunlight 24 hours a day, with the angle of incidence of the Sun's rays increasing until, at the summer solstice, they strike the pole vertically. Thus, over the year, the amount of solar radiation received by the poles depends on the obliquity, and increases with it. The same reasoning

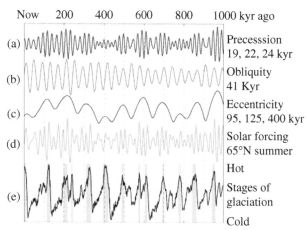

Fig. 14.1 The variation of the orbital parameters over the past million years, with their main periods: (a) precession of the axis of rotation (polar axis), expressed in the change in the Earth–Sun distance in June; (b) tilt of the axis of rotation; (c) eccentricity of the ellipse described by the Earth around the Sun; and (d) solar forcing at latitude 65°N in summer. The black curve in (e) illustrates the chronology of the ice-sheet volume deduced from the isotopic signal of benthic foraminifera (Lisiecki & Raymo 2005). High sea levels (in grey) correspond to warm stages (see the caption to Fig. 18.4). The numbering refers to the marine isotopic stages (odd numbers are interglacials or interstadials, and even numbers are glacials).

Source: Image created by Robert A. Rohde/Global Warming Art/CC-BY-SA-3.0/GFDL.

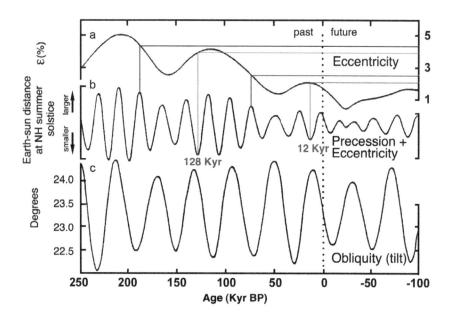

Fig. 14.2 The changes during the past 250,000 years and in the coming 100,000 years: (a) in the eccentricity of the Earth's orbit; (b) in the Earth–Sun distance at the NH summer solstice; and (c) in the tilt of the Earth's axis of rotation.

can be applied to the Equator: here again, the annual solar energy received at the Equator varies with the obliquity, but it *decreases* with increasing obliquity.

The increase in obliquity, which does not alter the total amount of energy received by the planet, nevertheless redistributes it at the different latitudes. The yearly energy received increases for latitudes situated above the tropics (middle and high latitudes), and decreases for those situated below (low latitudes), keeping in mind that the latitude of the tropics varies with the obliquity, by a few degrees.

High latitudes versus low latitudes

We come back to the middle and high latitudes. The greater the obliquity, the more pronounced are the seasons. The effect is increasingly marked at higher latitudes, as seasonal variations increase with latitude. High latitudes are thus most sensitive to variations in obliquity. The climate changes in these regions over tens of thousands of years are marked by this 41,000-year cycle, while in the tropical regions this cycle is much more attenuated.

14.4 Changes in the Earth's orbit and eccentricity: the impact on the Earth–Sun distance

Fig. 14.3 The Earth's orbit round the Sun is an ellipse, with one focus occupied by the Sun. The eccentricity of an ellipse is defined by the ratio c/a, where c is the distance between a focus and the centre, and a is the semi-major axis. The eccentricity of the Earth's orbit changes slowly in time, with a maximum value of 6%. Its present value is 1.67%.

The elliptical orbit of the Earth, in which the Sun occupies one of the foci, is defined by its eccentricity, ε (Fig. 14.3): ε is the ratio c/a, where c is the distance from the centre of the ellipse, O, to a focus (e.g. to the Sun), and where a is the semi-major axis of the ellipse (the distance OP or OA). The present-day value of ε is 0.0167, or 1.67%, which means that the current orbit of the Earth is almost circular.

Changes in eccentricity

Over the millennia, the value of the eccentricity changes. The eccentricity evolves in time with a principal period of 412,000 years and secondary periods (131,000, 123,000 and 99,000 years) that are clustered around 100,000 years.

Over the past million years, ε has fluctuated between a value close to zero (quasi-circular orbit) and its maximum value, 6%. These changes are shown qualitatively in Fig. 14.1. The numerical values are given in Fig. 14.2, for the past 250,000 years and for the coming 100,000 years. Every 412,000 years, the eccentricity thus passes through a minimum, and the orbit becomes almost circular. This was the situation about 800,000 years ago, and then 400,000 years ago. At present, ε is again small and will continue to decrease, reaching its minimum value close to zero in about 30,000 years. Superimposed on this 412,000-year periodicity is another periodicity of about

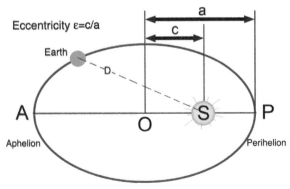

100,000 years that appears clearly in the figures. In particular, the three most recent maxima of ε occurred about 200,000, 100,000 and 10,000 years ago, respectively.

Eccentricity and irradiance

What is the impact of the variation in eccentricity on the annual solar flux received by the Earth? The Earth receives slightly more energy when the eccentricity increases. This variation in energy is very small, however. Over the past million years, it amounts to no more than 0.1%, which is too small to initiate the alternating climates considered here.

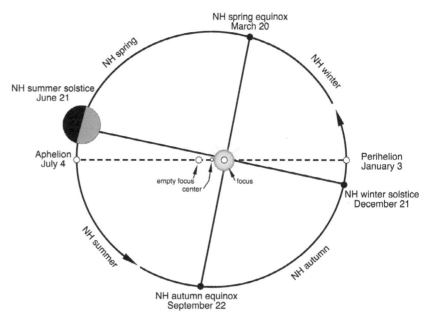

Fig. 14.4 At present, the Earth is at perihelion on 3 January and at aphelion on 4 July. The seasons shown refer to the Northern Hemisphere.

What is the effect of the variation in eccentricity on the solar energy received on Earth at different times of the year? Let us first evaluate the effect of eccentricity on the Earth–Sun distance D. In the course of the year, D varies. At perihelion P, the Earth is closest to the Sun, and at aphelion A, it is furthest away (Fig. 14.3). At present, Earth reaches perihelion on 3 January ($D = 147.1 \times 10^6$ km) and aphelion on 4 July ($D = 152.1 \times 10^6$ km) (Fig. 14.4). The difference between these extreme values is dD, and the relative difference is dD/D_m, where D_m is the mean distance between the centre of the ellipse and the Earth, at perihelion or at aphelion. Since the eccentricity is small, the value of D_m is very close to the yearly average of the Earth–Sun distance. The value of dD/D_m is equal to $2ε$ (Part III, Note 4). As already seen in Section 14.1, the irradiance E received on Earth depends on the distance D and varies as $1/D^2$. Hence the relative variation of irradiance during the year, dE/E is, to a good approximation, equal to $4ε$ (Part III, Note 5).

Currently, dE/E is equal to about 6.7%. This quantity is also equal to dE_s/E_s, where E_s is the average solar energy flux received per square metre at the Earth's surface. Since this flux is $340\,W/m^2$ (Chapter 4), the difference dE_s amounts to $23\,W/m^2$. *Thus, at the beginning of January, when the Earth is at perihelion, it receives 23 W/m² more than at the beginning of July at aphelion* (Part III, Note 6).

Precession of the orbit with respect to the stars

Lastly, the Earth's orbit exhibits a cyclic variation of a different order. The orientation of the ellipse, defined by the direction of its major axis in the ecliptic plane, is not constant, but rotates around the z-axis. It performs a complete rotation in about 135,000 years. This precession must be taken into account in order to evaluate the variation in solar flux arriving at different latitudes for each season.

14.5 Precession of the axis of rotation: the impact on the Earth–Sun distance at different seasons

The third parameter is the precession of the Earth's axis of rotation; that is, its polar axis. The direction of this axis, which remains practically constant for decades (Part III, Note 7), can be identified in the sky by any star that happens to lie on this axis. During the 24 hours of the Earth's rotation, this star appears to be immobile. At present, this is almost the situation of the pole star (Polaris).

Precession

The direction of this axis is not constant. The ellipsoidal shape of the Earth, which is due to the centrifugal force that broadens the Equator, makes the axis precess with a period of 25,772 years (a value adopted in 2000 by the International Astronomical Union). The rotational axis describes a cone around the z-axis of the ecliptic plane in the reference system of the stars. The axis thus traces out a circle in the celestial sphere (Fig. 14.5). Over the coming millennia, the pole star will gradually vacate its position, thus leaving its place to other rivals. This motion resembles that of a top that is spun on a table; the axis of rotation of the top gradually leans away from the vertical due to friction with the table and, as soon as the axis starts to tip, it begins to precess, describing a cone. In our case, the role of the top is played by the Earth, spinning on the ecliptic plane. Since friction is absent, the tilt of its rotational axis is constant and it precesses, describing a cone around the z-axis. This precession does not change the

Fig. 14.5 Due to precession, over a period of 25,772 years, the polar axis traces out a cone in the sky of about 23° around the axis Oz perpendicular to the ecliptic plane (Fig. 14.6). The blue points indicate positions of the celestial North Pole at various times during a precession cycle in the past and in the future. At present, the polar axis points practically directly towards Polaris (in yellow), in the constellation of Ursa Minor.
Source: Adapted from Berger (1992). Reproduced with permission of André Berger.

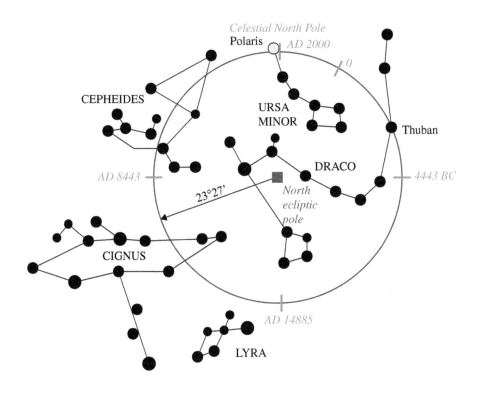

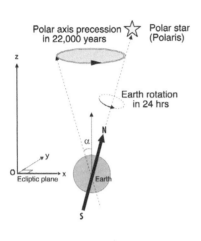

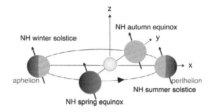

(a) Summer solstice in northern hemisphere occurs closer to Sun (at perihelion)

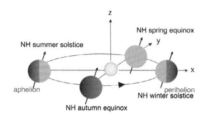

(b) 11,000 years later, the NH summer solstice occurs further from Sun (at aphelion)

Fig. 14.6 The axis of the Earth's rotation (polar axis) is tilted at an angle α with respect to the axis Oz normal to the ecliptic plane (left). Every 11,000 years, due to the precession of the polar axis around Oz, the position of the Earth at the summer solstice in the Northern Hemisphere is successively close to the Sun (top) and then far from the Sun (bottom).

obliquity. It simply reorients the axis differently with respect to the rest of the universe – notably with respect to the major axis of the Earth's elliptical orbit.

We have just seen, however, that the ellipse itself also precesses around the z-axis with respect to the stars, in approximately 135,000 years. These two movements (precession of the axis of rotation with a period of 25,772 years, and precession of the ellipse) around the z-axis combine to yield the following situation. If we place ourselves in the framework of the ellipse Oxyz (Fig. 14.6), in which the major axis Ox and the minor axis Oy are both stationary, the axis of rotation of the Earth appears to precess around Oz with a period of the order of 22,000 years (Fig. 14.6). In reality, this movement is composed of several periods (again expressed in kiloyears), 23.7 kyr, 22.4 kyr and 19.0 kyr, which, for simplicity, we reduce to a single period of 22 kyr. This precession is referred to as the *climatic precession*, or the *precession of the equinoxes*.

Precession and the solar flux received

What is the effect of the precession on the energy received from the Sun by the whole planet throughout the year? As precession does not modify the distance from the Earth to the Sun, it modifies neither the amount of energy received annually nor that received daily by the whole planet. This, however, is not true at a given latitude for a given season …

… or at a given latitude and in a given month? We consider, for example, the situation illustrated in Fig. 14.6, where the axis of rotation is inclined in such a way that during the year, the summer solstice in the NH occurs when the Earth is at perihelion, and 6 months later at the NH winter solstice, when the Earth is at aphelion. After half a cycle of precession, 11,000 years later, the situation is reversed. The NH summer solstice then occurs when the Earth is at aphelion and the NH winter solstice occurs at perihelion. Over the 11,000 years, the NH summer solstice gradually shifts around the ellipse. In the course of the 22,000-year cycle, perihelion will fall successively on each day of the year. If we now

consider the longer span of a month, rather than a day, in the course of 22,000 years, each month will successively take place at a given position on the ellipse (e.g. perihelion). Thus the separation between two consecutive months at this position on the ellipse is close to 2000 (22,000/12) years, since the whole of the ellipse is spanned by the full 12 months in 22,000 years.

We focus our attention on June, the month of the NH summer solstice. As will be seen in Part IV, the NH summer is a key factor in the climate oscillations that recur over tens of thousands of years. During a precession cycle, this NH solstice month falls alternately when the Earth is at perihelion, close to the Sun, and then, 11,000 years later, when the Earth is at its aphelion, far from the Sun. At middle and high latitudes in the NH, the summers alternate in being warmer, then cooler.

Changes in the Earth–Sun distance in June are a key parameter that will guide us through the epic of the ice ages and also the saga of the monsoons. These changes since 1 million years ago are illustrated in Fig. 14.1, and in greater detail in Fig. 14.2. We now examine two examples that will be used later in Part IV:

- During the present precession cycle, this distance passed through a minimum 12,000 years ago, followed by a maximum 1000 years ago. Between perihelion and aphelion, as noted above, the relative difference in the irradiance received on Earth is defined by the eccentricity ($dE/E = 4\varepsilon$). The value of ε was 1.96% 12,000 years ago, and it was 1.72% 1000 years ago (Fig. 14.2). It follows that the irradiance at the NH summer solstice decreased from $353\,(340+13)\,\text{W/m}^2$ when the Earth was at perihelion in June 12,000 years ago, to $328\,(340-12)\,\text{W/m}^2$ when the Earth was at aphelion in June 1000 years ago. This means that the energy received between these summer solstices fell by about 7.1%.
- In the past 200,000 years, the closest Earth–Sun distance in June occurred 128,000 years ago, at the beginning of the last interglacial period. At that time, the eccentricity was 4.05%, and at the NH summer solstice the solar flux at the Earth was $368\,(340+28)\,\text{W/m}^2$, an irradiance far greater than that of 12,000 years ago.

We have seen that the maxima and minima of the Earth–Sun distance D in June are determined by the eccentricity ε of the ellipse. If, for example, ε were zero, the Earth's orbit would become circular and the distance in June (or in any month) would be constant throughout the year, as well as throughout the 22,000-year cycle. The precession cycle would then no longer affect the irradiance in the different seasons. *It follows that precession affects the climate only through the eccentricity.* Since, over the millennia, the eccentricity changes, the amplitude of the 22,000-year oscillation in the Earth–Sun distance in June is therefore modulated by the 100,000-year cycle of the eccentricity (Fig. 14.2).

Summary

The impact of precession is revealed by the fact that the Earth travels in an elliptical orbit around the Sun, not a circle. The greater the eccentricity, the greater is the effect of precession. Precession does not modify the amount of solar energy received annually at each latitude, but it does modify its distribution during the different seasons.

For a given latitude above the tropics, and for a given hemisphere, the winter and summer seasons become increasingly pronounced in the course of a precession cycle as the summer solstice in that hemisphere approaches perihelion, and then progressively less marked 11,000 years later as the summer solstice moves close to aphelion. Seasonal variation is stronger closer to the polar regions. During a precession cycle, if the seasonal character is enhanced in one hemisphere, it is attenuated in the other, since the seasons are out of phase.

We shall see that these modifications generate feedback mechanisms that can change the total energy balance of the planet and drive climate changes.

14.6 Changes in irradiance

In summary, all the information we need to estimate the variation in irradiance as a function of time, for each latitude and for any given month in the year, is contained in just two curves (Figs. 14.2b & 14.2c):

- The change in the Earth–Sun distance in the month under consideration (June is generally selected because it is the month of the summer solstice in the NH – as we shall see, this period of the year plays an essential role in climate change). The change in the Earth–Sun distance for June gives the variation of the total energy received by the Earth in that month.
- The change in obliquity. The variation in obliquity indicates how the distribution of this energy varies at the different latitudes: the greater the obliquity, the stronger is the irradiance in summer at middle and high latitudes.

The succession of climates over the past million years can be understood qualitatively by reading these two curves simultaneously (Part IV). One may ask, for example, *when will summer irradiance be at a maximum in the NH at middle and high latitudes?* Answer: when the two effects reinforce each other and enhance irradiance; in other words, a small Earth–Sun distance in June combined with high obliquity. *When will this irradiance be at a minimum?* Answer: when the Earth–Sun distance in June is large and the obliquity is small.

What is the identifying property of the changes in irradiance that take place over thousands of years, for a given latitude and a given month? The variations of irradiance with time are a composite of three different cycles: eccentricity, obliquity and precession. When the combination of these cycles is analysed over several hundred thousand years, a signal emerges that contains a series of frequencies containing periods of about 100,000 years (eccentricity), 41,000 years (obliquity), and a cluster of periods of about 23,000 years, 22,000 years and 19,000 years (precession).

Chapter 15
The composition of the atmosphere

15.1 The effect on the climate: the mechanism

As we have already seen, the composition of the atmosphere governs three main functions:

- reflection of solar radiation, which, by restricting the heat arriving at the surface, cools it (this function is fulfilled by clouds, dust and aerosols and scattering by the air);
- absorption of solar radiation – this also deprives the surface of heat and cools it (it involves certain gases, dust, aerosols and clouds); and
- absorption of radiation emitted by the surface of the Earth – this warms the surface of the planet through the greenhouse effect (contributions from greenhouse gases, clouds, dust and aerosols).

Finally, in the atmosphere the water cycle plays a crucial transverse role in the climate:

- in the energy transferred by evaporation/condensation from the Earth's surface to the atmosphere;
- through water vapour, the largest contributor to the greenhouse effect; and
- by the clouds, which contribute both to cooling and heating the planet (solar reflection and greenhouse effect).

Climate Change: Past, Present and Future, First Edition. Marie-Antoinette Mélières and Chloé Maréchal.
© 2015 John Wiley & Sons, Ltd. Published 2015 by John Wiley & Sons, Ltd.
Companion website: www.wiley.com\go\melieres\climatechange

Changes in some constituents of the atmosphere can thus generate either cooling through reflection or absorption of solar radiation (e.g. volcanic aerosols) or heating (greenhouse gases) by absorption of radiation from the surface of the Earth (terrestrial radiation), or both simultaneously (clouds etc.).

15.2 How the composition has changed, and why

To retrace the principal changes in composition of the atmosphere since the birth of the planet would require more than one chapter. Following the rapid disappearance of the Early Atmosphere, a new secondary atmosphere was created from the outgassing of the Earth's crust by volcanoes, mostly with CO_2 and H_2O. In the first few billions of years, the CO_2 content gradually decreased. Under the action of the water cycle, it was progressively stored in the form of carbonate, and is found in the great calcareous mountain ranges of the planet. Dioxygen (O_2) emerged, mainly due to biological activity. Over the past billion years (the Cambrian, Carboniferous etc.), the CO_2 level is believed to have undergone large fluctuations. These are some of the stages that are most often invoked to explain the temperate climate and the existence of liquid water on Earth. The composition of the main components of the atmosphere (O_2 and N_2) appears to have been stable for many millions of years. *What about other gases?* For the past million years, the main features of the atmospheric composition have been reconstituted from the contents of air bubbles trapped in ice sheets. Since the middle of the 20th century, this composition has been determined by direct atmospheric measurements.

15.2.1 GREENHOUSE GASES AND DUST

Ice records yield the composition of the atmosphere over the past 800,000 years. The precision of modern analysis reveals the changes in trace gases in the atmosphere (CO_2, CH_4 and N_2O), as well as the dust content. In particular, these records enable us to appreciate the devastating change that the atmosphere has undergone under the impact of human activity in the past two centuries.

The greenhouse gas (GHG) concentrations (restricting ourselves to carbon dioxide, CO_2, and methane, CH_4) vary on different timescales. At a given location, their concentrations vary slightly over the year according to the season, because their emission from the Earth's surface (and their absorption) is related both to photosynthesis and to the temperature of the soil and the ocean. From the beginning of the present warm interglacial period 12,000 years ago until the Industrial Revolution (18th century), the global averaged GHG concentration hardly varied (CO_2 varied between 260 and 280 ppm, and CH_4 between 580 and 720 ppb). *And what happened before that?* For several hundred thousand years, the concentration fluctuated between two values. The higher value is associated with interglacial (warm) periods; the lower value is representative of glacial periods. For CO_2, the difference is about 30%, and for CH_4, a factor of two. These glacial–interglacial variations are the consequence of the difference in climate, and the difference in ocean circulation.

Similarly, the dust concentration in the atmosphere varies strongly, increasing in glacial periods. As the atmosphere is drier and colder during these periods, dust is removed much less efficiently. Moreover, with the decrease in vegetation cover and lower sea level, new sources of dust emerge.

15.2.2 VOLCANIC AEROSOLS AND WATER VAPOUR

Strong explosive volcanic eruptions inject aerosols into the stratosphere. In that stable stratified region of the high atmosphere, the water cycle is insignificant. This means that volcanic aerosols that are injected reside there for 2–3 years before falling back into the troposphere (the lower part of the atmosphere, lying between zero and roughly 12 km), where they are washed out. This residence time is long enough for the aerosols to disperse all around the Earth, thereby guaranteeing a worldwide impact on the climate. They cool the surface of the Earth mainly by screening the incoming solar radiation. The cooling effect of strong eruptions is on average a few tenths of a degree, but their effect is limited in time to about a year (Box 15.1).

BOX 15.1 VOLCANIC ERUPTIONS

During eruptions, volcanoes eject dust particles and gases (mainly SO_2) into the atmosphere. The dust particles generate spectacular plumes (Fig. B15.1) that disrupt the surrounding regions. Their mass, however, causes them to fall into the troposphere within a few months, where the water cycle quickly brings them back to the Earth's surface. In very powerful eruptions, the expelled gas reaches the stratosphere, where the SO_2 condenses into sulphated aerosols in about 35 days. Only the strongest explosive eruptions are able to inject a large amount of aerosols into the stratosphere and perturb the climate worldwide. Thus, for example, the 1980 Mount St Helens eruption, albeit large, was not sufficiently powerful to project aerosols to such a height, and consequently had no incidence on the global mean temperature. Conversely, Mount Pinatubo, which appears to be the largest eruption in the 20th century, clearly perturbed the climate with its layer of aerosols that accumulated in the stratosphere: the annual temperature, averaged over the surface of the Earth, fell by 0.2°C.

In this region of the stratosphere (about 15 km above the ground) where the volcanic aerosol layers form, the air is stable and the water cycle is negligible. Aerosols remain confined for 1–2 years to the altitude at which they arrive (with a decay time of 14 months). During this period, transported by stratospheric circulation, the particles spread around the whole planet, forming a veil that substantially modifies the radiation balance of the planet (Fig. B15.2). In creating marvellous red sunsets, the increased

Fig. B15.1 The eruption of Mount Pinatubo in the Philippines (15 June 1991) was the largest explosive volcanic eruption in the 20th century. It injected 15 million tonnes of sulphur dioxide into the stratosphere, where it was transformed into sulphated aerosol.

Source: NASA/GFSC.

scattering of the Sun's light by this veil is an inspiration to painters. The watercolours of Turner pay homage in this way to the awe-inspiring eruption of Mount Tambora in 1815.

Aerosols modify the energy balance of the planet. They interact with the incident solar radiation, partly by absorption and partly by scattering. Absorption heats the aerosols and thus the stratosphere, but deprives the Earth's surface of some of its heat, and causes it to cool (the 'anti-greenhouse effect': see Chapter 8). Scattering of the Sun's radiation increases its reflection and thereby also cooling. The latter mechanism, which is far stronger, produces net cooling of the Earth's surface. Observations show that 4–6 W m^{-2} of the solar radiation were reflected in this way in the 1991 Mount Pinatubo eruption. The effect of warming in the stratosphere and cooling on the ground is illustrated in detail in Fig. 23.2, where the changes in temperature over the past few decades are highlighted.

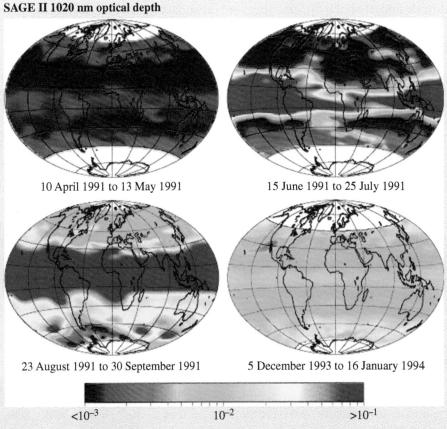

SAGE II 1020 nm optical depth

10 April 1991 to 13 May 1991   15 June 1991 to 25 July 1991

23 August 1991 to 30 September 1991   5 December 1993 to 16 January 1994

$<10^{-3}$ 10^{-2} $>10^{-1}$

Fig. B15.2 Changes with time in the veil of sulphated aerosol injected into the stratosphere in the tropics by the Mount Pinatubo eruption of 15 June 1991. These images were recorded by the Stratospheric Aerosol and Gas Experiment (SAGE), on board the ERBS satellite. The false colours show the optical depth of the stratosphere, which is directly related to the amount of aerosol present. The optical depth of the stratosphere increased by a factor of 10 or 100 after the eruption and reduced the global mean temperature by several tenths of a degree in the ensuing 15 months.

Source: NASA Earth Observatory.

Because of its situation in the atmosphere, the water cycle plays a fundamental role in the climate. The cycle is governed principally by the temperature at the surface of the Earth. Thus, in the troposphere, the concentration of the most important greenhouse gas, water vapour, is determined not by physico-chemical processes, but essentially by the temperature. Its role is that of an amplifier of climate change: an increase in temperature increases the amount of water vapour that can be stored in a given volume of air, and hence further increases the temperature through the greenhouse effect. Another associated key ingredient is cloud coverage. Clouds contribute both to warming the surface by their strong greenhouse effect and to cooling it by reflecting solar radiation. As these two opposing effects are of the same order, their outcome is hard to evaluate. Cloud cover is made up of different types of cloud, with different consequences for the radiation: high clouds (cirrus) warm the surface, while low clouds (stratus, cumulus) cool it. Cloud cover, which is related both to the water cycle and to the distribution of condensation nuclei, is also complex and hard to track back into the past. Mathematical modelling methods are the principal tool used to tackle this subject.

15.2.3 RECENT CHANGES

This section cannot be brought to an end without mentioning the main protagonist in the recent changes to the atmosphere, namely humankind. Compared to the millions of years that went before, the recent change appears as a sudden cataclysm, given the speed at which it has invaded the atmosphere and other compartments of the planet (oceans, soils, ecosystems etc.). Human activity has perturbed the equilibrium of the whole planet, especially in the most recent decades (earlier, the disturbance was less obtrusive). It is a completely new source of change in the atmosphere, mainly through the increase in greenhouse gases and different aerosols (black carbon and sulphates), all of which affect the climate. This aspect will be discussed in Part V.

15.3 What are the consequences?

Over the billions of years since the formation of the planet, one of the most remarkable changes in the atmosphere was undoubtedly the progressive reduction in its CO_2 content (along with its strong greenhouse effect) resulting from the water cycle. This mechanism ensured the transition from an atmosphere that was initially rich in CO_2 to one that was greatly depleted. The depletion gradually reduced the initial greenhouse effect, providing conditions of temperature and pressure in which liquid water can exist and maintain an active water cycle. Such changes failed to occur on our neighbouring planet, Venus. Smaller variations in the CO_2 content (always the same gas!) during the past hundred million years or so are more difficult to establish with certainty (Chapter 24). Nevertheless, they offer a possible mechanism that is often invoked to explain climate change. In the past few hundred thousand years, the alternation between interglacial and glacial periods was accompanied by an atmosphere that was alternately richer and poorer in greenhouse gases (Chapter 19). We shall see that this climate alternation is not triggered but only amplified by the atmospheric GHG changes.

Lastly, we come to the disruption that has now been taking place for two centuries (Chapter 25), subsequent to the increase of greenhouse gases in the atmosphere. *What are the fast changes that we sometimes dub 'catastrophic'?* It depends on the timescale over which the observation is made. Something that appears to us to be gradual in our lifetime, such as the change in greenhouse gas content of the atmosphere over the past century, could very well appear to a future observer hundreds of years from now as a violent accident. One recently reported investigation of undersea sediments revealed the detailed history of an 'accident' that happened 55 million years ago, the Paleocene–Eocene Thermal Maximum (PETM). The signal record exhibited an enormous outflow of greenhouse gases that was released within less than a thousand years (the resolution of the sediment recordings). This caused a sudden increase in mean surface temperature by about 5°C, which sharply reduced biodiversity and marked the end of a geological division (the Paleocene–Eocene transition; Chapter 24). Other accidents have studded the long history of the Earth and its atmosphere, branding the living planet and causing mass extinctions. Enormous emissions of gas in the wake of sudden volcanic activity, collisions with meteorites and so on – the violence of such disruptions on the climate and on the environment have deeply perturbed life cycles.

Chapter 16
Heat transfer from the Equator to the poles

16.1 The impact on the climate: the mechanism

The source of the heat that is continuously transferred from the Equator to the poles is the solar radiation gathered mainly in the tropics. This transfer warms the polar regions and cools the tropics. As already noted, the heat is carried by the circulation of the two fluids, the atmosphere and the ocean, together with the water cycle.

The water cycle involves a change of state of the water (liquid/vapour) between the two fluid reservoirs. Energy is transported by latent heat; that is, heat related to the change of state of the water. Evaporation of the water cools the ocean and the land where evaporation can take place, and the heat thus extracted from the surface returns to the atmosphere when the water vapour condenses in the form of clouds. The circulation of the air masses from the tropics to the poles redistributes this heat in the high latitudes.

16.2 How and why can the transfer vary?

Any obstacle placed in the path of the air or ocean currents between the Equator and the poles will hinder heat transfer between these regions. Over long timescales, in the ocean, this transport is influenced mainly by the distribution of the continents, which help or hinder marine currents at key passages between the tropics and the poles. Over much shorter time

Climate Change: Past, Present and Future, First Edition. Marie-Antoinette Mélières and Chloé Maréchal.
© 2015 John Wiley & Sons, Ltd. Published 2015 by John Wiley & Sons, Ltd.
Companion website: www.wiley.com\go\melieres\climatechange

periods, the transport will vary if the deep ocean currents are modified. Deep ocean currents are subject to the physico-chemical conditions that prevail in areas of deep-water formation, and these can vary abruptly. For the atmosphere, transport is hindered by mountain barriers, which influence the path of the air masses and their humidity. Mountain barriers change with uplift, which is governed by the slow process of plate tectonics, as are also the positions of the continents. Such changes take place over millions of years (the separation between America and Europe, for example, is increasing at a rate of 2 cm/yr). In summary, the relief, the size and the distribution of the continents all play a role in the transfer of heat and humidity from the Equator to the poles.

16.3 What are the consequences?

Let us imagine that the transfer of heat from the Equator to the poles were to slow down. The poles would become colder and the Equator warmer. But as the surface area occupied by the poles is much smaller than that of the tropics, the effects of the slowing down would be amplified in the polar regions, since the shortfall in energy is confined to a smaller area. (We recall that half of the surface of the planet extends between 30°N and 30°S, and the other half between 30° and the poles.) This implies strong cooling in the polar regions and weak warming in the tropics.

The slow cooling that the planet has experienced over the past few tens of millions of years (see Chapter 18) may be interpreted at least in part by this scenario. The substantial global cooling over the past 50 million years, albeit slow, gave birth in turn to ice sheets in the polar regions (first in the Antarctic and then, about 3 million years ago, in the Northern Hemisphere). During that time, the continents and the surface of the oceans in the tropics did not experience significant cooling. Nonetheless, during the past tens of millions of years, major changes have occurred both in the position of the continents and in the uplift of great mountain ranges (Chapter 18).

Chapter 17
Oscillations due to ocean–atmosphere interactions

17.1 The impact on the climate: the mechanism

Here the timescale is much shorter, that of climate fluctuations ranging from *inter-annual* (i.e. from one year to the next) to *decadal variations*. The atmosphere and ocean interact constantly, which means that they act and react with each other, giving rise to fluctuations around their average state.

The circulation of the atmosphere is directly related to the distribution of air pressure over the Earth's surface. This distribution is not uniform: it fluctuates in time, but large centres exist where the pressure is almost permanently high or low (Chapter 5).

These centres of high and low pressure, separated by several thousand kilometres and occupying an average position that hardly varies, nevertheless vary in intensity. Their fluctuations affect the strength and direction of the winds and the storm tracks, as well as the marine surface currents. The large and permanent high- and low-pressure centres alternate around the Earth's surface (Fig. 5.4). Some of them, associated in pairs, are particularly powerful, forming a high- and low-pressure tandem. The strength of the wind that a given pair controls is a function of the pressure difference between the two centres: the variation of this difference is known as the *index*. The average difference of pressure between the two centres defines the zero value of this index. It is positive when the pressure difference increases and negative in the opposite case; a large positive index thus indicates that the pressure of the

Climate Change: Past, Present and Future, First Edition. Marie-Antoinette Mélières and Chloé Maréchal.
© 2015 John Wiley & Sons, Ltd. Published 2015 by John Wiley & Sons, Ltd.
Companion website: www.wiley.com\go\melieres\climatechange

anticyclone is exceptionally high and that the low pressure is particularly low. Fluctuations in the index are the source of climate fluctuations, which often extend far beyond the region of the high- and low-pressure pair. Furthermore, they tend to persist throughout the year, or even for several years.

Among these pairs, one is especially influential because it is in the tropics, the warmest region of the planet. It is located in the Pacific Ocean, which possesses the longest east–west stretch of tropical ocean, where winds can develop unhindered by obstacles. This zone is a source of climate fluctuations. Here, water masses overheat and the water cycle has free rein. Climate oscillations that are generated here give rise to two types of events, El Niño and La Niña, discussed in the next section. Their consequences are felt over a large part of the globe.

Other high-/low-pressure pairs exist elsewhere on the planet. These generate oscillations, but with more modest climate perturbations, and the consequences are more regional. The examples that we discuss are the North Atlantic Oscillation, which has repercussions on the Europe–Mediterranean zone, and the Arctic Oscillation, which affects a wider area.

17.2 The El Niño Southern Oscillation and trade wind fluctuations

The largest fluctuation is the *El Niño Southern Oscillation (ENSO)*, which can affect the climate over a large part of the globe for up to a year. It is coupled to fluctuations of the trade winds in the tropical zone of the Pacific Ocean. The strength of the trade winds is controlled by the permanent high- and low-pressure pair that they cross in the tropical zone of the Pacific. The low-pressure region of the pair is centred near the Equator, approximately at Darwin (Australia), and the high-pressure region is in the east of the Pacific in the tropics, on average over Tahiti. The index describing the pressure difference between these two centres is the *Southern Oscillation Index (SOI)*. If the pressure difference between these centres decreases (negative index), the trade winds weaken in this region of the planet, and give rise to a climate perturbation called El Niño. Conversely, if they strengthen, the opposite pattern, called La Niña, takes its place. *The SOI index is thus also a measure of the strength of the trade winds in the equatorial Pacific.* El Niño events occur at irregular intervals every 3–7 years, and alternate with La Niña.

To grasp the consequences of these perturbations, we follow the path of the normal trade winds from the Atlantic to Asia (Fig. 5.4). The trade winds, arriving over the South American continent from the Atlantic, bring water-laden air masses that progressively release rain into the Amazon basin. On reaching the barrier of the Andes mountain range, which rises to more than 6000 m, the air masses unload their last remaining rain, thereby feeding the tropical glaciers of the Andes. Once over the Andes barrier, these dry air masses have no choice other than to contemplate and pass over desert regions: the Pacific coast is a true desert, because this strip of land is stuck between the Andes range, from where only dry winds blow, and the ocean, where the cold Humboldt Current discourages evaporation – and precipitation. This cold current is the consequence of *upwelling* due to the combined effects of the trade winds and the winds from the high-pressure zone over Tahiti. These winds drag

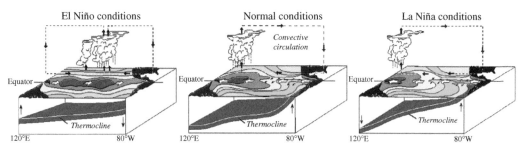

Fig. 17.1 El Niño and La Niña events. In normal conditions (centre), the warmest waters of the globe accumulate in the western equatorial Pacific Ocean, driven west by the trade winds. In an El Niño event, the trade winds weaken. This lets the warm waters move east and reduces upwelling at the eastern edge of the Pacific. The thermocline, defined by the bottom of the warm water mixing layer, rises in the west and sinks in the east. The region of strong atmospheric convection and heavy rainfall then moves from the west to the centre of the Pacific. The reverse situation of a La Niña event (where the trade winds strengthen) brings about the opposite results (warmer waters move west and upwelling is intensified).
Source: National Oceanic and Atmospheric Administration Paleoclimatology Program.

the ocean surface water westwards, causing the cold deeper water, which is rich in nutrients, to rise (Fig. 17.1). The stronger the trade winds, the more intense is the upwelling and the richer life becomes there.

The trade winds, blowing from east to west, continue their route over the tropical Pacific, pulling the warm water on the surface westwards to the frontiers of Asia. This is where the warm water accumulates and constitutes the warmest ocean region in the world, the so-called 'warm pool', which pushes the thermocline downwards. This is a zone of intense convection in the atmosphere, with abundant rainfall.

Fluctuations in the trade winds

The Tropical Ocean Global Atmosphere programme (TOGA; see http://www.pmel.noaa. gov/tao/elnino/toga-insitu.html) extended its observation system between 1985 and 1994. Its record of the thermal profile of the surface layer (to 200 m depth) in the tropical Pacific Ocean illustrates the impact of the seasonal fluctuation of the trade winds. Every year around Christmas, the trade winds along the coast of Peru weaken as the thermal Equator moves south. The upwelling of cold waters then diminishes, the surface water becomes warmer and the mixing layer becomes thicker (i.e. the thermocline, which separates the warm surface mixing layer from the colder lower layers, plunges lower). This annual phenomenon is clearly visible in Fig. 17.2, which shows the seasonal oscillations in the thermal profile of the surface layer (to 200 m depth) of the ocean between 95°W and 105°W; that is, off the coast of Peru. At first, it was given the name El Niño, after the infant Jesus. By extension, this name was later applied not to the seasonal phenomenon, but to a major weakening of the trade winds lasting several months. Such a situation arises whenever the SOI index becomes negative (the pressure difference decreases). Conversely, when the index becomes positive, the trade winds strengthen and the situation is reversed. La Niña then sets in, together with an increase in the thickness of the warm surface layer in the east Pacific.

Eastern equatorial Pacific (1979–93)

Temperature, 0–250 m

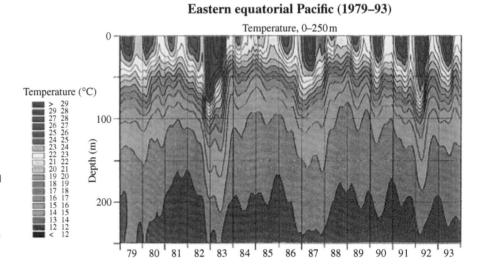

Fig. 17.2 Seasonal changes in the thermal profile of the first 250 m in the equatorial Pacific Ocean waters. Measurements made between 1979 and 1993 in two characteristic zones: top, in the east Pacific (between 95°W and 105°W), where upwelling of cold water is particularly active in October (seasonal upwelling); bottom, in the western Pacific (between 150°E and 160°E), the site of the warmest waters of the planet. Due to the trade winds, the thermocline is generally higher in the east (about 50 m) and deeper in the west (between 150 and 200 m). When the trade winds slacken in an El Niño episode, this difference diminishes. The El Niño episodes of 1983, 1987 and 1992 are particularly conspicuous, as is also the La Niña of 1988. These measurements were carried out in the Tropical Ocean Global Atmosphere programme (TOGA).

Source: Reproduced with permission of Joel Picault/TOGA.

Western equatorial Pacific (1979–93)

Temperature, 0–250 m

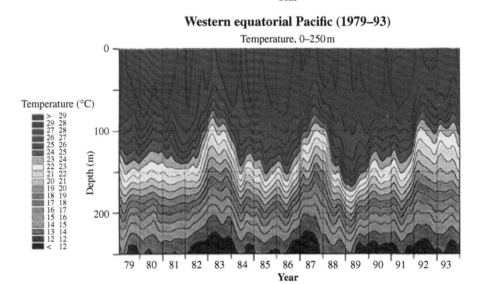

The impacts of these two events on the temperature distribution in the surface waters, as well as the changes in the thermocline, are outlined in Fig. 17.1. Observations of the two situations are illustrated in Fig. 17.3. This shows the changes in the ocean surface temperature at three different times: (i) in 1993 (the normal situation); (ii) in 1997, during the El Niño event, with the warm pool moving east; and (iii) in 1998, during the La Niña event, where, on the contrary, the warm pool moves back westwards. This impact is also visible in the thermal profile of the ocean between 1979 and 1993 (Fig. 17.2). In the east equatorial region (between 95°W and 105°W) the thermocline, which is usually situated at a depth of about 50 m, descends to about 100 m during an El Niño event (in 1983, 1987 and 1992). In the same years in the west equatorial region (between 150°E and 160°E), the thermocline rose from its normal depth of approximately 150 m to about 100 m.

El Niño and its consequences

When the trade winds slacken, what are the consequences of El Niño on the climate? Let us look back over the path of the trade winds coming from the Atlantic, discussed above. A decrease in their intensity brings in turn: less rain to eastern Brazil, a smaller supply of water to the Amazon basin, less snow on the Andean tropical glaciers, reduced upwelling along the Peru–Chile coast, accompanied by a drastic loss of marine production in this zone (young and mature fish, and birds, are all affected and the fishing industry is particularly badly hit) and rising temperatures in the coastal areas as a result of the warmer sea surface, accompa-

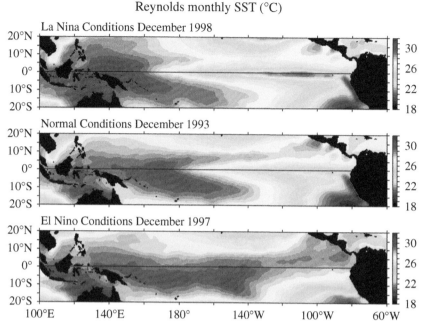

Reynolds monthly SST (°C)

Fig. 17.3 The difference in temperature of the surface waters in the Pacific between the normal situation of 1993 (centre), the El Niño episode in December 1997 (bottom) and that of La Niña in December 1998 (top), illustrating the diagram in Fig. 17.1. **Source:** National Oceanic and Atmospheric Administration Paleoclimatology Program.

nied by violent rainstorms in Ecuador and Peru (with the explosion of sporadic vegetation in this desert region, and deadly landslides of unconsolidated desert soils loosened by the violent rain). Further west, in the Pacific and in South-East Asia, the migration of the world's heat reservoir, the warm pool and its attendant rains from the west of the Pacific towards the central region reduces rainfall in South-East Asia and in Australia, and increases periods of drought and forest fires.

Although the impact of this oscillation is most immediately visible in the tropical Pacific zone, it also affects more distant parts (Asia, Australia, North and South America and Africa), sometimes reaching regions at mid-latitudes. Figure 17.4 summarizes the consequences of an El Niño event on the different regions of the world. These events affect regions as distant as the Southern Ocean (not shown in this figure), where the sea-ice distribution around the Antarctic is modified. By contrast, they have little effect on the Atlantic Ocean at mid-latitudes in the Northern Hemisphere: these latitudes are partly under the influence of the North Atlantic Oscillation (NAO).

In brief, El Niño events are a major perturbation for the climate over the whole of the globe. They last for several months and mark the year. The warming effect generated in the tropical zone is such that the global mean temperature of the planet rises by as much as +0.2 to +0.3°C. Moreover, they strongly perturb the water cycle with the increase of water vapour in the Pacific tropical zone (+4% humidity during the 1997–8 El Niño). On average, less water is stored on the continents, and the global mean sea level is consequently higher. They also affect the carbon cycle, with increased atmospheric CO_2 (see Fig. 29.2), probably from forest fires in South-East Asia. The impact is visible in the annual averages of these quantities, with repercussions on certain regions at medium latitudes. La Niña episodes, by contrast, have the opposite consequences.

Warm episode relationships, December–February

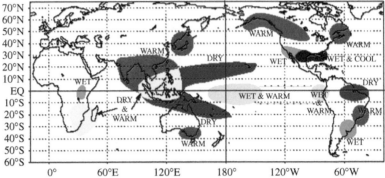

Warm episode relationships, June–August

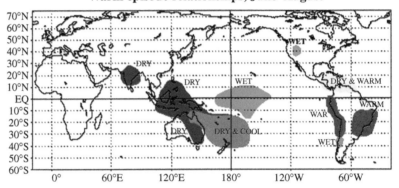

Fig. 17.4 The regional impact of El Niño episodes. The change in rainfall and temperature: top, boreal winter; bottom, boreal summer. **Source:** National Oceanic and Atmospheric Administration Paleoclimatology Program.

17.3 The North Atlantic and Arctic Oscillations

17.3.1 THE NORTH ATLANTIC OSCILLATION

Besides the ENSO, oscillations exist between other low- and high-pressure centres, but the effects are more regional. Here, we mention only the North Atlantic Oscillation (NAO), which affects primarily the climate of Europe, the Mediterranean region and the Middle East. In the North Atlantic, two major centres influence the eastward atmospheric circulation at mid-latitudes to which these countries are subjected. The two centres are the high pressure over the Azores (tropical high pressure) and the low pressure in the vicinity of Iceland (Fig. 5.4). Their relative strength influences the path taken by storms and the amount of rain delivered to Europe and the Middle East (Fig. 17.5). Their relative intensity is defined by the NAO index, which is principally employed over the winter period. A strong anticyclone combined with a deep depression thus corresponds to a positive NAO index (NAO+), and brings winter rains and storms preferentially to Northern Europe, while leaving the southern part of Europe, the Mediterranean region and the Middle East relatively dry. This is illustrated by the changes in the rate of flow of large rivers in the Middle East (Cullen & deMenocal 2000), where weak flow rates are associated with a positive NAO index (Fig. 17.6). A negative index

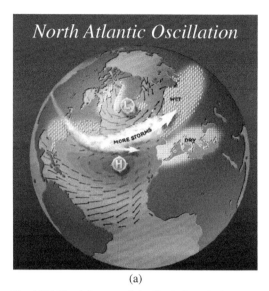
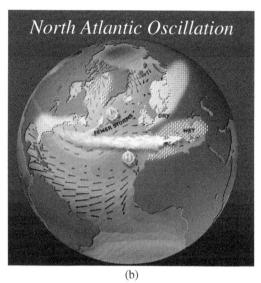

(a)   (b)

Fig. 17.5 The influence of the North Atlantic Oscillation (NAO) on the trajectory of air masses for (a) a positive NAO index and (b) a negative index. The index is positive when the difference in pressure between the Azores anticyclone and the depression over Iceland is reinforced, and negative when it weakens. Depending on this index, wet air masses are directed preferentially either to Northern Europe or to Southern Europe and the Mediterranean.

Source: Figure courtesy of Martin Visbeck. Reproduced with permission.

(NAO–) produces the opposite pattern: reduced rainfall in the north and reinforced rainfall in the south. Here too, the index changes from year to year and can increase over several years.

17.3.2 THE ARCTIC OSCILLATION

The Arctic Oscillation (AO) is a pattern in which the atmospheric pressure at polar and mid-latitudes fluctuates between negative and positive phases. Its phase strongly affects the weather in the northern regions. The AO is often in phase with the North Atlantic Oscillation (NAO).

The negative phase stems from higher-than-normal pressure over the polar region and lower-than-normal surface pressure at mid-latitudes. The polar low-pressure system (the *polar vortex*; Section 5.2.2) over the Arctic is weaker, and the upper-level winds (the *westerlies*) are consequently weaker. In the negative phase, this weaker belt becomes more distorted, allowing (i) colder, arctic air masses to penetrate more easily southward, bringing increased

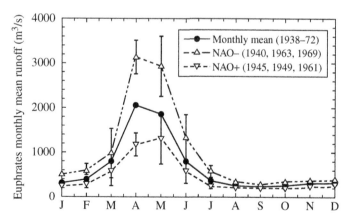

Fig. 17.6 The monthly average flow rate of the Euphrates at Keban, Turkey: (a) from 1938 to 1972; (b) for the three years with the lowest NAO index during this period (NAO–); and (c) for the three years with the highest NAO index (NOA+). A clear anti-correlation exists between the flow rate of the river and NAO index.

Source: Cullen and deMenocal (2000). Reproduced with permission of John Wiley & Sons, Ltd.

storminess to the mid-latitudes, and (ii) easier northward penetration of warm air to the high latitudes. A strongly negative phase brings cold, stormy weather to the more temperate regions. This is illustrated by very cold winters in the western United States and in Europe, as in 2009–10 and 2012–13, when the AO index reached a deeply negative value.

The positive phase brings the opposite conditions. The polar circulation is stronger and the polar vortex is reinforced. The ring of strong winds confines colder air within the polar regions. It brings the ocean storms further north, making the weather wetter in Alaska, Scotland and Scandinavia and drier in the western United States and in the Mediterranean region. Europe and Asia enjoy warmer than usual temperatures, while Greenland and northern Canada are colder than normal.

PART III SUMMARY

In Part III, the different causes of climate change are reviewed by tracing the path of energy radiated from the Sun until it is finally absorbed at the surface of the Earth. Since the absorbed energy is what keeps the average climate constant, any change along its trajectory will affect the mean climate. This description, however, is insufficient. Changes in the conditions that affect the distribution of this energy at the Earth's surface must also be considered.

The Sun is the starting point. The energy it radiates depends on its activity, which varies over many timescales. Over the past hundreds of millions of years, the Sun's activity has nevertheless remained fairly stable. On the timescale of the past 10,000 years, its fluctuations have been limited and their effects on the climate are moderate, not unlike what occurred in the Little Ice Age, the cold spell that marked the centuries of the recent past.

The distance between the Earth and the Sun governs the flux of solar energy arriving on Earth. Over the past few million years, the annual average of this distance has been practically stable. However, over the millennia it varies in each given season, owing to periodic changes in the orbital parameters of the Earth (mainly precession and obliquity, but also the eccentricity of the planet's elliptical orbit around the Sun). The relevant periodicities of these parameters are all measured in tens of thousands of years. It follows that their effects on the climate are a function of the latitude, modifying the monsoon at low latitudes and creating alternations between cold and warm climates at high latitudes (see Part IV).

The next role is that played by the atmosphere. Its composition controls the flux of radiation that passes through it on its journey both from the Sun and from the Earth, and calls into play many different mechanisms (Parts I and II). Warming from the greenhouse effect and cooling due to reflection are among the foremost mechanisms in the atmosphere that, combined with the water cycle, govern the energy balance on the planet. Variations in the composition of the atmosphere are therefore one of the main causes of climate change. Such modifications can take place on very different timescales, ranging from 1 year to millions, or even thousands of millions of years. The most recent example of change is that due to humankind, whose activities have mightily modified (and continue to do so) the greenhouse gas content of the atmosphere (Parts V and VI).

The final role is that played by the distribution of energy at the surface of the Earth. Heat transport from the Equator towards the poles is ensured by atmospheric and ocean circulation, and also by the water cycle. Transport through the ocean is helped or hindered by the relative positions of the continents. Movement of the air masses is affected by mountain barriers, which interfere with their flow. The position of the continents and the uplift of mountain ranges are both driven by plate tectonics, and take place on a timescale much longer than the past million years. On much shorter timescales, the distribution of heat and of rainfall is modulated by oscillations that develop between the atmosphere and the ocean, and which are in constant interaction. The timescale of these oscillations ranges from a year to ten or even several tens of years. Although year-long climatic fluctuations, such as those related to El Niño events, have been clearly identified, those that last for several decades are more difficult to apprehend and are still being investigated.

This approach will help us to understand the climate changes of the past as well as the present trend of the climate.

PART III NOTES

1. A variation of $1.5\,W/m^2$ in the solar energy flux directly striking each square metre (see Fig. 4.1) corresponds to a value four times weaker received per square metre of the Earth's total surface (Part I, Note 3). As the albedo (the reflection coefficient for solar radiation) of the planet is about 30%, the variation in the absorbed solar flux is about 70% of this amount; that is, approximately $0.25\,W/m^2$.

2. Although the half-life of ^{14}C is 5730 ± 40 years, in accordance with international conventions that were adopted in 1951 by the radiocarbon scientific community to avoid confusion, the period used is that found by Willard Libby, namely 5568 ± 30 years. This value is still used for the calculation of age.

3. The total flux crossing the surface of a sphere of radius D, namely $E \times 4\pi D^2$, where E is the solar flux arriving at distance D (Fig. 4.1), is equal to that crossing the surface of the Sun, $E_0 \times 4\pi R_S^2$, where R_S is the radius of the Sun and E_0 is the flux emitted by its surface. The irradiance E received by the Earth can therefore be expressed as a function of distance by $E = \text{constant} \times (1/D^2)$.

4. The eccentricity of an ellipse is the ratio of the distance c between a focus and the centre (distance OS in Fig. 14.3, where the Sun is at S and the centre is O) and the semi-major axis a of the ellipse (distance OA or OP). Thus $\varepsilon = c/a = \text{OS/OA}$. The difference dD between the largest value of D, SA (Earth at aphelion, A), and the smallest, SP (Earth at perihelion, P), is $\text{SA} - \text{SP} = 2c$. The relative difference is dD/D_m, where the mean distance D_m is OA or OP. Thus dD/D_m is $(\text{SA} - \text{SP})/\text{OA}$, which is equal to $2c/a = 2\varepsilon$. Therefore $dD/D_m = 2\varepsilon$.

5. The irradiance E received by the Earth, $E = \text{constant} \times (1/D^2)$, can be expressed by its logarithm: $\log E = \text{constant} - 2\log D$. The derivative of this expression is, in absolute value, $dE/E = 2dD/D$ and, for small variations in D, $dE/E = 2dD/D = 4\varepsilon$.

6. Judging from the difference in winter sunshine received over a year in each of the hemispheres (NH winter at perihelion, SH winter at aphelion), it might be concluded that winters in the Northern Hemisphere are currently less cold than in the Southern Hemisphere. This, however, is not true. The difference in irradiance received is offset by the impact of the large ocean surface in the Southern Hemisphere, which attenuates seasonal variations.

7. On the timescale of tens of years, the axis of the Earth's rotation displays a very slight oscillation, called *nutation*, due to the attraction of the Moon and the Sun. The period of this oscillation is 19.6 years and its amplitude is 9". This weak oscillation appears to exert no influence on the climate.

PART III FURTHER READING

Houghton, J. (2009) *Global Change – The Complete Briefing*, 4th edn. Cambridge University Press, Cambridge, UK; see Chapter 3.

Neelin, J.D. (2010) *Climate Change and Climate Modeling*. Cambridge University Press, Cambridge, UK; see Chapter 4.

Ruddiman, W.F. (2008) *Earth's Climate – Past and Future*, 2nd edn. W.H. Freeman, New York; see Chapter 8.

PART IV

LEARNING FROM THE PAST ...

Part IV retraces the broad outlines of the natural climate changes that the Earth has known in the past. In this part, we therefore omit the very recent period that begins with the Industrial Revolution, when the growing impact of human activity started to modify the environment substantially. Climate changes that occurred in this period will be examined in Part V. This return to the past will help us to appreciate the magnitude of the climate changes that are anticipated during the 21st century. We concentrate on examples that will serve as benchmarks for future changes of climate.

The two following questions arise:

- *How does the warming of the most recent decades compare with natural climate variations in the past?* In Part V, we shall see that the recent rise in mean global temperature (i.e. the annual average over the whole of the Earth's surface) is close to 1°C.
- *How do future changes compare with the natural climate variations of the past?* The average warming anticipated in the 21st century depends on the economic scenario. In the current state of modelling, it ranges between about +2°C and +5°C with respect to the 19th century, as we shall see in Part VI.

Such comparisons with the past underscore the urgency of the threat from man-made emissions of greenhouse gases.

First, we briefly overview the climate that the Earth has known since its beginnings, 4.56 billion (thousand million) years ago. Then we concentrate on the past few million years – which, on geological timescales, is 'recent'. Although evolution of the different animal and plant species continued throughout this 'recent' period, the major lines of species living at the beginning of this period at that time were broadly comparable with those of the present. Likewise, the distribution of the oceans and continents was similar to that of today. This time window is therefore pertinent for comparisons with the present and the near future.

Chapter 18
Memory of the distant past

18.1 Over billions of years …

Hydrogen and oxygen are two of the most abundant elements in the universe. Combined, they produce the molecule H_2O, which is prevalent everywhere in space. Depending on the prevailing conditions of temperature and pressure, water can, or cannot, exist in the liquid state. Its existence in the liquid state is an essential factor for life as we know it to develop on Earth. In the solar system, only on our planet are the conditions met to maintain large reserves of liquid water and for a water cycle to exist (a return ticket from one to another of the three states of matter: solid, liquid and gas). Over the 4.56 billion years or so since the Earth was formed, the sediment records tell us that this reservoir of liquid water was present practically from the start (the oldest data go back to a few hundred million years after the formation of the Earth). These data also show that the temperature on Earth varied within a range that, from 2.3 billion years ago (the date of the first ice age), allowed large ice sheets to exist inter-mittently: these periods are usually called glaciations. After their appearance at that early date (the *Huronian glaciation* in the *Palaeoproterozoic era*) and their subsequent disappearance, ice sheets reappear in a cluster of several successive major phases between 850 and 550 million years ago. One hypothesis still under debate assumes that during these glaciations the Earth was entirely covered with ice (the Snowball Earth hypothesis). Albeit attractive, this hypothe-sis is controversial, but still remains a subject of scientific dicussion.

Climate Change: Past, Present and Future, First Edition. Marie-Antoinette Mélières and Chloé Maréchal.
© 2015 John Wiley & Sons, Ltd. Published 2015 by John Wiley & Sons, Ltd.
Companion website: www.wiley.com\go\melieres\climatechange

During the primary era, which started 540 million years ago, glaciations continued to make their appearance in the life of the planet. The ice sheets, centred on the poles, took over once more for a brief period 440 million years ago (the *Ordovician Period*), and then reappeared between 330 and 260 million years ago (in the *Permo-Carboniferous Period*). Finally, after the warm *Cretaceous Period* (145–65 million years ago), cooling set in, ushering in a new glaciation that began 35 million years ago. It first affected the coldest continent, the polar continent of the Antarctic in the Southern Hemisphere (SH), and then, 3 million years ago, extended to the high latitudes of the Northern Hemisphere (NH).

This brief panorama shows how climate conditions on Earth stayed within a range that allowed the coming and going of polar ice sheets to continue over several billion years. Throughout all this time, life developed and evolved, even strongly modifying the physical and chemical conditions of the planet. Among the most spectacular consequences of this were the gradual decrease of carbon dioxide (CO_2) in the atmosphere and the enrichment of oxygen (O_2) about 2.3 billion years ago. This change enhanced the oxidizing capacity of the atmosphere.

18.2 The past tens of millions of years: slow cooling

The Cretaceous is a period of relatively stable climate that was much warmer than now (at the end of the Cretaceous, the global mean temperature was roughly ten degrees higher than at present). It ended at the Cretaceous/Tertiary boundary (65 million years ago), with a major crisis that disrupted the biosphere and caused large-scale mass extinction on land and at sea. At least 75% of species disappeared, including non-avian dinosaurs. This marks the end of the Secondary Period.

Two possible causes of this crisis are generally put forward. One, of extraterrestrial nature, involves the impact of a meteorite, the crater of which is located at Chixculub on the coast of the Yucatan Peninsula in Mexico. The other is volcanic, and is related to the formation of the Deccan Traps in India. These two events could well have taken place together, producing an even more disruptive effect. It is difficult to imagine that events such as those could have happened without repercussions on the climate. And such climate changes would only intensify the overall crisis.

Shortly after the dinosaurs disappeared, at the beginnning of the Tertiary Period the surface of the Earth was about 25°C (ten degrees warmer than now). The average temperature (+25°C compared to the present +15°C) resembled that of the Cretaceous. During the Early Eocene Climatic Optimum (54–48 million years ago), the world became even warmer, and the average temperature reached about +29°C. This warm world not only held sway at the surface but also extended to the ocean depths. The mass of water in the oceans was similar to that of the present day, which is enormous – its surface area covers two thirds of the planet, with an average depth of about 3750 m. At that earlier time, all this water was much warmer: at the bottom of the ocean it was about ten degrees higher, both in the tropics and at high latitudes. Over the past 50 million years, the mass of water has gradually cooled, in fits and starts, ending up with a global deep ocean temperature close to 0°C. The surface waters, by contrast, have cooled unevenly: at the Equator, for example, their temperature has hardly changed since then. In the polar regions, however, it has fallen to freezing point. Figure 18.1

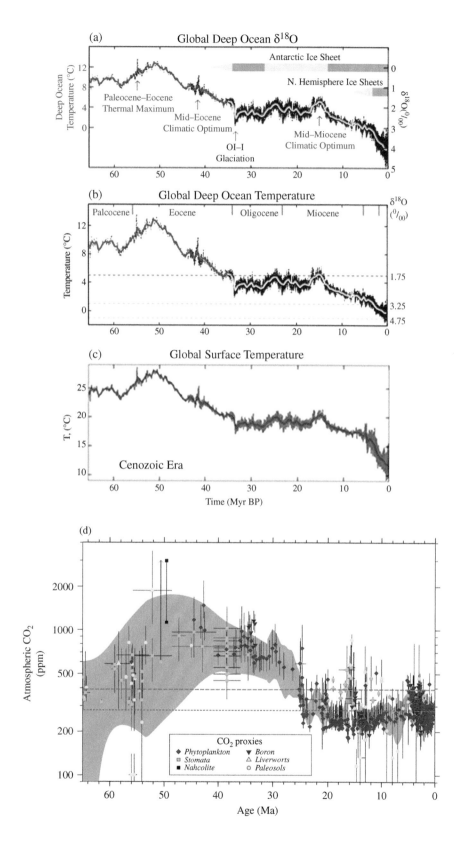

Fig. 18.1 (a) Variation in the $\delta^{18}O$ isotopic signal since 65 Myr BP from deep ocean zooplankton (benthic foraminifera) buried in the sediments (Zachos et al. 2008). This signal can be used to determine the deep ocean temperature (red curve) provided that ice does not accumulate on the continents. Since 35 Myr BP, the signal (blue curve) has been affected by polar ice cap build-up (light blue/blue horizontal bars). Coarse temporal sampling reduces the amplitude of glacial–interglacial oscillations in the intervals 7–17, 35–42 and 44–65 Myr BP. (b) The estimated mean deep ocean temperature and (c) the estimated mean surface temperature, from Hansen et al. (2013a). (d) The concentration of atmospheric CO_2 in the last 65 Myr, reconstructed from marine and terrestrial proxies. Individual proxy methods are colour coded. Light blue shading is the one-standard deviation uncertainty band. **Source:** (d) IPCC 2013. *Climate Change 2013: The Physical Science Basis.* Working Group I Contribution to the Fifth Assessment Report of the Intergovernmental Panel on Climate Change, Figure 5.2, bottom. Cambridge University Press.

shows how the average temperature of the deep ocean water and that of the Earth's surface have evolved with time.

This remarkable cooling, which lasted tens of millions of years, was punctuated by two particularly important events. As the temperature dropped, ice sheets began to build up on continents in the coldest parts of the Earth, first in the south, then in the north. The first episode occurred about 35 million years ago. This episode appears to be contemporaneous with some decrease in the CO_2 content of the atmosphere (Fig. 18.1). This period marks the establishment of ice sheets on the coldest continent of the planet, Antarctica. This continent had been located at the South Pole for tens of millions of years, but it gradually became separated and isolated from the others as a result of tectonic plate movement (Fig. 18.2). Since then, after a period of warming between roughly 25 and 15 million years ago, cooling has resumed. From about 15 million years ago, the East Antarctica ice sheet has been permanently established and the volume of ice increased until 11–10 million years ago. The Antarctic polar ice sheet then underwent large fluctuations, before reaching its maximum size at the end of the Miocene, between 6 and 5 million years ago, when the West Antarctica ice sheet set in. An incipient, or at least seasonal, ice cover on the Arctic Ocean and Alaska may date from this period. In the Pliocene (5.3–2.6 million years ago), the average climate was warmer by about +2 to +3°C than the present interglacial warm period. The Antarctic Ice Sheet shrank significantly and the sea level was higher than now by at least 15–20 m.

The second episode is situated in the past three million years. Gradual cooling of the climate is attested by changes of the $\delta^{18}O$ isotopic signal in marine sediments (Part IV, Note 1). About 2.6–2.3 million years ago, the continents at high latitudes in the NH had cooled sufficiently for snow to accumulate each year, despite the warm season, leading to the formation of large ice sheets. This marks the beginning of the glaciations in the NH. Unlike the Antarctic, however, the ice sheets did not become permanently established. Their distance from the

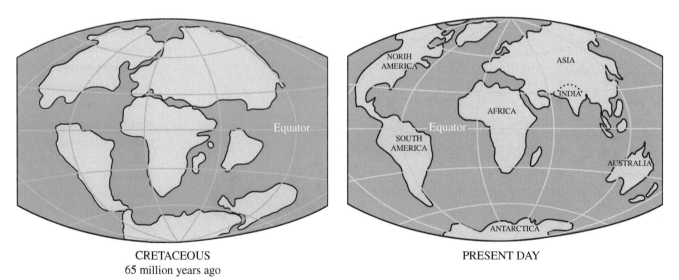

CRETACEOUS
65 million years ago

PRESENT DAY

Fig. 18.2 Changes in the positions of the continents between 65 million years BP and now. The plate tectonics that drive continental drift have substantially modified heat transfer from the Equator to the polar regions.

Source: Department of the Interior/USGS.

North Pole made them more vulnerable to the cyclic changes of the Earth with respect to the Sun: since that time, they have periodically grown and then almost disappeared, bringing about huge variations in sea level.

Over the long slow cooling period that started 50 million years ago, ocean circulation has changed in response to the decrease in temperature of the polar regions. Near the poles, the surface water, on becoming cold and dense, can penetrate into the mass of the deep water and sink to the ocean floor, cooling and oxygenating the bottom. This mechanism, which became predominant in the North Atlantic, started there about 12 million years ago. The present deep-sea circulation was gradually set in motion at that time. This circulation is known as *thermohaline*, because the temperature and the salinity determine the density of the surface water and allow it to sink, or not, to the ocean floor (Chapter 5).

In summary, what were the characteristics of the climate some 50 million years ago, in the early Eocene Climatic Optimum? Palaeoclimatologists have determined the orders of magnitude for that much warmer climate:

- a global mean surface temperature of about 29°C (+15°C higher than at present), and higher in the polar regions by about +20°C in NH, and by about +40°C in SH (where ice sheets were absent).
- ocean waters warmer than now (temperature at the bottom higher by about 12 degrees);
- ocean circulation very different from now, most likely with stratified layers and oxygen-poor deep water owing to limited penetration by surface water; and
- flora and fauna largely different from those we know at present.

Causes of cooling

The slow cooling that took place during the Tertiary could have several causes, of which two are particularly notable: (1) the progressive depletion of the CO_2 content of the atmosphere, causing the temperature to fall; and (2) the drift of the continents and changes in their relief, modifying the heat transfer from the Equator to the poles.

The first hypothesis is based on reconstructions of the atmospheric CO_2. These reconstructions contain uncertainties that depend on the method employed (Fig. 18.1). Over the past 25 million years, there has been no notable difference in atmospheric CO_2 from its present value. Before that, in spite of the large uncertainty, the CO_2 content was higher, but a distinct reduction took place about 35 million years ago, very probably in connection with the onset of the Antarctic Ice Sheet.

The second hypothesis involves migration of the continents and changes in their relief, which modified the dynamics of the atmosphere and the oceans, as well as the water cycle. For example, the thermal isolation of the Antarctic continent was reinforced when the Australian continent broke away, migrated northwards, and, when the Drake passage opened, allowed the Antarctic circumpolar current to develop. Since then, the winds and sea currents have constituted an efficient barrier to transport of heat from the Equator to the South Pole, thereby isolating the Antarctic continent, bringing on its gradual cooling and the trend towards a permanent ice sheet. Other events doubtlessly also contributed to alter the heat transfer between low and high latitudes: for instance, uplift of the Himalayas and the establishment of the Indian monsoon; the uplift of the Andes and the Rocky Mountains; the formation of the Panama isthmus; the modification of the Fram Strait between Greenland and Spitzbergen; and so on.

During this slow disruption in the Tertiary, biological evolution led to the development of mammals, and these appropriated the ecological niche left by the dinosaurs. Note that the ability of mammals to keep their body temperature constant prepared them better for the cooling climate. This period also witnessed the appearance of prairies and their subsequent extension beyond the tropics (about 35–40 million years ago), as well as the explosion of C4 vegetation, which was more suited for survival in warm dry environments and more efficient in capturing CO_2 (Part IV, Note 2). The fauna adapted to the changes in vegetation and climate, and in particular to running over steppe and open savannah.

Thus, at the beginning of the Tertiary, several tens of millions of years ago, the prevailing climate was much warmer than now. The environment was very different – no ice sheets, totally different ocean circulation, other plant and animal families, etc. During the following tens of millions of years, the climate cooled severely, and evolution gradually gave rise to the flora and fauna that we know today. The changes in temperature over the globe are shown in Fig. 18.3 for three periods in the past, the Early Eocene Climate Optimum (54–48 million years ago), the Miocene Pliocene Warm Period (3.3–3.0 million years ago) and the Last Glacial Maximum 21,000 years (i.e. 21 kyr) ago. These past variations are compared with a fourth period, the widely studied end-of-the-21st-century scenario RCP8.5 for 2081–100 (see Part VI).

18.3 The entry of Northern Hemisphere glaciations

The past few million years have seen the appearance of climate oscillations. Modest at first, they have become increasingly pronounced. They are characterized by temporary ice sheets at high latitudes in the NH, small at first, but then of increasingly greater volume. The isotopic signal of seawater (Lisiecki & Raymo 2005) records in its main features the changes in volume of these ice sheets (Fig. 18.4). A gradual increase in the average volume of ice took place between 3.6 and 2.4 million years ago, with the appearance of regular oscillations in the ice-sheet volume around 2.7 million years ago. These were accompanied by fluctuations in the sea level of about 50 m.

The changes involved both hemispheres. Between 2.9 and 2.7 million years ago, the marine sediments contain a deposit of ice-rafted debris transported by icebergs both in the Antarctic Peninsula and in the region to the south-east of Greenland, demonstrating that the event was common to both polar regions. In New Zealand, which is situated at mid-latitudes in the SH, traces of glaciation appear around 2.6 million years ago. On the continents in the NH, in central Asia, this climate transition is equally spectacular. Large deposits of loess appear at that time, the sign of a transition from a warm and humid environment to that of continental steppes. Finally, it was at that time that the modern features of the deep ocean circulation became established.

A contributory factor to the build-up of the large ice sheets in the NH was almost certainly the closure of the Panama isthmus, which prevented exchange of tropical waters between the Atlantic and the Pacific Ocean and strongly modified the North Atlantic circulation. This tongue of dry land connecting South to North America gradually built up between 5 and

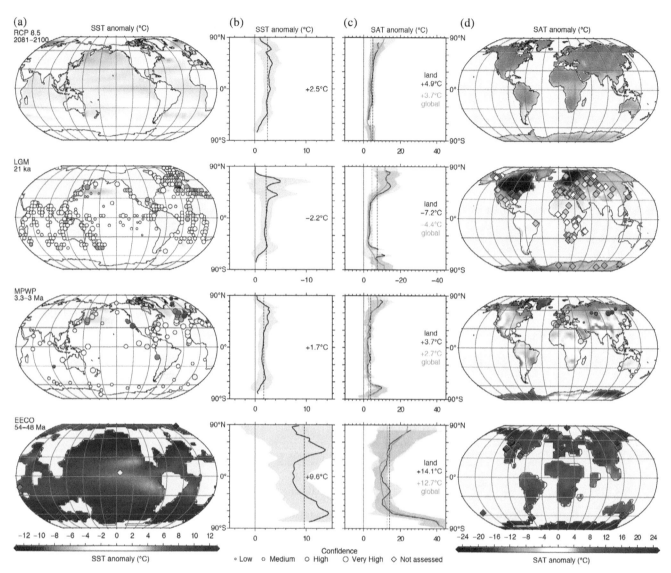

Fig. 18.3 Comparison of data (circles) and simulations for four time periods, showing (a) sea surface temperature (SST) anomalies, (b) zonally averaged SST anomalies, (c) zonally averaged global (green) and land (grey) surface air temperature anomalies (SAT) and (d) surface air temperature anomalies (SAT). The time periods are 2081–2100 (top row) for scenario RCP8.5, Last Glacial Maximum (LGM, 21 kyr ago, second row), Mid Pliocene Warm Period (MPWP, 3.3–3 Myr ago, third row) and Early Eocene Climatic Optimum (EECO, 54–48 Myr ago, bottom row). Model temperature anomalies are calculated with respect to the pre-industrial value; site-specific temperature anomalies estimated from proxy data (circles) are calculated with respect to present site temperatures. Shaded bands in (b, c) indicate two standard deviations. Note difference in temperature scale for sea and continents.

Source: IPCC 2013. *Climate Change 2013: The Physical Science Basis.* Working Group I Contribution to the Fifth Assessment Report of the Intergovernmental Panel on Climate Change, Box 5.1, Figure 1. Cambridge University Press.

3 million years ago. Previously, the Atlantic and the Pacific Oceans communicated, with the east–west equatorial current transferring water masses from the Atlantic towards the Pacific. The formation of the isthmus brought about two major changes to the ocean circulation: (i) the establishment of the Gulf Stream in the North Atlantic, and (ii) the increased salinity

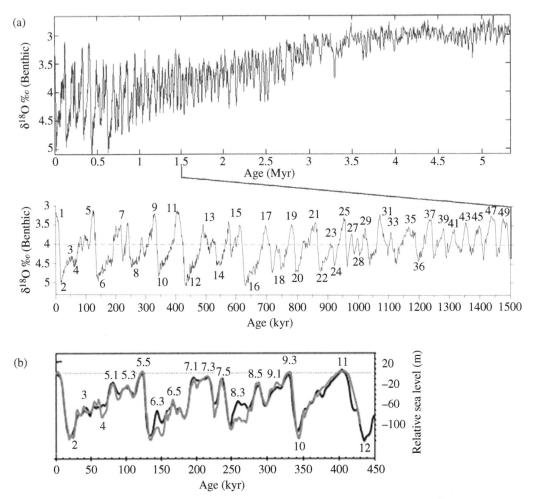

Figure 18.4 (a) The change in $\delta^{18}O$ of carbonates from ocean-bed zooplankton tests (benthic foraminifera) over the past 5 million years. This signal arises from changes in the total ice-sheet volume and in the deep ocean temperature. This composite curve was constructed by Lisiecki and Raymo (2005), by analysing marine cores from 57 sites located mostly in the Atlantic Ocean. The numbering of the different isotopic stages is indicated on the lower curve (even numbers, glacials; odd numbers, interglacials and interstadial stages). (b) The reconstructed sea level based on benthic foraminifera isotopic records from the North Atlantic (black line) and from the Pacific (grey line) (Waelbroeck et al. 2002). The change in deep ocean temperature during the glacial–interglacial cycle is the cause of the non-linear relationship between these two curves.

Source: Figure 18.4(a) reproduced with permission of Lorraine Lisiecki (b) Waelbroeck et al. 2002. Reproduced with permission of Elsevier..

of the water in the Atlantic compared to that of the Pacific. Numerous scenarios have been put forward to explain how with the help of feedback mechanisms, these modifications could have contributed to establish ice sheets in the north. Other factors, however, have certainly played a large part in the changes that have occurred over the past ten million years – such as, for example, the uplift of the Tibetan plateau and its repercussions on atmospheric circulation.

The transition that took place around 2.6 million years ago thus opens into a world that is dominated by a succession of alternating cold and warm climates, in which ice sheets on the continents in the NH intermittently advance and then recede. This date was recently designated to define the Tertiary–Quaternary transition. The oscillation involves two states of the planet: (i) glacial stages with a cold climate (with more or less extended ice sheets in the NH); and (ii) interglacial stages with a much warmer climate similar to that of the present, in which the ice sheets in the NH either almost disappear or are absent. The to-and-fro between the two climates is reflected in the oscillation of the sea level, since the growth of the ice sheets is supplied by the water in the oceans. These changes are global and practically synchronous in the two hemispheres. We have entered what might be called a 'glaciation dance', the tempo of which is set by astronomy.

Chapter 19

Since 2.6 million years ago: the dance of glaciations

19.1 The archives of the dance

Various *proxies*, or indirect means, can be used to reconstruct climate oscillations. Many proxies exist but, notably, two of them have revolutionized our understanding of the past. Both are based on changes with time of the ratio of the oxygen isotopes ($^{18}O/^{16}O$). It was first shown in the 1960s that marine sediments contain continuous records of the volume of ice stored on the continents. Later came the records from ice cores, which give access to changes in the atmosphere over several hundred thousand years, together with local temperatures. These two types of record have the outstanding advantage of being deposited continuously over long periods of time, thus giving access to a continuous reading of the past. With their help, we can go back in time in great detail, in the first case over millions of years, and in the second over almost a million years.

19.1.1 MARINE SEDIMENT RECORDS: THE ISOTOPE SIGNAL AND THE NOMENCLATURE OF THE CLIMATE STAGES

The volume of ice stored in the polar ice sheets is related to the oxygen isotope ratio $\delta^{18}O$ in seawater, which is recorded in marine sediments. *How? The isotopic content depends on the amount of water evaporated from the ocean in building up the ice sheets.* Snow that accumulates on the ice sheets comes from water vapour that evaporates from the ocean reservoir. This water

Climate Change: Past, Present and Future, First Edition. Marie-Antoinette Mélières and Chloé Maréchal.
© 2015 John Wiley & Sons, Ltd. Published 2015 by John Wiley & Sons, Ltd.
Companion website: www.wiley.com/go/melieres/climatechange

vapour is enriched in the light isotope, giving $H_2{}^{16}O$. The more water that evaporates from the ocean to construct the ice sheets, the more the ocean becomes enriched with the heavy isotopic form ($H_2{}^{18}O$). It follows that the heavy isotope content of the seawater increases when the ice sheets grow, and vice versa. These changes in content are recorded in the calcium carbonate that is formed in the *test* (hard shell) of living marine organisms such as foraminifera. A fall in sea level of about 120 m between an interglacial and a glacial stage increases the isotopic ratio $\delta^{18}O$ of the seawater by about 1‰. This difference is recorded in foraminifera tests. After death, the tests fall into the sediment, carrying with them their precious message. A complication arises, however, when the message is decoded: the $\delta^{18}O$ of the carbonate tests is also a function of the temperature of the seawater in which they were synthesized. In glacial and in interglacial stages, this temperature is different. To reduce the effect of temperature, measurements are made on benthic foraminifera. These live on the sea floor, where the temperature varies less than in the surface water. The glacial–interglacial change in the deep-water temperature is typically about 4°C in the Atlantic Ocean and 1.5°C in the Pacific Ocean. A correction for the deep water temperature, which depends on the coring location, transforms the variation in the isotopic signal into that of the sea level. An increase by 1‰ of $\delta^{18}O$ in seawater induces a change of about +2‰ in the benthic foraminifera tests. Since the first coring expedition in 1947–8, a large number of sediment cores have been retrieved in every ocean of the world. The collation of these data carried out in the 1970s provides a continuous record of the sea-level changes over the past million years. More recently, a set of 57 marine cores was selected for their quality. The resulting new 5.3 million year benthic $\delta^{18}O$ stack was published by Lisiecki and Raymo (2005). The stack signal is shown in Fig. 18.4, together with the reconstructed sea level over the past 450,000 years (Waelbroeck et al. 2002). In such curves, which are the result of assembling several cores, fluctuations are smoothed and extremes attenuated.

The different glacial and interglacial stages have been identified in this way and numbered on the basis of the isotopic ratio $\delta^{18}O$ in the carbonate tests; for this reason, they are labelled MIS (Marine Isotopic Stage). Even numbers are attributed to the glacial stages (low sea levels) and odd numbers to the interglacials (high sea levels). These stages are indicated on the curve of Fig. 18.4. The present warm period, starting about 12,000 years ago, is the most recent interglacial, and is designated MIS 1. The preceding glaciation (i.e. the glacial stage with its maximum about 20,000 years ago), is MIS 2 and so on. During these stages, principally in the warm interglacials, secondary oscillations appear. These are numbered according to the same principle, by a second numeral.

We take the example of the previous interglacial, the warm MIS 5 that began about 130,000 years ago. This can be divided into three warm periods. The first was the warm interglacial 5.5 (the Eemian period), where the sea level was 5–7 m higher than now. This period was followed by two cooler periods, interstadials 5.3 and 5.1, which were nevertheless warm but the sea level was lower. These are separated by MIS 5.4 and 5.2, during which the sea level was much lower. After correction for the deep-water temperature, the difference in sea level between stage 5.5 and the two interstadials 5.3 and 5.1 amounts to a fall of about 20 m. This example illustrates the importance of temperature correction, because if a direct reading were made from the marine isotope curve in Fig. 18.4 (i.e. neglecting the correction), it would suggest a fall in the sea level of about 40 m. Another example is the preceding warm period, MIS 7, which displays three warm interstadials: MIS 7.5, 7.3 and 7.1. The two interstadials 7.3 and 7.1, which are well resolved in the ice-core records, are very close. In the marine record, however, where time resolution is poorer, they often appear as a single warm stage that stretched from about 220,000 to 190,000 years ago. The consequences of this poor resolution will be illustrated in Chapter 21. These last two examples underline the importance of time resolution in the different records and the degree of caution with which they must be interpreted.

19.1.2 ICE RECORDS: ISOTOPES AND AIR BUBBLES

Ice on Earth is stored in different compartments. Its ability to archive the memory of the past, however, is not the same in all places (Box 19.1).

BOX 19.1 ICE SHEETS, GLACIERS, ICE FLOES, ICE EVERYWHERE

The three great ice reservoirs have in common that successive layers of ice are formed over each other (or underneath in the case of an ice floe), giving rise to compact accumulations of ice. There the similarities end. All three reservoirs differ in their surface area, in their total volume and in their effect on the sea level.

Glaciers and ice sheets grow by accumulating snow. When it is sufficiently thick (between 60 and 100 m for the polar ice sheets), the snow gradually turns into ice. Ice sheets form at high latitudes, in surroundings sufficiently cold for well-developed glaciers to grow and connect with each other over thousands of kilometres. A cover of ice as thick as 3000 m gradually builds up that can persist for thousands of years (such is the case for the ice sheets that appear periodically at high latitudes in the Northern Hemisphere (NH)), or even millions of years (as is the case in Antarctica). At the present time, there are two ice sheets, one on Greenland and the other on the Antarctic continent. If both were to melt completely, the sea level would rise by 7 m and by 58 m, respectively. The ice sheets permanently flow into the sea, shedding icebergs hundreds of metres high, which then drift towards lower latitudes. The friction of the ice on the rock bed rips off gravel and rocks and imprisons it in the ice. The ice sheet transports this debris towards the ocean, where it is carried off by the icebergs. When these melt, the debris is released into the ocean and ends up in the marine sediment. The rock and gravel in the marine sediments, the so-called ice-rafted debris (IRD), constitutes a record of ancient iceberg melts and of their geographical range. In recent years, the annual balance of each of these ice sheets has become negative, which means that they are losing ice and are decreasing.

Glaciers are much smaller structures. In the tropics they form at high altitudes, while at high latitudes they descend to sea level. The equilibrium line of the glacier, which demarcates the upper zone of accumulation of snow from the lower zone of ablation (Part V), is located at an altitude of about 5250 m in the tropical Andes, at around 3000 m in the European Alps, while at high latitudes it drops to sea level. If all the glaciers that lie outside the vicinity of the Greenland and the Antarctic ice caps were to melt completely, the sea level would rise by an amount that is estimated to be between 20 and 40 cm. Present observations indicate that glaciers are in retreat over practically the whole of the planet.

Sea ice, or pack ice, for its part, starts to grow on seawater when it cools to a temperature between −1.7 and −1.9°C. A film forms at the surface of the sea, on the underside of which, in contact with the ice/liquid interface, ice progressively accretes. The average thickness of sea ice is about 3 m, which means that only a few tens of centimetres emerge from the surface of the sea. Movement due to storms, however, can cause blocks of pack ice to overlap and increase the thickness, which can be as much as a dozen metres. When seawater freezes, salt is expelled into the seawater, because the solid consists only of fresh water. The surrounding water into which the salt is expelled becomes denser and sinks to a lower level. In certain zones of the planet this water can sink to the bottom, and feeds the deep-water circulation (Chapter 5). The melting, or formation, of sea ice in no way modifies the sea level, just as the ice cube in a glass of whisky does not change the level of the liquid when it melts. In the past few decades, the area of the arctic ice floe has been shrinking regularly, especially in summer, and its thickness is decreasing strongly. That of the southern ice floe has remained stable.

The extremes of climate and its details are most easily perceived in the ice records through the changes of the $\delta^{18}O$ isotopic signal and δD of the ice. The cause of their variation is quite different from that in the marine sediments. In ice sheets, the isotopic signal depends on the history of the temperature of the air masses between the point of evaporation and where the snow crystals are deposited, while in marine sediments $\delta^{18}O$ corresponds to a global change in the seawater after the ice sheets are formed. The $\delta^{18}O$ signal in the ice is determined principally by the local temperature at the point at which the snow is deposited: cooling is recorded as a decrease in $\delta^{18}O$. This means that during the glaciation cycle this signal varies in the opposite direction to that of ocean water. Taking into account the change of $\delta^{18}O$ in the seawater, transitions from an interglacial to a glacial stage appear in the ice of the Antarctic central plateau as a decrease in $\delta^{18}O$ of about 5‰, in which the signal decreases from −51‰ to −56‰. In Greenland ice, this decrease is greater: for instance at the North Grip site, at the centre of the ice sheet, the isotopic signal decreases from about −35‰ to −44‰. The isotopic ratio in ice is therefore not a *global* property, since, in contrast to that of marine sediments, it represents a change at a given location. But it offers the priceless advantage of being continuous with a far better time resolution. The continuous deposit of snow provides a chronological basis that is independent of that of marine sediments. Fine resolution and continuous dating make ice cores a choice tool for studying climate mechanisms.

The number of ice cores that have been extracted from the Antarctic and Greenland is much smaller than the number of marine cores. They are obtained under extremely difficult conditions (Box 19.2).

BOX 19.2 ANTARCTIC ICE-SHEET DRILLING AT VOSTOK

From 1970 onwards, Soviet teams carried out the first deep drilling of the Antarctic ice cap. The first ice core, obtained in 1982 after 12 consecutive years of drilling, contained 2000 m of glacial records and spanned the last climatic cycle (130,000 years of archives). A second ice core, of length 3350 m and spanning the last four climatic cycles – that is, 420,000 years of climate history – was extracted in 1996 after 6 years of drilling. The drill had then reached the ice that covers the subterranean lake under the Vostok site. To preserve the water of the lake from pollution, the ice layer was not penetrated.

Drilling conditions were not easy: the Vostok station is situated at an altitude of 3488 m and boasts an average annual temperature of −55°C (the lowest value, −89°C, was recorded in July 1982). Drilling continues throughout the year. Matters were not helped by the difficulties that confronted the Soviet Union during that period, causing interruptions of supplies at that end of the world. It can be readily understood that, for the Soviet groups, those years of drilling were not exactly equivalent to living in a five-star hotel! The collaboration that developed with the French teams working at the site led to the analyses of the ice being performed in two French laboratories (LGGE and LSCE) (Petit et al. 1999).

Since that time, the drilling campaign (1996–2004) carried out in the framework of the European Project for Ice Coring in Antarctica (EPICA) in the centre of the Antarctic ice cap at Dome C has produced an ice core 3270 m long that contains climate records going back 800,000 years.

The reproducibility of the isotope signal obtained from the different drilling locations both in Antarctica and Greenland is remarkable, with similar details at different drilling sites separated by several hundred kilometres. This indicates that the temperature changes are, to first order, similar over the whole of the ice sheet.

There is, however, a difference between the time resolution of the ice records in the Greenland and the Antarctic ice sheets. The resolution in Greenland is far better because more snow falls there than in the Antarctic, where the temperature is much lower. At the centre of the Greenland Ice Sheet, the annual accumulation is about 50–60 cm of snow, corresponding to roughly 20 cm of liquid water. This is 10 times greater than at the centre of Antarctica, where the atmosphere is colder and drier. As the two ice sheets are of comparable thickness, the Greenland record covers a shorter period. The drilling sites GRIP (GRIP Project Members 1993; North Greenland Ice Core Project Members 2004) and GISP (Mayewski & Bender 1995; Stuiver & Grootes 2000) in Greenland cover about 100,000 years, and that of Dome C (EPICA 2004; Jouzel et al. 2007) in Antarctica 800,000 years. The interpretation of the isotopic signal in terms of the temperature is, however, more complex in Greenland than in the Antarctic. While the isotopic signal in the Antarctic can be related directly to the temperature change between glacial and interglacial periods, additional techniques are needed to obtain a quantitative evaluation of this change (Part IV, Note 3).

Another piece of information contained in the ice is that delivered by the air bubbles trapped inside (Fig. 19.1). Air is imprisoned as the snow is deposited and compacted, and it gradually becomes cut off from the atmosphere. These air bubbles close at a depth of about 100 m in the central regions of the Antarctic and at about 60 m in Greenland. The difference in age of a trapped air bubble and that of the ice that entraps it is thus substantial, and is a source of complication in determining the chronology. The difference is of the order of thousands of years in the Antarctic and hundreds of years in Greenland. The bubbles constitute a continuous sampling of past atmospheres that only the glacial records are able to deliver.

In this way, it was recently possible to reconstruct the changes in annual temperature together with the changes in the atmospheric contents of greenhouse gases over 800,000 years from the EPICA ice-core drilling at the centre of Antarctica (Dome C) (Jouzel et al. 2007). Figure 19.2 displays the variations in carbon dioxide, CO_2, and methane, CH_4, as well as temperature over the past 600,000 years. Also indicated in this figure is the $\delta^{18}O$ isotopic signal from marine sediments, a proxy for the volume of the ice sheets. The marine and ice records both indicate the successive glacial and interglacial stages. In Part V, which focuses on changes in the atmosphere during the past century, we shall see how these reconstructions put into perspective the sheer magnitude of human impact.

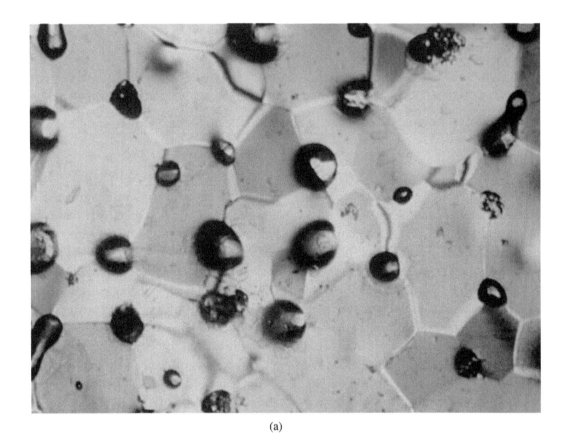

(a)

Fig. 19.1 (a) An Antarctic ice core taken from a depth of about 100 m (false colour photograph). Air bubbles imprisoned in the ice are located at grain boundaries, and appear black. (b) An ice core drilled at Dome Concordia, Antarctica. The diameter is about 10 cm and its length is no more than 2 m. **Source:** (a) P. Duval (LGGE-UJF-Grenoble/CNRS). (b) © CNRS Photolibrary/Laurent Augustin.

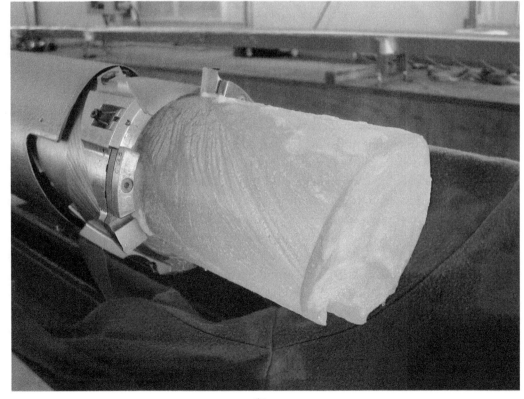

(b)

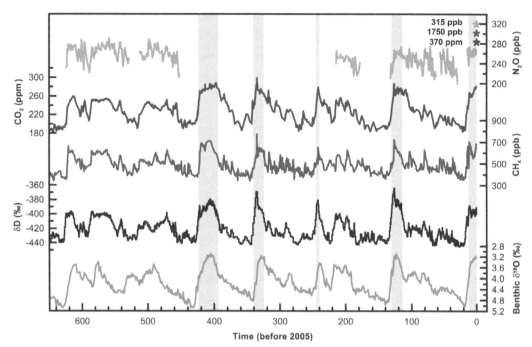

Fig. 19.2 Variation in the deuterium (δD) proxy for local temperature, and the atmospheric concentrations of the greenhouse gases CO_2, CH_4 and nitrous oxide (N_2O) since 600,000 years ago. These curves are obtained from analysis of the ice cores from Dome C (centre of Antarctica). The annual temperature increased from about –65°C during the Last Glacial Maximum (~20 kyr ago) to about –55°C during the present interglacial. The gas concentrations are obtained by analysing air bubbles trapped in the ice. Asterisks indicate the 2000 values of these gases; in 2013, it reached 395.4 ppm (CO_2), 1893–1762 ppb (CH_4) and 326–324 ppb (N_2O) respectively (CDIAC data). The lower curve (d), also shown in Fig. 18.4, traces the chronology of the volume of the ice caps deduced from the isotopic signal of benthic foraminifera. The grey bars mark interglacial stages that were comparable to the present Holocene period. Note that the shaded vertical bars are based on the ice-core age model. The marine record is plotted in its original timescale, based on tuning to the orbital parameters. The five warm stages (grey) correspond to elevated sea levels, as deduced from marine sediments.

Source: IPCC 2007. *Climate Change 2007: The Physical Science Basis.* Working Group II Contribution to the Fourth Assessment Report of the Intergovernmental Panel on Climate Change, Figure 6.3. Cambridge University Press.

19.1.3 CONTINENTAL RECORDS

On the continents, there are numerous indicators that also record glacial and interglacial cycles. These indicators are less straightforward to decipher because they can be strongly affected by the local context, but they have the outstanding advantage of revealing the conditions in which the flora and fauna developed on land. There are three main types of record:

- Lake and peatbog sediments, which record changes in the vegetation (notably in the pollen deposit) and in the fauna.
- Calcareous deposits, particularly from *speleothems* (stalagmites and stalactites that form in caves). These alone can provide absolute dating.

- Long sequences of soil, notably in certain regions in Central Asia where *loess deposits* (windborne sediments deposited during glacial periods) alternate with *palaeosols*, which are richer in organic matter and are generally deposited during interglacials. Such sequences have provided regional records over several millions of years.

19.2 The glacial–interglacial cycles

19.2.1 SINCE 2.6 MILLION YEARS AGO: THE VARIOUS RELEVANT PERIODS

The Pliocene–Pleistocene stack sediment record of Lisiecki and Raymo (2005), of globally distributed benthic $\delta^{18}O$ records, illustrates the changes in volume of the ice sheets (Fig. 18.4). These observations show that the alternations between glacials and interglacials are cyclic. They clearly appeared 2.6 million years ago, with a *period* of the order of 40 kyr; that is, 40,000 years (about 40 glaciations took place between 2.6 and 1 million years ago). When the ice sheets developed, the sea level fell by roughly 50 m. Around 1.2 million years ago, the amplitude of the oscillations increased, and a new cycle of about 100 kyr gradually started. Over the past 700,000 years, the oscillations have entered a new regime. The amplitude of the glacial–interglacial cycle increased and the 100 kyr cycle has become dominant. The changes in sea level, by as much as 120–130 m, reflect the development of much thicker ice sheets.

This evaluation is corroborated by frequency analysis of the isotopic signals. From 1.5 million years ago to the present, two frequencies always occur, with one period close to 20 kyr and the other about 40 kyr. The 100 kyr period, which appeared discretely about 1.2 million years ago, has become more pronounced over the past 700,000 years. These three periods are those of the three orbital parameters of the Earth (Chapter 14), namely: (i) the precession of the rotational axis of the Earth (19, 22 and 23 kyr); (ii) the obliquity of this axis (41 kyr); and (iii) the eccentricity of the ellipse described by the Earth around the Sun (~100 and ~400 kyr).

This remarkable result from frequency analysis of sediments has been progressively established by the palaeoclimatology community since the 1970s. *This is the key that reveals the underlying cause of the glacial–interglacial oscillations: the primary cause is indeed astronomical, modulating the solar radiation received by the Earth.* How this irradiance can induce such oscillations will be addressed in Chapter 20.

19.2.2 THE DURATION OF GLACIALS AND INTERGLACIALS

Over the past 2 million years, warm periods (interglacials) have not been on an equal footing with cold periods (glacials). The record of events displays a climate that is on average more glacial than interglacial. A very rough estimate indicates that over the last climate cycle, warm periods have occupied about 10% of the time that has elapsed. *Does this mean that for the rest of the time only glaciations have existed, with a sea level 120 m lower than now?* Certainly not. The gradual advances of the ice sheets were interspersed with phases of marginal warming and pauses in the descent, or even a slight rise, of the sea level. At times the rise was more pronounced and, while not reverting to a full interglacial, nevertheless produced a relatively warm climate. Such climate phases are denoted *interstadials* (Fig. 18.4). This is the case, for example, for the interstadials 5.3 and 5.1 (~100,000 and ~80,000 years ago, respectively) that followed MIS 5.5, with a lower sea level.

The average duration of an interglacial has been of the order of 10 kyr. This was the case for the preceding interglacial, MIS 5.5 (the Eemian), which lasted from about 128,000 to 118,000 years ago. Nonetheless, the duration could be as short as a few thousand years, as in MIS 7.5 (~245,000 years ago) or, exceptionally, as long as 30 kyr, as in MIS 11 (~420,000 years ago). As will be seen in Chapter 20, these different durations are determined by the conjunction of astronomical parameters.

19.2.3 ASYMMETRIC TRANSITIONS

Recovery from a glacial period, where the ice sheet melts and the sea level rises by about 120 m, lasts less than 10,000 years; that is, one half of a precession period. This is much shorter than the opposite process of large ice-sheet build-up. The build-up of ice sheets, which is marked by a decrease in sea level, is generally gradual, advancing in fits and starts over several precession cycles (period ~22 kyr). All these steps are initiated by the irradiance from the Sun. The rate depends on the shortfall of irradiance in summer in the NH, which in turn depends on the orbital parameters (Chapter 20).

19.3 Glacials and interglacials: very different climate stages

In what way are these two climates so different from each other? The last glacial stage (MIS 2) and the present interglacial (MIS 1) illustrate these differences.

19.3.1 THE GLACIAL STAGE

The last glacial maximum, which peaked about 20,000 years ago, is one of the most pronounced glaciations to have occurred in the past 2 million years. Its principal features are as follows:

- a global mean temperature lower by about 5°C than the present value;
- two large ice sheets in the NH, one of which descended as far as the mid-latitudes of North America, while the other is located over Finnoscandinavia, and one ice sheet in the Southern Hemisphere (SH), covering the Antarctic continent;
- a sea level about 120 m lower than now;
- completely different vegetation cover at the middle and high latitudes; and
- very different animal populations in these same latitudes.

What does the surface of the Earth look like during a glaciation? In the Northern Hemisphere … Large ice sheets, as much as 3000 m thick, cover the continents of North America and Eurasia (principally Northern Europe). Greenland seems fatter than now (Fig. 19.3;

see also the extent of glaciation and chronology maps at the website http://booksite.elsevier.com/9780444534477/index.php). In Europe, the southern part of the ice sheet reaches Berlin, and almost touches London (Fig. 19.4). With the sea level lower by about 120 m, much of the continental shelf around the continents emerges, thereby modifying land connections and allowing the migration of various fauna. About 20,000 years ago, Europe was physically united and hunter–gatherers could travel on foot from Paris to London (although they would have had to cross the palaeo-Channel, into which flowed the Thames, the Seine and the Rhine). They could also reach North America from Asia on dry land across the Bering Straits or, for example, Papua New Guinea from the Australian continent. In the European Alps, glacier fronts advanced by several hundred kilometres,

Fig. 19.3 The ice-sheet coverage in the Northern Hemisphere during the Last Glacial Maximum, 20,000 years ago. Purple, sea ice; grey, emerged land.
Source: Image reproduced from Ehlers and Gibbard (2004) by permission of the authors.

far beyond their present limits, and many alpine valleys lay under more than a 1000 m of ice.

In the Southern Hemisphere … As in the NH, the lower sea level caused the continental shelf to emerge. On the eastern coast of South America, in the latitude band where the winds are violent, the large area of newly exposed land became a source of dust in the atmosphere. *Was the Antarctic Ice Sheet much bigger? Would that even have been possible?* During a glacial period, the mean annual temperature at the centre of the Antarctic continent is about 10°C lower than during an interglacial. This reduces the snowfall by half. At the Vostok site near the centre of Antarctica, the annual mean temperature fluctuates between −55°C in the present warm period and −65°C in glacial periods. In a climate that is already cold, where the drop in temperature further reinforces the aridity, this difference is too small to change the ice sheet perceptibly. Moreover, in the present interglacial, the ice sheet already covers almost the whole of the Antarctic continent. Thus, during a glacial period it can hardly grow any larger, since it is completely surrounded by the ocean, and further extension is prevented by the absence of solid bedrock. In some places, however, the ice can become thicker. Modelling of the dynamics of this ice sheet has led to the conclusion that during the last glaciation its volume was larger than now, contributing by about 15 m to the drop in the sea level.

At high latitudes, therefore, the ice field extends during glacials, the temperature decreases and the amount of stored ice increases. In the low and mid-latitudes of both hemispheres, glaciers advance strongly. In the tropics and at the Equator (the Andes,

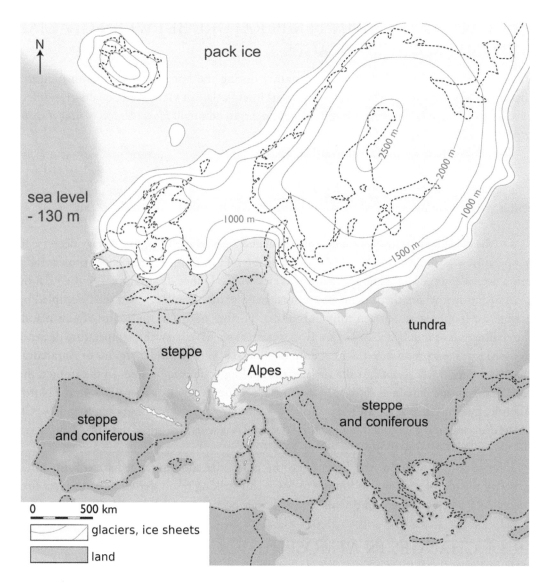

Figure 19.4 The ice coverage in Europe during the Last Glacial Maximum and the types of vegetation present at that time. Note that deciduous trees also took refuge on the Mediterranean coast. **Source:** Reproduced by permission of S. Coutterand.

Papua New Guinea, Kenya etc.) glacier fronts advanced, descending to 1000–2000 m lower than now.

19.3.2 THE INTERGLACIAL STAGE

The principal features of the present warm period are its high sea level, which has remained practically stable for about 6000 years, and the absence of polar ice sheets over the north of America and Eurasia. The sole vestige of the ice sheets that remains from the last glaciation in the NH is the small Greenland Ice Sheet. Its volume is about 4% of all ice that once covered the Earth during the previous glacial stage (Part IV, Note 4). The present warm period, the Holocene, is typical of the interglacial stages that preceded it. It is described in detail in Chapter 22.

19.3.3 THE CHANGE IN TEMPERATURE BETWEEN GLACIAL AND INTERGLACIAL STAGES

Reconstruction of the global mean temperature during the last glacial maximum indicates that it was lower by about 5°C than the present interglacial. In view of the qualitative difference between these two stages, this difference appears to be small! *How was this cooling distributed in the different regions of the planet?*

The cooling has two main features (Fig. 18.3):

- it is more intense close to the poles; and
- it is much more pronounced on the continents than on the surface of the ocean.

On the continents at low latitudes (equatorial and tropical regions), the average decrease in temperature was about 5°C. At mid-latitudes in the NH in Europe, a variety of proxies indicate a decrease in the annual temperature of at least 10°C. France, for example, on the Atlantic coast of Europe, where the present-day annual temperature is +12°C, was then occupied by permafrost (permanently frozen soil) and tundra. Further north, in the centre of Greenland, the cooling was as much as 20°C or so. By contrast, as we have seen, the temperature decrease in the SH was less pronounced, with a drop of about 10°C in the polar regions of Antarctica. *What happened in the oceans?* At the surface, the cooling was more limited. In the tropics, the water temperature remained close to its present-day value, falling by a mere 2°C. In the North Atlantic, by contrast, the fall in temperature was closer to ten degrees (between 6 and 10°C). In the deep water, the recorded temperature decrease was several degrees: recent evaluations in the Great Southern Ocean indicate a drop of 5°C.

On the continents from the Equator to the poles, the cooling ranged from 5 to 20°C, although the global cooling was limited to 5°C. This striking result derives from the fact that oceans cool much less than continents, and that they occupy two thirds of the Earth's surface.

19.3.4 CHANGES IN ATMOSPHERIC COMPOSITION BETWEEN GLACIAL AND INTERGLACIAL STAGES

The variations in the atmospheric composition during the past 800,000 years are found by analysing air bubbles trapped in the Antarctic ice. Reconstructions of the atmospheric content of the greenhouse gases CO_2, CH_4 and N_2O show that they have fluctuated between two limiting values, low during glacial periods and high in interglacials (Fig. 19.2). As deglaciation begins, the carbon dioxide (CO_2) content of the atmosphere increases, rising from about 180 ppm in a glacial period to close to 280 ppm in an interglacial. The methane (CH_4) content also increases (from 0.3 to 0.7 ppm), as does that of N_2O (from 0.22 to 0.27 ppm). The increase in CO_2 appears to stem principally from modifications of the ocean reservoir, a point that is discussed more fully in connection with the carbon cycle (Chapter 29). Warming of surface water releases CO_2 (warm water dissolves less gas) and accounts for about 20% of the increase in atmospheric CO_2. The rest of the increase is probably a consequence of the deep ocean current, which is more powerful during interglacials, thus shortening the transfer and storage times in the various ocean reservoirs. It should be borne in mind that interglacial oceans contain 60 times more CO_2 in the form of dissolved gas, carbonate ions and organic carbon than is found in the atmosphere (Fig. 5.12). Thus, even a minor change in ocean circulation can affect the equilibrium of the partial pressure of CO_2 between the atmosphere and the ocean. With the other two greenhouse gases, their increase stems from the enlarged wet zones at both low and high latitudes, and also from the warming of the continental surfaces. Return to glaciation is accompanied by these transformations in the reverse direction.

19.4 Glacials and interglacials: similar but never identical

In the glacial–interglacial cycles over the past 2 million years, the climates were never identical. Each glacial and interglacial possessed its own characteristic properties, whether in its rate of onset, its duration, its progression or its temperature, because the three orbital parameters that determine the irradiance vary independently. Their combination governs the way in which the climate unfolds, sometimes causing the ice sheets to grow, sometimes making them disappear. Their combined effect is never exactly the same at the beginning, or during or at the end of each stage. This point is discussed in detail in Chapter 20.

19.4.1 GLACIAL STAGES

The detailed reconstruction of the glacial cycles during the Quaternary that was started in the 1960s, based on isotopic techniques in marine sediment cores, has shed new light on the number of glaciations in the past. An earlier view, according to which only six major glaciations had occurred in the Quaternary (going back in time, in Europe, these were the Würm, Riss, Mindel, Günz, Donau and Biber, respectively), gave way to the discovery of about a dozen glaciations during the past million years. Moreover, between 2.6 and 1 million years ago, there were 40 or so oscillations with rather modest glaciations (Fig. 18.4).

Which, then, were the major glaciations? Extremely low sea levels, which are the counterpart of large ice sheets, can enlighten us (Figs 18.4 and 19.2). There can be no doubt that the last glaciation, which started about 70,000 years ago and reached its maximum about 20,000 years ago, was the Würm glaciation. This period thus spans the different stages MIS 4, 3 and 2. There is no ambiguity either in attributing to stage MIS 6 the preceding Riss glaciation, which was even more pronounced than that of the Würm. That glacial stage began 190,000 years ago and reached its maximum about 130,000 years ago. For the earlier two glaciations, the Mindel and the Günz, the two lowest sea levels appear to have occurred in MIS 12 (between about 460,000 and 430,000 years ago) and MIS 16 (between about 660,000 and 630,000 years ago), and could be attributed, respectively, to these glaciations. These four glaciations stand out as the most pronounced in the past 700,000 years, in which the Finnoscandinavian Ice Sheet extended south of the Baltic Sea. These stages correspond to the thickest deposits of loess in Central Europe.

And what happened before that, in the 2 million years prior to this period? During that earlier period, where glaciations begin and slowly amplify, two particularly severe cooling spells took place, the first at about 1.2 million years ago (MIS 36) and the second (including the pair MIS 22 at about 0.9 million years ago and MIS 20 at about 0.8 million years ago) point to a gradual worsening in the climate. They show up as two very large deposits of loess in Central Asia, the sign of an exceptionally cold and dry climate. The second of these two phases is accompanied by a marked decrease in deep-water production coming from the north of the Atlantic Ocean.

19.4.2 INTERGLACIAL STAGES

The high sea levels reconstructed from the marine sediment cores, as well as the local temperatures recorded over 800,000 years in the Antarctic ice (Fig. 19.2), consistently show that the past five interglacials (MIS 11, MIS 9, MIS 7, MIS 5 and MIS 1) were particularly warm, among the warmest since the beginning of the glacial oscillations. The present interglacial is not the warmest: the previous interglacial (the Eemian MIS 5.5, 125,000 years ago) was substantially warmer, as is attested by the flora and fauna of the period. The average temperature at that time was about 2°C higher than the average temperature of the Holocene. According to current evaluations, at the beginning of the Eemian the temperature was higher by about 4°C in the Antarctic and by about 5°C in Greenland. This interglacial, along with MIS11 and, for a short period, MIS 9, stands out as having been one of the warmest during the Quaternary.

Interglacials also differ in their duration. While the average interglacial lasts about 10,000 years, some are shorter (e.g. a few thousand years for MIS 7.1) and others are delayed until the next 22 kyr cycle, giving a total length of about 30,000 years. This rare event took place about 400,000 years ago, in the interglacial stage MIS 11.

In Chapter 20, where these different examples are taken up again, we shall see how the length and intensity of the different interglacials are governed by astronomical parameters. *And what would be the 'natural' length of the present interglacial, the Holocene?* Astronomical theory suggests that the present interglacial will naturally (i.e. in the absence of human impact) last even longer than MIS 11 (Chapter 21).

19.5 Abrupt climate changes in the last climate cycle

19.5.1 CHAOTIC CLIMATE AND ICEBERG MELTING IN THE LAST GLACIATION

The progress of the last glacial period, which began about 70,000 years ago, with its maximum 20,000 years ago, has been reconstructed in detail, mainly from the information recorded in the ice cores of Greenland. This period was interspersed with two kinds of abrupt climate events, the Heinrich and the Dansgaard–Oeschger events (Fig. 19.5).

The first of these, the Heinrich events (named after the sedimentologist who discovered the evidence for them in 1988), are recorded in the marine sediments and correspond to armadas of icebergs released into the North Atlantic (Bond & Lotti 1995). At least six of these events have occurred in the past 60,000 years. The events leave a trace in the marine sediments in the form of the gravel and rocks transported by the iceberg (IRD), which are released when the

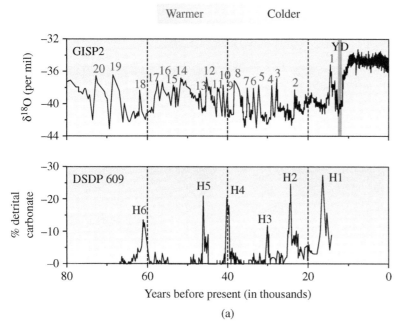

Warmer   Colder

(a)

Fig. 19.5 (a) Top: the isotopic signal $\delta^{18}O$ of ice is related to the local temperature. Its variation over the past 80,000 years in Greenland ice (drilling campaign GISP2) reveals that the temperature stability was greater in the past 10,000 years than during the glacial period. The $\delta^{18}O$ record shows 20 of the 25 rapid fluctuations (Dansgaard–Oeschger events) that occurred during the last glacial period (Grootes et al. 1993). Bottom: the record of detrital carbonate in sediment, a proxy for the ice-shelf break-up episodes (Heinrich events), from a deep-sea core in the North Atlantic (Bond and Lotti 1995). The abrupt cooling of the Younger Dryas is indicated by a blue bar. (b) A comparison between climate signals from high latitudes in the Northern Hemisphere ($\delta^{18}O$ in Greenland ice, green curve) and from the Southern Hemisphere (temperature in Antarctica at Dome C, red curve). The fluctuations in Greenland during glacial periods are more pronounced and the glacial–interglacial temperature differences are stronger. Note the reversed timescale.

Sources: (a) Grootes et al. (1993). Reproduced with permission of the Nature Publishing Group. (b) Adapted from Jouzel et al. (2007) by A. Landais and V. Masson-Delmotte. Reproduced with permission of V. Masson-Delmotte.

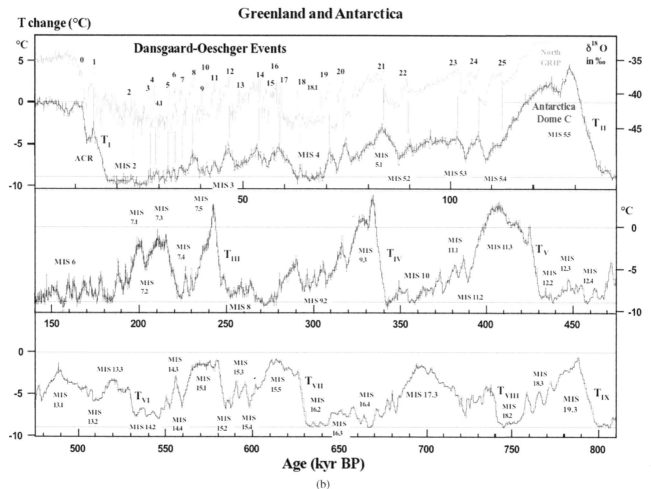

(b)

Figure 19.6 An example of different marine sediment core proxies used to reconstruct climate variations and their effect on vegetation in the last glacial period: pollens (top), coarse sediment released by icebergs (ice-rafted debris, or IRD; bottom left), and polar plankton foraminifera (bottom right). **Source:** Reproduced with permission of M.F. Sanchez Goni.

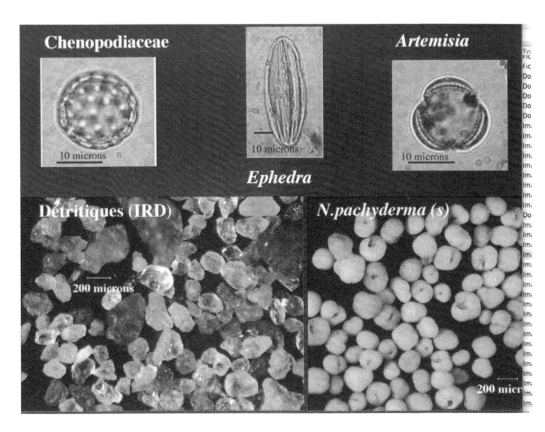

iceberg melts (Fig. 19.6). Cartography shows that the icebergs descended as far as 45°N. Such events are clearly related to changes in the ice sheets on the Atlantic coast in the NH, mainly the Laurentide Ice Sheet (North America). Periodically the ice sheets partly collapse, spawning numerous icebergs that invade the Atlantic Ocean and supply its surface with cold fresh water. The arrival of the cold water is corroborated by the presence of polar planktonic foraminifera (zooplankton living in surface waters near the poles) in the sediments. Each melting event is accompanied by abrupt cooling of the surface water by about 5°C, and by a reduction of $\delta^{13}C$ in the carbonate tests of deep-sea foraminifera. The reduction indicates a slowing of the deep ocean currents and reduced ventilation in the Atlantic deep water. The effects of these events on the temperature appear to be limited mainly to the North Atlantic regions and to the surrounding continents.

The Dansgaard–Oeschger (or D–O) events (named after the two glaciologists who identified them in 1980) stand out particularly clearly in the isotopic record from Greenland ice cores (Grootes et al. 1993). These events, which happen on the timescale of a millennium, occurred about 20 times during the last glaciation (Fig. 19.5). The climate follows a zig-zag pattern: abrupt warming, where the temperature in Greenland could increase by about 10°C within less than 100 years, and then, a few centuries or millennia later, the cold gradually returns. Each event is accompanied by an increase in atmospheric methane content, which is believed to be a consequence of thawing of the permafrost surrounding the ice sheets in the NH.

These rapid climate fluctuations are associated with changes of plant coverage in Europe. For Southern Europe, these changes were reconstructed from marine sediment cores extracted close to Spain, both on the Atlantic and on the Mediterranean coast. The period associated with cold interstadials in Greenland corresponds to the spread of desert/steppe-type vegetation. Warm periods (warm interstadials in Greenland) are associated with wet temperate climates with open forests (pine, green oak, olive trees, pistachios etc.). Likewise, the cooling due to the Heinrich events corresponds to a return to desert/steppe vegetation.

How far did these fluctuations extend? It has been known for several years that the Dansgaard–Oeschger events had a planet-wide effect, reaching as far as Asia, South America and Antarctica. They are clearly recorded in the Antarctic ice cores (Fig. 19.5b). The ice core record suggests that the Dansgaard–Oeschger events are related to the so-called Antarctic Isotope Maxima through climate coupling between the two hemispheres, the 'bipolar seesaw' (Blunier & Brook 2001; Stocker & Johnsen 2003). On a regional scale, the D-O temperature changes in Greenland could be caused by changes in the sea-ice extent in the North Altantic (Li et al. 2005).

Although the cause of the Heinrich events appears to be well understood, that of the D-O events still belongs to the realm of research. The cause of these glacial events is still subject to debate. Some interpretations attribute them to solar fluctuations, while others invoke possible internal mechanisms in the climate system.

19.5.2 THE END OF THE LAST GLACIATION

The continents

The NH enjoyed a hesitant warming period after the glacial maximum of 20,000 years ago. The first major thaw, however, occurred abruptly about 14,700 years ago, with a temperature increase that marked the arrival of the warm Bølling–Allerød stage. Its rapidity is recorded in the ice cores from Greenland (Fig. 19.5). The warm climate, which started about 14,700 ago, was subjected for about 2000 years to repeated assaults of cold that brought gradual cooling. The particularly warm Bølling stage was interrupted 700 years later by one of these cooling events; this was then followed by the slightly less warm Allerød stage, which lasted a little more than a thousand years. Then, suddenly, 12,800 years ago, the cold event of the Younger Dryas (YD) set in, which lasted for 1100 years and effectively interrupted the thawing process. The YD event ended within a few decades, as suddenly as it started. These sudden changes appear to be related to rapid changes in the deep ocean circulation. Since then, the warm Holocene climate has become firmly established.

In the SH, a continuous record of the warming that marked the end of the last glaciation exists in the Antarctic ice core (Fig. 19.5). The warming started hesitantly about 18,000–19,000 years ago and developed, less chaotically than in the North Atlantic region, until about 12,000 years BP; that is, *12,000 years before present*, where the reference date is 1950 (other dating conventions on this timescale are indicated in Part IV, Note 5). This warming did not continue unimpeded: 14,000 years ago it was interrupted for about 2000 years by the Antarctic Cold Reversal (ACR), during which the temperature was almost constant. This period coincides with that in which the climate in the NH cooled fitfully.

BOX 19.3 THE SEVERE COLD SNAP OF THE YOUNGER DRYAS

The Younger Dryas (YD) was a sudden dramatic event that occurred between 12,800 and 11,500 years ago, drastically cooling the climate on a regional scale. It is named after the alpine tundra wild flower (*Dryas octopetala*), which became common in Europe during this time. This event is recorded in various types of archive, among which are speleothems. These have the advantage of providing absolute U/Th chronology. Figure B19.1 displays the YD event recorded in speleothems of different regions (Genty et al. 2006). The impact of this event extended mainly over the North Atlantic and its surrounding regions: Greenland, Europe, North America and North Africa. Beyond that range, its effect decreased and it is not recorded in the climate of the Antarctic.

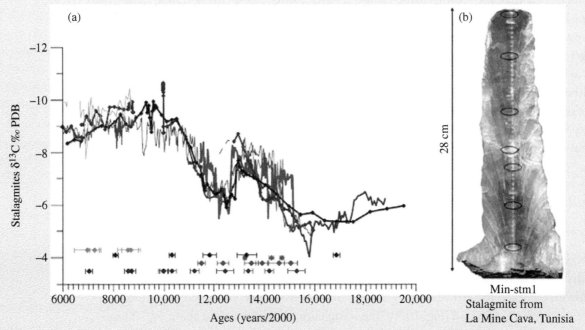

Fig. B19.1 (a) The abrupt Younger Dryas event (YD) as recorded on continents in stalagmites. Stalagmites have the singular advantage of yielding absolute chronologies. Here, the $\delta^{13}C$ profiles at two Western European sites are shown (blue and green lines, Villars Cave, South-West France; red and orange lines, Chauvet Cave, South of France) and one in North Africa (black line, La Mine Cave, Tunisia). The decrease in the $\delta^{13}C$ signal during the Holocene is the result of cooling. The abrupt YD cooling is clearly recorded in North Africa, together with the recovery of the warming phase. The same figure also shows two $\delta^{13}C$ records from stalagmites in China (purple line, Hulu cave, from Wang et al. 2001; light purple line, Dongge Cave, from Dykoski et al. 2005). These observations suggest that the YD event was extensive in the Northern Hemisphere. (b) A polished section of the stalagmite (height 12 cm) from 'La Mine' in Tunisia.

Sources: (a) Genty et al. (2006). Reproduced with permission of Elsevier. (b) Photograph by and reproduced with permission of D. Genty.

Within a few decades, the prevailing Böllering–Allerød warm period was interrupted. In Western Europe, the annual temperature dropped by about 5°C. A cold dry climate set in once again. The continents became arid, and the forests, which had invaded the land, suddenly retreated: steppe returned. But ... 1300 years later, this cold snap came to an end as suddenly as it had begun, and forests quickly regained control of the territory. In Poland, for example, where a precise local reconstruction was established from the varved clay sediments of Lake Gosciaz, the cold steppe vegetation (including, notably, wormwood and juniper) gave way in less than 50 years to a deciduous forest of elm, poplar and birch.

The sudden cold snap changed the surface of the continents to such an extent that the level of methane in the atmosphere fell quickly. As with each cold period, the reduction of methane is the consequence of a modification of the soil and constitutes an important proxy for cooling. The records of two gases in the Antarctic ice show that, contrary to methane, the CO_2 content of the atmosphere was unaffected by the event. This means that the CO_2 cycle was not modified. The increase in CO_2, which started about 18,000 years ago, marked a pause between 14,700 and 12,800 years BP, and then continued on its course as before.

What caused the YD? The marine sediments conserve the trace of a violent and intense discharge of fresh glacial water that spread over the surface of the Atlantic Ocean at that time. This water, released by the gradually melting ice sheets, had progressively accumulated on the North American continent, forming an immense lake, *Lake Agassiz*. A change in the topography (the rupture of a dam, or the lowering of a threshold) opened a passage for it to flow out, and let the water suddenly pour into the North Atlantic Ocean. The reason for the rupture is still a matter of discussion. It could have been the result of a change in the landscape due to the continuous melting of the ice sheet. But it could also have been the aftermath of a meteorite that fell on to North America when the YD set in. This hypothesis is based on deposits that contain trace elements characteristic of a meteor impact, overlaid by a thick organic deposit that reveals disruption of the ecosystems in the region. Moreover, the numerous extinctions among American megafauna, as well as that of the Clovis civilization, appear to date from this period. That hypothesis is still under debate. A scenario can be constructed from the sudden decrease in temperature and in salinity of the surface water in the North Atlantic. The lower density of this water hindered its downwelling in the North Atlantic. It therefore reduced the heat flow carried by the Gulf Stream from the tropical regions to the mid-latitudes. A thousand years later, when the cause ceased, the ocean current started again as fast as it had stopped. The progression towards the warm interglacial climate, predetermined by the Earth's orbit round the Sun, could then start again. Its arrival was merely delayed by the accident of the YD episode. This is an illustration of the frightening speed at which climate can change on a regional scale within a few decades, disrupting the vegetation from the Arctic to North Africa for several centuries.

The sea level

By monitoring the rise in sea level from ocean sediments and coral records, we can gain access to the chronology of the melting of the ice sheets (Fig. 19.7). The rise in sea level began around 17,000 years ago, then advanced steadily until about 14,700 years ago, when it abruptly accelerated at the beginning of the Bølling warming. This fast rise, exceeding 4 cm/yr, increased the sea level by about 16 m in less than 350 years (Deschamps et al. 2012). It is known as the *Meltwater pulse 1A*. *What caused this extraordinary ice melt?* One scenario involves the contribution from the *West Antarctic Ice Sheet* (WAIS), in addition to the melting of the northern ice sheets. The marine portion of the WAIS, with bedrock below sea level, is potentially unstable. It could have become partly destabilized by the rapid increase in sea level, thus reinforcing it. After this abrupt episode, the rise in sea level resumed its previous rate, and then gradually slowed until the sea level practically stabilized at its present height about 6000 years ago. The average rise in the sea level would have been 50 m in 5000 years (1 cm/yr), roughly three times faster than the rise in sea level that has been recorded over the past 10 years.

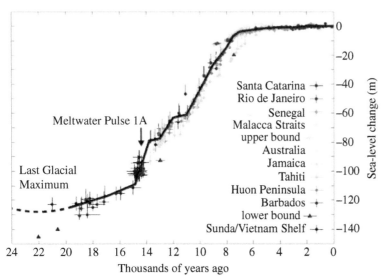

Figure 19.7 The rise in sea level since the end of the last glacial period. **Sources:** Fleming et al. (1998), Fleming (2000) and Milne et al. (2005). Image created by Robert A. Rohde/Global Warming Art/ CC-BY-SA-3.0/GFDL.

The 120 m rise in sea level changed the coastal geography. It opened the Bering Straits and the straits in the Sea of Japan; it gave birth to the Arafura Sea, which separates Australia from New Guinea, and flooded the Black Sea with water from the Mediterranean, as well as the Persian Gulf and so on. These episodes, which were punctuated by abrupt variations from melting of the ice sheets, must have marked human memory deeply, whether it was in the Pacific Ocean, where atolls barely emerge from the water, or in the Middle East, where written legends of the Deluge survive. The loss of land connections undoubtedly raised formidable barriers to the migration of species, including humans.

The duration of the transition

The last glacial–interglacial transition extended over about 10,000 years. During that period, living species, both animal and plant, had either to adapt or to migrate. At middle and high latitudes, this gave rise to plant coverage and animal populations that were very different from those preceding it (Chapter 21), even though the two physical environments differed in average temperature by only 5°C. This is the order of magnitude that we must bear in mind when we contemplate a 5°C anthropogenic warming in the future (Part VI).

Chapter 20
Glacial–interglacial cycles and the Milankovitch theory

The astronomical theory of palaeoclimates was developed in the middle of the 20th century by the Serbian astronomer Milutin Milankovitch (see Milankovitch 1941). In this theory, the sequence of glacial and interglacial periods is attributed to variations in the basic astronomical parameters and their influence on solar irradiance. Milankovitch showed how the three principal periods of these astronomical parameters drive the glacial–interglacial cycles. The theory was adapted and further enriched by the Belgian astronomer André Berger in 1988. In Chapter 14, we outlined the relevant astronomical parameters and their different cycles. We now give a simplified account of how, in the light of the Milankovitch theory, glacial–interglacial cycles arise.

Glacial–interglacial cycles are the consequence of periodic variations in seasonal irradiance at middle to high latitudes. As noted in Chapter 14, irradiance and how it varies are determined by a combination of three orbital parameters (the eccentricity of the Earth's elliptical orbit round the Sun; the obliquity, or the tilt of the polar axis; and the precession of the axis). The different cycles of the orbital parameters have always existed. It was only 2.6 million years ago, however, that this irradiance began to act as a pacemaker for glacial periods, because, after a long cooling period lasting several tens of millions of years, it was then that the Earth's climate was cool enough for ice sheets to become established on land in the Northern Hemisphere (NH). From that moment on, after the ice sheets started to build up, they were affected by the cyclic fluctuations of the seasonal irradiance and changes were driven by the orbital parameter periods.

Climate Change: Past, Present and Future, First Edition. Marie-Antoinette Mélières and Chloé Maréchal.
© 2015 John Wiley & Sons, Ltd. Published 2015 by John Wiley & Sons, Ltd.
Companion website: www.wiley.com\go\melieres\climatechange

20.1 The leading role of the Northern Hemisphere

In the Southern Hemisphere (SH), the Antarctic continent is permanently covered by an ice sheet. Centred at the South Pole, it stretches as far north as 60°S and is surrounded by an ocean that extends practically to the tropics. In the NH, the reverse situation holds: the Arctic Ocean, centred at the North Pole, descends to 70°N. There, it is almost completely surrounded by continents that cover a large fraction of the mid-latitudes. Ice sheets cannot take hold on the surface of an ocean, only on a continental base. This is why ice sheets in the NH build up when the continents lying in the latitudes 45–70°N are cold enough for the snow cover to persist throughout the year and for ice to become permanently established. In the course of tens of thousands of years, these cooling conditions oscillate. In the south, the situation is completely different, since the Antarctic Ice Sheet is exposed to a global climate that is sufficiently cold for it to remain stable permanently, and where, between latitudes 45°S and 70°S, the ocean does not allow an ice sheet to build up.

It is therefore the climate conditions in the high NH latitudes and their land masses that drive the glaciation dance. Meanwhile, the SH endures these major climate changes, oscillating between glacial and interglacial conditions, in which the ice sheets expand or retreat. Of course, secondary interactions exist that couple the climates of the two hemispheres. This topic, however, will not be dwelt upon here: we only seek the reasons for these huge changes in direction of the climate vessel. For this, we turn our attention to variations of solar heating at high latitudes in the NH.

Seasonal irradiance at 65°N?

Instead of looking only at the irradiance at 65°N in the summer solstice month of June (Fig. 14.1d), as is usually done (this involves the three astronomical parameters, but is limited to one particular latitude), we examine two parameters: the Earth–Sun distance in June (governed by changes in precession and eccentricity) and the obliquity. We can then see directly how the combination of these two parameters, each of which affects the entire planet, defines the tempo of the dance. *The advantage of this approach is that two different climate oscillations can be distinguished. One is the glacial–interglacial cycles, driven by the irradiance at high latitudes. The other is the tropical monsoon cycles, driven by the irradiance at low latitudes. Although both are driven by the irradiance, these two types of oscillation are not always identical, as we shall see in Chapter 22.*

20.2 Seasonal irradiance, the key parameter in Quaternary glaciations

In Chapter 14, we saw that as a consequence of the precession period of approximately 22 kyr (i.e. 22,000 years), at the NH summer solstice in June, the position of the Earth is alternately closest to (perihelion) and furthest from (aphelion) the Sun every 11,000 years (see Fig. 14.6). Every 11,000 years, the summer days at NH middle and high latitudes are thus alternately warmer, then cooler, and the ground is accordingly heated to a greater or lesser degree. We consider a warm (interglacial) stage, starting when the Earth is at perihelion in June. To simplify greatly, we may say that the ground in summer at middle and high NH latitudes then cools gradually over the following several thousand years as the summer solstice advances in its 11,000-year journey from perihelion to aphelion. Seasonal snow mantles linger increasingly later each year. The cooling process starts at the highest latitudes and, over the course of several millennia, the ice remains, potentially triggering the build-up of ice sheets ('potentially', because it depends on how intense the cooling is, and that is governed by the orbital parameters). In this way, the next glacial period is insidiously prepared during an interglacial period in the half-cycle of the precession. This sequence of events continues, although it is still warm and the sea level remains practically constant, since the build-up of ice sheets at high latitudes has not yet begun. Conversely, when ice sheets already exist, increasingly warmer summers gradually enhance the summer melting and, depending on the orbital parameters, reduce the ice sheets or cause them to disappear within a few thousand years.

The magnitude of this effect depends both on the coupled eccentricity and precession parameters, which modulate the Earth–Sun distance in June (Chapter 14), and on the obliquity, which determines seasonality (increased eccentricity brings the perihelion in June closer to the Sun; increased obliquity increases the irradiance in June). The irradiance in June therefore depends cyclically on the three parameters – precession, eccentricity and obliquity – each of which has a different frequency (Figs. 14.2 and 14.3). According to this basic hypothesis, it is principally the irradiance in summer that governs the fluctuations of the NH ice sheets. Precipitation is involved to a lesser degree (Part IV, Note 6). This hypothesis is justified by its ability to replicate the observations *a posteriori*.

Evidence

From the 1960s onwards, scientists could access the continuous records of glaciation history retrieved from marine sediments. They were thus able to establish a precise time frame for these variations and to detect the different frequencies involved. The numerous marine cores, which span the past million years, reveal unequivocally the existence of the three principal frequencies of the irradiance: obliquity, with a period of 41 kyr; precession, with periods of 19 kyr and 23 kyr; and eccentricity, with periods of ~100 kyr and ~400 kyr. The results unambiguously confirm the key role of the irradiance in the Quaternary climate cycles (Hays et al. 1976).

20.3 Two types of configuration

As a first approximation, the dynamics of glaciations appears to be governed by two situations at NH high latitudes. We call these states 'a' and 'b', and they alternate during the precession. The 'a' situations are those in which the NH summer takes place at perihelion, which stimulates thawing of the ice sheets (if present). Such situations are reinforced when eccentricity is high (summers are warmer because perihelion is closer to the Sun), giving situation 'a1'. They are also strengthened if the tilt of the polar axis is greater (summer in the middle and high altitudes is warmer when obliquity is high): this is situation 'a2'.

Conversely, 'b' situations are those periods in which the NH summer occurs far from the Sun (June at aphelion), giving rise to cool summers with weaker irradiance in June. These conditions are reinforced when the eccentricity is large (summers are cooler because aphelion is further from the Sun): this is situation 'b1'. They are also strengthened if the tilt of the polar axis is low (summer at middle and high latitudes is cooler when obliquity is low): this is situation 'b2'.

With this starting point, we analyse the climate over the past 250,000 years, in terms of the variation of the Earth–Sun distance in June together with that of the obliquity (Fig. 14.2). This analysis will help us to understand the sequence of glaciations and deglaciations that occurred in the last two climate cycles, and also what the future holds in store for us.

20.4 The climate in the past 250,000 years

20.4.1 PAST RECORDS

Changes in volume of the ice are recorded in the marine sediments (Fig. 18.4). Figure 19.2 plots these changes for the past 600,000 years, together with the changes in temperature in the Antarctic Ice Sheet at Dome C. In this figure, the interglacial stages (high sea level) are marked in grey. We rely on the ice-sheet record because of its good time resolution (Fig. 20.1). It must be borne in mind, however, that neither of these records, marine or ice, faithfully reflects the changes in sea level; that is, the volume of the ice sheets. In particular, in the ice record, interstadials MIS 5.3 (105 kyr BP) and MIS 5.1 (85 kyr BP) appear to be strongly attenuated, although during these interstadials the sea level fell by only about 20 m from the previous interglacial MIS 5.5 (Fig. 19.7).

The past 250,000 years exhibit sequences of warm stages, grouped into three major warm stages: MIS 7, MIS 5 and MIS 1. In the following, the dates are accurate to within 5 kyr. Each group starts with the deglaciation that terminates the preceding glacial stage (MIS 8, MIS 6 and MIS 4–3–2), thus causing the sea to rise by about 120 m. These three warm stages begin at the terminations 8/7, 6/5 and 2/1. The dates attributed to these terminations, according to Lisiecki and Raymo (2005), are 243 kyr BP, 130 kyr BP and 14 kyr BP. Note that the first two of these groups are each decomposed into three warm stages, interrupted by cooling phases: stages 7.5, 7.3, 7.1 and the stages 5.5 (Eemian), 5.3 and 5.1.

Each of the three major warm stages, MIS 7, MIS 5 and MIS 1, is different. MIS 7 began at about 240 kyr BP with interglacial 7.5, which was unusually short. It lasted only a few thousand years, which means that extremely high resolution would be required to detect the corresponding high sea level in the marine core record. As such resolution is rarely at hand, however, the intensity of this interglacial is sometimes underrated. Its properties are nevertheless clearly revealed in the ice cores, the time resolution of which is excellent. This interglacial was interrupted by an intense cooling phase, which led to the establishment of stadial MIS 7.4, a cold period that lasted

Fig. 20.1 Top: marine isotopic stages (MIS) corresponding to warm stages (interglacials and interstadials) are marked in red; glacial stages are marked in blue. Bottom: glacial and interglacial stages recorded at the Vostok site (Antarctica): local temperature variations (black curve) are calculated from the isotopic signal of the ice.

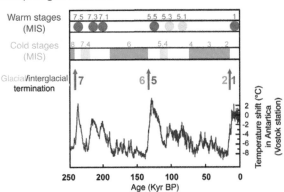

for about 15,000 years and was accompanied by low sea level. Two warm stages followed, MIS 7.3 (centred at ~215 kyr BP) and MIS 7.1 (200 kyr BP), separated by a period of moderate cooling. Since these two warm stages were separated by a short time interval, they are rarely distinguished in marine cores of moderate resolution, and frequently appear as a single long warm stage. The next warm period, MIS 5.5, occurs after the long Riss glaciation (glacial stage MIS 6). It begins with the Eemian interglacial (MIS 5.5), which lasted about 10,000 years and ended with an intensely cold spell that, within a few thousand years, caused a strong drop in the sea level. This in turn was followed by the two warm interstadials, MIS 5.3 and MIS 5.1. The three warm periods (5.5, 5.3 and 5.1) are centred successively at approximately 125 kyr BP, 105 kyr BP and 85 kyr BP. After that, the long Würm glaciation began, and lasted for several tens of thousands of years, until the present Holocene interglacial settled in permanently about 12,000 years ago.

The two last glaciations of the Riss (MIS 6) and the Würm (MIS 4–3–2) each progressed in a different way. The beginning of stage MIS 6, which occurred at about 195 kyr BP, brought an end to the interglacial MIS 7.1. The sea level fell rapidly, by about 70 m, and at about 180 kyr BP, the ice-sheet volume increased enormously. From then until the end of this glacial stage at about 130 kyr BP (termination 6/5), the sea level continued to fall gradually until it reached about –120 m. Such was not the case with the Würm glaciation, which spanned the three stages MIS 4, MIS 3 and MIS 2. The onset of the Würm glaciation occurred at about 70 kyr BP, with a drop in sea level by about 60 m (Fig. 19.7). A phase of intermediate climate, MIS 3 (which may be considered as an interstadial), then intervened, during which the decrease in the sea level was halted, and even displayed a slight increase. A second phase of ice-sheet growth then started at about 30 kyr BP, with a rapid decrease in sea level by about 50 m. Finally, 20,000 years ago, the sea level reached its lowest point, 120 m below its present position. That glaciation came to an end at the 2/1 termination.

Fig. 20.2 The interglacial stages in the past 250,000 years, and the conditions that give warm summers at middle and high latitudes in the Northern Hemisphere. These are defined by the configurations 'a1', where the Earth–Sun distance is at a minimum, and 'a2', where obliquity is high (see text).

20.4.2 THE ONSET AND DEVELOPMENT OF WARM PHASES

Let us look at configurations in the past that gave rise to warm stages. For these to happen, warm NH summers (strong irradiance in the NH summer) are needed. In Fig. 20.2, the pink circles placed at the minima of the curve of the Earth–Sun distance in June indicate when this distance is shortest during a precession cycle. These positions correspond to the situation 'a1', when warm summers prevail in the NH. In principle, such configurations either bring on deglaciation or halt the growth

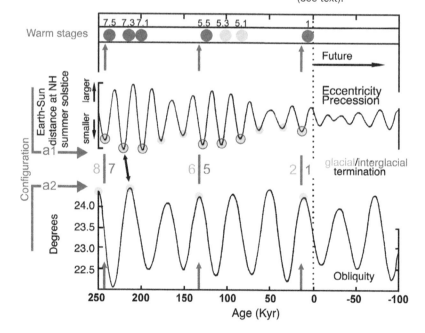

of ice sheets at middle and high NH latitudes. Some minima, however, are deeper than others, and the most prominent are denoted by an extra circle around the pink symbol. In some, the warming effect of the minimum is reinforced if the maximum of obliquity (situation 'a2') is simultaneously very large. These are indicated by a yellow circle on the obliquity curve. Of the seven cases of maximum obliquity that have occurred since 240,000 years ago, this reinforcement has occurred only four times.

The timing of the three glacial terminations is indicated by the three sets of vertical red arrows and bars: these are the glacial–interglacial transitions 8/7, 6/5 and 2/1. As in Fig. 20.1, interglacials (MIS 7.5, 7.3, 7.1, 5.5 and 1) are marked by red circles at the top of the figure, and interstadials (MIS 5.3 and 5.1) are shown in pink. Interglacial 7.5 lasted a few thousand years, and interglacial 5.5 about 10,000 years. The third interglacial, MIS 1, is still with us.

What do we see? The first two terminations, 8/7 and 6/5, occur when the Earth–Sun distance D in June is particularly small (configuration 'a1') and the obliquity is very large (configuration 'a2'). The recent termination 2/1 stands out from its predecessors in that D, albeit at a minimum, seems not to be particularly small. The configuration of 12,000 years ago was in fact a rare occurrence in the last few hundred thousand years. When 'a1' and 'a2' situations happen close to each other in time, they are rarely in phase – except for 12,000 years ago, when the modest minimum in D coincided perfectly with the obliquity maximum. The effects were therefore reinforced.

After each of the three deglaciations, the interglacials MIS 7.5, 5.5 and 1 come to a maximum, owing to the strong irradiance in June. The three prominent minima of D at approximately 245, 220 and 200 kyr BP produced the three warm stages of MIS 7. Interglacial 7.5 was cut short by the intense cooling that began at about 235 kyr BP. The two subsequent interglacials (MIS 7.3 and 7.1) are related to the unusually short Earth–Sun distances at approximately 220 and 200 kyr BP, but warm stage 7.3 was delayed until *after* the June minimum of D, because the strong maximum in obliquity appeared only later, at about 215 kyr BP. In other words, the maximum irradiance in June at high NH latitudes, which is the consequence both of a short Earth–Sun distance and strong obliquity, was delayed with respect to the minimum in D. As a result, interstadials 7.3 and 7.1 occurred close together in time. For this reason, in the marine core records, where time resolution is poor, these events sometimes appear as one. We shall see below that this exceptional configuration also gave rise to the extraordinary situation of enhanced tropical monsoon during a cold stage.

The strong minimum of D at about 130 kyr BP is remarkable in that it was practically in phase with an obliquity maximum. This led to the outstandingly warm Eemian interglacial 5.5. Neither of the two succeeding minima of D (~105 kyr BP and 85 kyr BP) was reinforced by strong obliquity. The two latter configurations yielded interstadials 5.3 and 5.1, during which the sea level remained lower by about 20 m than its present height.

Finally, 12,000 years ago, with a minimum Earth–Sun distance in June and close to a maximum obliquity, summer irradiance reached its maximum. In spite of the small eccentricity, the simultaneity of the maxima of summer irradiance and obliquity was sufficient to cause almost complete melting of the NH polar ice sheets, leaving Greenland as the only survivor.

20.4.3 THE ONSET AND DEVELOPMENT OF GLACIAL PHASES

We now concentrate on configurations in the past that favoured the build-up of ice sheets in the NH. For this, cold summers are needed, with weak irradiance in NH summers. In Fig. 20.3, the blue markers indicate when the Earth–Sun distance at the NH summer solstice is greatest during the precession cycle: these correspond to the situation 'b1', with cool summers. Some maxima in D are much more pronounced than others. In some, the cooling effect is reinforced if, at the same time, the minimum obliquity is very small (situation 'b2'), as indicated by blue markers on the obliquity curve. These correlations are marked by blue vertical lines. Over the past 250,000 years, five occasions were particularly promising for triggering or reinforcing an ice age: 235, 190, 115, 75 and 30 kyr BP. The precision of these estimates is about 5 kyr.

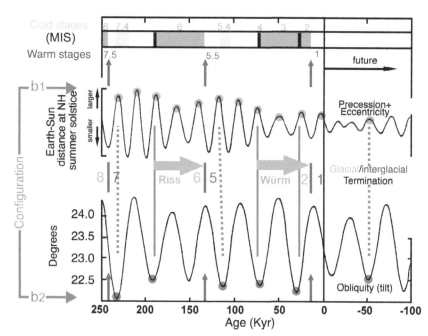

Figure 20.3 The glacial stages in the past 250,000 years, and the conditions that give cool summers at middle and high latitudes in the Northern Hemisphere. These are defined by the configurations 'b1', where the Earth–Sun distance in June is at a maximum, and 'b2', when the obliquity is small (see text).

What is the impact on climate change?

The following description, although somewhat detailed, will enable us to understand an extraordinary climate situation that will be discussed in Section 22.5.4, where the Earth underwent a glacial period MIS 7.4 at high latitudes at the same time as an intense monsoon cycle in the tropics.

First, we inspect the two situations denoted by dotted blue vertical lines at around 235 and 115 kyr BP. These put an end to the preceding interglacials (MIS 7.5 and 5.5) and, within one precession cycle, decreased the sea level by several tens of metres. They greatly increased the volume of the polar ice sheet, and ushered in the cold stadials 7.4 and 5.4. Very soon after 240 kyr BP, interglacial 7.5 was interrupted in this way by the combined effects of a large Earth–Sun distance in June and extremely low obliquity. The NH summer received so little irradiance that interglacial 7.5 lasted only a few thousand years. The decrease in summer irradiance over half a period of the precession cycle (~11 kyr) was the largest in the past 400,000 years and was associated with an exceptionally short interglacial. As a result, the ice sheets built up rapidly and the sea level decreased tremendously within a few millennia. The trend, however, was quickly counteracted by the strong minimum of the Earth–Sun distance in June in the following precession cycle. A warm phase then emerged, in which the warm interstadials 7.3 and 5.3 were established. In the case of interstadial 5.3, where the sea level is well documented, the warming effect was not strong enough completely to melt the ice sheets that had built up during the cooling event of stadial 5.4. The sea level remained about 20 m lower than in the Eemian (interglacial 5.5).

The glaciations that started so briskly towards 235 and 115 kyr BP were stopped in their tracks in the half-cycle of the following precession, when conditions began to turn to vigorous deglaciation. This was completely different from the great cooling periods that began at around 190 and 75 kyr BP. The first intense cooling period started the Riss glaciation, during which ice sheets developed sporadically over about 60,000 years. The sea level began first by falling rapidly, by approximately 70 m, within one half-cycle of the precession, then more gradually, to reach its lowest level at around 130 kyr BP, when the ice sheets were most extended. This extension was halted by the especially favourable conditions of irradiance in June that began to appear, and which established the warm Eemian phase (interglacial 5.5). The duration of this glaciation is indicated in Fig. 20.3 by the thick horizontal blue arrow, and its cut-off is denoted by the vertical red line, at the 6/5 termination.

In a similar but not identical manner, at about 75 kyr BP the second cooling event set the foundation for the Würm glaciation. It lasted for about 50,000 years. By 65 kyr BP, the sea level had decreased substantially, by about 60 m within half a precession cycle, accompanied by rapid expansion of the ice sheets. This was the first phase of the Würm glaciation (glacial stage MIS4). In the following tens of thousands of years, the conditions for efficient deglaciation were never met; eccentricity slowly decreased and no maximum of obliquity reinforced the warmth of the NH summer when the Earth was close to the Sun in June. The timid maxima in the summer irradiance produced only modest variations in the sea level, by about 20 m. It was only towards 30 kyr BP that a new configuration set in that strongly favoured ice-sheet expansion (the blue vertical line in the figure). A second phase of ice expansion then took place, that within half a precession period decreased the sea level by a further 50 m. About 20,000 years ago, the sea reached its lowest level (–120 m), corresponding to the maximum extension of the ice sheets (glacial stage MIS 2).

The end of the glaciation started in the next half-cycle of the precession, where, as already seen, the NH summer irradiance reached its maximum 12,000 years ago. The duration of the glaciation is indicated in the figure by the thick horizontal blue arrow, and its cut-off by the vertical red line (the 2/1 termination).

20.5 Glacials and interglacials: similar situations, never identical

The above account explains why the different glaciations never develop in the same way and why the last two great glaciations did not have the same dynamics. The Riss glaciation, after a powerful beginning, then advanced quite slowly. By contrast, in the last glaciation (the Würm), two occasions arose in which ice sheets could extend rapidly. These appear in the form of two separate glacial stages (MIS 4 and 2) and two major falls in sea level that marked the beginning of the stages.

This analysis also illustrates why the features of the last interglacial (the Eemian, MIS 5.5) and of the present Holocene (MIS 1) are quite different. The orbital eccentricity of the Earth during the Eemian period was greater (Fig. 14.3). Although the mean annual irradiance ($340\,W/m^2$) was the same 128,000 years ago, the maximum irradiance in June was then greater by about 15 W/m^2 than 12,000 years ago (Section 14.5). During the Eemian, this situation produced particularly hot summers in the NH, with a mean global temperature about 2°C higher than in the Holocene, a sea level that was about 7 m higher and a diminished Greenland Ice Sheet. The climate conditions affected the flora and fauna. In Europe, for example, they were typical of a much warmer climate.

The story is different for each interglacial period, and for each glaciation: although they are roughly comparable, they do not progress identically or have the same features, because the combinations of the Earth–Sun distance in a given season and of the obliquity are never repeated identically. This happens because the periodicities of the three basic orbital parameters (eccentricity, obliquity and precession) are completely independent.

20.6 The energy budget: radiative forcing and feedback

The glacial–interglacial cycles are driven by the periodic variations of the seasonal irradiance at high latitudes. During these variations, however, the global solar energy received annually by the Earth remains constant. *How is it, then, that the change in the energy balance at the surface of the planet can increase the average temperature of the Earth by 5°C between a glacial stage such as the Last Glacial Maximum (LGM) and an interglacial such as the Holocene? What are the parameters that create such a difference in the energy balance?*

When deglaciation begins, a large number of characteristic features of the planet gradually begin to change:

- Modification of the cryosphere and shrinking or disappearance of ice caps; reduction of ice floes; a rise in the sea level and flooding of emergent continental shelves; modifications of the vegetation cover; the spread of forests at middle and high latitudes; shrinking of steppes and tundra; and so on.
- Changes in the deep ocean circulation as it reactivates; changes in the carbon cycle and in the biosphere, both on land and in the ocean; changes in the composition of the atmosphere, with increasing greenhouse gas content and decreasing airborne dust; changes in the atmospheric circulation and in the water cycle as a result of the reduced temperature difference between low and high latitudes and the smaller height of the ice caps; rising temperatures on the land and in the oceans that shift the physico-chemical equilibrium conditions; and so on.

All these changes modify the parameters that control the global energy balance, and introduce radiative forcing mechanisms that bring their own rapid feedback processes. A new climate is established that imposes a new global mean temperature on the Earth. The underlying cause of these changes, namely the irradiance, imposes its own rhythm and the different climate conditions adjust accordingly.

How is the energy balance altered? Let us return to the simplified picture of Fig. 4.5, which shows the annual amount of solar energy received by the planet, the albedo, which determines how much solar energy is reflected, and the composition of the atmosphere. The solar energy received each year does not vary. The two other components, on the contrary, do change. The albedo of the planet decreases, mainly because of shrinking ice and snow cover, and greenhouse gases build up in the atmosphere. The resulting radiative forcing (Section 10.1) increases the average temperature of the planet.

Attempts have been made to estimate the radiative forcing due to the increase in greenhouse gases between the LGM and the Holocene. The extra radiative forcing is estimated at about $+2$–$3\,W/m^2$, of which the major contribution is that of CO_2. The radiative forcing due to the change in albedo at the beginning of the Holocene warm period has numerous components. To estimate it, the glacial conditions must be known precisely, but only orders of magnitude are available. Ice-sheet melting and the rise in sea level produce a forcing of about $+3\,W/m^2$; more than $+1\,W/m^2$ could be caused by changes in the vegetation and reduction of the dust content in the air. Between the LGM and the Holocene, the final net estimate of the forcing due to albedo and greenhouse gases is often given as 7–$8\,W/m^2$.

Warming generated by this radiative forcing is amplified by fast feedback mechanisms (Chapter 9). When these mechanisms are taken into account, we saw that with our current climate conditions a radiative forcing of $1\,W/m^2$ increases the temperature by approximately $0.8\,K$ (0.4–$1.2\,K$). Studies of palaeoclimatology suggest that this relationship has remained approximately unchanged over the past few million years (PALEOSENS Project Members 2012). The above estimate of 7–$8\,W/m^2$ for glacial/interglacial forcing therefore implies a temperature rise of about 5–$6°C$, in agreement with the palaeoclimate reconstructions. Moreover, the estimates of the two radiative factors show that the increase in greenhouse gases is only a secondary factor in the final warming from glacial to interglacial: the greater part is changes of albedo.

Chapter 21
The glaciation dance: consequences and lessons

21.1 The impact on life of glacial–interglacial cycles

Here we touch on the critical aspect of this book, the effect of climate change on life. We have just defined the magnitude of the climate variations in the Quaternary: how did these disturb the equilibrium of ecosystems, of the vegetation, of the animal kingdom? For ultimately, this is the only question: how will living beings, and the resources on Earth, be affected in a future climate?

21.1.1 CHANGE IN HABITAT FROM GLACIALS TO INTERGLACIALS

The severe drop in temperature that took place when the first ice sheets started to develop in the Northern Hemisphere (NH), coupled with the oscillations between glacial and interglacial climates, caused major modifications in the habitats from the tropics to the poles. Transitions that take place between two such different climates in a few thousand years challenge the adaptive capacities of vegetation and fauna. Regions of tundra, steppe and permafrost extend around the edges of the ice sheets that resemble those in modern-day Russia and Canada, but with temperatures that are lower by about 10°C than those of today. In general terms, when glaciation sets in, the areas of vegetation migrate southwards by about 2000 km. At

Climate Change: Past, Present and Future, First Edition. Marie-Antoinette Mélières and Chloé Maréchal.
© 2015 John Wiley & Sons, Ltd. Published 2015 by John Wiley & Sons, Ltd.
Companion website: www.wiley.com\go\melieres\climatechange

mid-latitudes, cold steppe and taiga are the dominant forms of vegetation. Rodents, reindeer, bison and other animals that can tolerate severe cold roam these regions. Adaptations in physiology, feeding habits or morphology (involving, for example, blood circulation, hibernation, growth of woolly hair etc.) give rise to species that are typical of extremely cold environments.

The brief warm periods of the interglacials between the glacial stages allow the regrowth of forest that is inhabited by very different, temperate wildlife, such as deer, auroch, wild boar, lynx and so on. The fauna recorded, for example, in France between the last ice age and the Holocene illustrate these great changes, with, however, an important difference with respect to previous warm periods: in the Holocene, domestic animals were introduced. Certain species that are more adaptable to ecology and climate may find themselves in different ecosystems at the same time (notably carnivores, which depend only indirectly on plant production). To summarize, such climate switches cause a major geographical redistribution of plant species. Over the whole of Eurasia, for example, forest taxa migrate south during a cooling phase, while steppe with mammoths develops in the north. This process is accompanied by evolutionary factors, of which the most remarkable is a trend towards gigantism in the taxa at the end of the Pleistocene (the period that stretched from the beginning of the Quaternary to the beginning of the Holocene).

The tropics, for their part, undergo alternating phases of drought and humidity. The history of the Sahara provides a perfect illustration: during phases of drought, the inhospitable deserts expand, while during wet phases, a richer variety of life forms returns. Human beings can then find a means of subsistence. During the Holocene, they were even able to establish a pastoral lifestyle.

As in all major climate changes, the equatorial zone is the least affected by the oscillations, because the temperature variations in this belt of latitude are the weakest, and rainfall is always available. The habitats do, however, undergo large modifications as a result of the drop in temperature by several degrees and the periodic regional desertification. Refuges nevertheless exist. These enable flora and fauna to survive the perturbations without excessive hardship.

21.1.2 THE IMPACT ON BIODIVERSITY: TAXA, 'CLIMATE REFUGEES' AND SO ON

The flora and fauna of the Tertiary were profoundly affected everywhere on Earth by the repeated cooling events that struck in the course of successive glaciations. At about 2.7 million years BP in Europe, for example, the nature of the vegetation changed from subtropical to arctic. In northern Eurasia, a major change took place at about 2.6–2.2 million years BP with the appearance of large modern ruminants (Cervidae, Bovidae) and Equidae. This change was triggered both by the increasing aridity and by the overall cooling, which opened up extended spaces (plains). The emergence of the species *Homo* in Africa is situated in the same period, between 2.6 and 2.45 million years BP. Its appearance is thought to be related to the major climate change that intervened at this epoch (cooling and the onset of glacial–interglacial oscillations) at the same time as the appearance of other new species, notably ungulates. Finally, between 1 million and 0.8 million years BP, a major reorganization of the flora and fauna (large and small mammals) is evident in Eurasia. The majority of animals that then remained were the direct ancestors of our present-day species. Particularly severe glacial periods caused the extinction of numerous species, both on the continents and in the oceans. In the oceans, for example, the intense cooling phases that occurred in glacial stages MIS 24 (0.92 million years BP) and MIS 22 (0.88 million years BP) (Fig. 18.1), during which the deep ocean current was sharply reduced, brought extinction to some 50 deep-sea foraminifera species. On the continent – for example, in Greece – the extreme glacial phases MIS 22 and MIS 16 (0.63 million years BP) were responsible for the disappearance of numerous residual taxa. Owing to its geographical position at the southern end of Europe, the Greek peninsula is frequently a refuge for taxa that migrate south during glacial stages. Should an even more intense cold period occur, then migration of these climate refugees would be thwarted by the barrier of the Mediterranean Sea: all that is left for them is to die out. Extinctions of thermo-philic (warmth-loving) plant taxa were not confined to extreme glacial phases: they took place gradually throughout the Quaternary, earlier in the colder more northerly territories than in Southern Europe and around the Mediterranean Sea.

In the equatorial zone as well, the continents were exposed to the counterblows from the ebb and flow of these two very different climates. Unlike the rest of the planet, however, the variations in temperature were greatly attenuated, not exceeding 5°C, and rainfall remained abundant. In privileged, relatively accessible places, islands of refuge persisted that allowed the huge variety of existing species to subsist during glacial phases. In this way, the flora and fauna could retain their heritage of biodiversity. From one swing of the climate pendulum to the next, the equatorial zone and its virgin forests could transmit the fabulous inheritance of tens of millions of years of evolution.

For the mid-latitudes, however, this was not the case. We shall see below that these latitudes permanently undergo successive recolonizations by very different flora and fauna, during which the sole survivors are the species that are most able to migrate and adapt. The physical appearance of the forests is thus very different from that of equatorial forests. During the warm climate of the interglacials in Europe, whole forests grow in which the number of dominant species is often to be counted on the fingers of one or two hands, with perhaps 10 species per hectare. This situation is vastly different from tropical rain forests where, typically, each neighbouring tree belongs to a different species, and as many as 150–250 different species of tree thrive in each hectare.

21.1.3 EACH INTERGLACIAL: A SUCCESSION OF SIMILAR VEGETATION

Pollens deposited in peat (soil rich in organic matter) have made it possible to reconstruct the succession of types of vegetation that give way to each other in the course of the different interglacials. A regular feature at mid-latitudes is that when the warm interglacial sets in at the end of a glacial period, extensive grasslands are gradually replaced by lignaceous species. Then, when the pendulum swings back, grasslands again reconquer the territory. Ecological equilibrium is renewed almost permanently. The NH mid-latitudes have been the site of some of the most drastic inter-climate changes in vegetation, passing from frozen soil occupied by tundra to deciduous tree forest covering. For several hundred thousand years, history has been incessantly repeating itself in these regions, as vegetation reconquers the land.

The record of this history is found, for example, in pollen trapped in the sediments of the crater lakes in Le Velay (Massif Central, France) (de Beaulieu 2006). The sediments go back more than 400,000 years, covering not only the present Holocene but also the four preceding interglacials. Figure 21.1 shows the changes in the vegetation during each of these interglacials. As soon as warming starts, the different stages of vegetation follow each other in almost exactly the same sequence from one climate cycle to another, even though the warm phases are separated by a hundred thousand years. There were several small differences, however. For example, during the last warm period 125,000 years ago (the Eemian, MIS 5.5), beech (*Fagus*) failed to return. This tree species appeared regularly during the interglacials MIS 11 (about 420,000 years ago) and MIS 9.5 (about 330,000 years ago), and was timidly present in MIS 7.5 (~240,000 years ago). It was absent during the Eemian (MIS 5.5), but made a brilliant comeback in the present interglacial MIS 1, the Holocene.

What is striking in these records is the practically identical succession of the various plant species. This constant repetition demonstrates how similar are the climates of the different interglacials at our latitudes, with a few small variations: the vegetation responded unfailingly, with the same succession of plant cover. Selection takes place gradually among the different species, with the progressive disappearance of those that do not survive climate extremes. The only ones to subsist are those that have had the time and ability either to adapt or to find a refuge, and then to migrate back when conditions again became favourable. In temperate latitudes, this dynamic led to a vegetation cover of forest that is dominated by a limited number of species.

What is the effect of natural barriers in hindering migrations? In Europe, the east–west span of the Mediterranean Sea barrier limits available refuges for southward migration, and increases the loss of species during the different glaciations. Similarly, the east–west barrier of the Alps and the Pyrenees on the European continent was without doubt a formidable barrier to the north–south (and south–north) migration that takes place at each climate change. These obstacles to north–south migration could only limit the number of species that were able to adapt to these imposed migrations, thus reducing the diversity of flora and fauna in Europe. Such handicaps did not exist on the North American continent, which enjoys a large continuity of territory between 60°N and 30°N, with mountain ranges oriented mainly in a north–south direction (Appalachians, Rockies). This may be the reason for the greater variety of species in North America today. Records there indicate 253 species of trees and about 650–700 species of birds: in Europe, these numbers are only about 124 and around 500, respectively (figures relating to tree species from Latham & Ricklefs 1993; figures relating to bird species from private communications in 2012 with J. Blondel & S. Gillihan, executive director at the American Ornithologists Union). Although reduced biodiversity could be one of the consequences of climate change, another factor may well have affected the two continents, modifying the flora and fauna differently: man. Nonetheless, although the hand of man could conceivably affect animal resources, it is hard to believe that it could have decreased the number of tree species by a factor of two. It thus seems plausible that the difference between the two continents is a heritage of the glacial–interglacial alternations, and that it illustrates the impact of large climate changes on biodiversity.

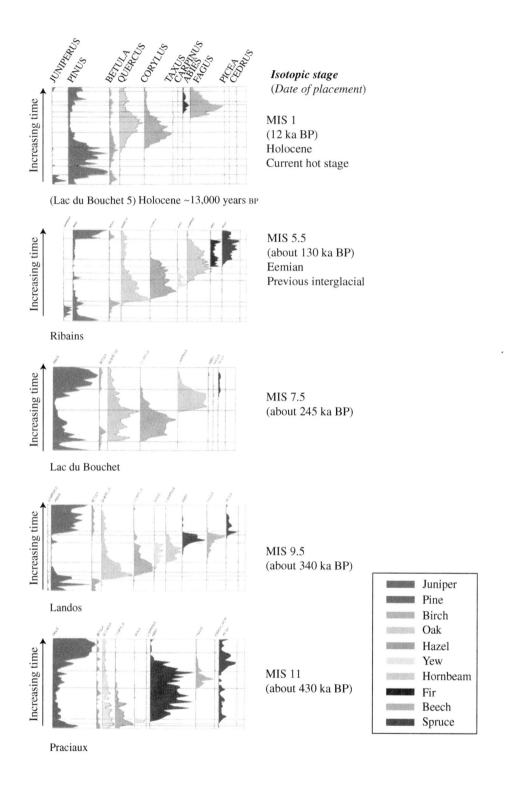

Isotopic stage
(*Date of placement*)

MIS 1
(12 ka BP)
Holocene
Current hot stage

(Lac du Bouchet 5) Holocene ~13,000 years BP

MIS 5.5
(about 130 ka BP)
Eemian
Previous interglacial

Ribains

MIS 7.5
(about 245 ka BP)

Lac du Bouchet

MIS 9.5
(about 340 ka BP)

Landos

MIS 11
(about 430 ka BP)

Praciaux

	Juniper
	Pine
	Birch
	Oak
	Hazel
	Yew
	Hornbeam
	Fir
	Beech
	Spruce

Fig. 21.1 The last five interglacials: successive changes in plant cover in the Massif Central in France during each of the last five interglacial stages (warm periods), reconstructed from pollen abundance (*x*-axis) in lake sediments of the Velay crater. The vertical axis is the depth of the sediment; that is, the date of deposit. Each figure scans the deposit during the interglacial. The horizontal axis is the pollen abundance for each listed species. The similarity between the different interglacials is striking. Primary colonization, first by pine and subsequently by birch, is followed by a phase of expansion of oak and hazel that may be shorter or longer. Then come deciduous trees and conifers that are less thermophilic and/or spread more slowly. The boreal forest comes last. Except for this last stage, which has not been reached, the same dynamics can be observed in the Holocene.
Source: de Beaulieu (2006). Reproduced with permission from Jacques-Louis de Beaulieu.

21.2 Lessons to be drawn

The characteristics of the different glacial and interglacial stages enable us to put into perspective the scale of present and future climate changes. We will confine ourselves to the following points.

21.2.1 THE SIGNIFICANCE OF AN 'AVERAGE' DEGREE

The glacial periods of the past illustrate the meaning of a change in temperature of 1°C when it refers to the global mean temperature; that is, averaged over a whole year and over the whole of the surface of the Earth. Between two climates as profoundly different as a cold period (glacial stage) and a warm period (interglacial), the average temperature on Earth varies by 'merely' about 5°C. This small difference in 'average degrees' involves substantial changes at low latitudes, disruption of the flora and fauna at mid-latitudes and complete transformation of the high latitudes, which are overrun by ice sheets. It is salutary, for example, to remember that during the last glacial stage, in places where deciduous forests now grow there was bare tundra and permafrost.

What lesson can be drawn? Among the scenarios for climate change that are projected for the end of the 21st century, a 5°C increase in global mean temperature is far from unlikely. This change is similar to what occurred in the transition from glacial to interglacial conditions. But this time round, the projected warming is in the context of the current warm period, which, to within 1 or 2°C, is the warmest period that the Earth has ever known in the past few million years. Within less than a century, we are in danger of heading towards a totally new climate. The risk depends on the choices that we make.

21.2.2 DISPARITY BETWEEN LATITUDES

Climate change depends strongly on latitude: temperature changes are enhanced by a factor of two or three on going from the Equator to the polar regions. During the last glaciation, on land, the temperature in the tropical regions fell by about 5°C, while at the NH mid-latitudes the temperature fell by 10°C and, in Greenland by 20°C. *What is the cause of this amplification?* One of the principal factors is albedo feedback. The change of albedo increases with increasing latitude as the area of snow-covered land and ice floe increases. Reflection of solar energy increases, and heating of the surface falls.

What lesson can be drawn? The polar regions, which receive the least energy, are those that react most strongly to change (increase or decrease) in the total energy available at the surface of the Earth; that is, to changes in the average temperature.

21.2.3 THE DIFFERENCE BETWEEN CONTINENT AND OCEAN

At the same latitude, variations in temperature on the surface of a continent are more pronounced than on the surface of an ocean. In the transition from a glacial to an interglacial stage, the temperature of the surface water in tropical oceans rises by only about 2°C, while on the continents at the same latitude it rises by 5°C.

This difference derives from a combination of several factors. First, solar radiation penetrates deeper into the ocean than it does into land. The same solar flux is thus absorbed by a greater amount of matter in the ocean than on a continent, and the increase in temperature is accordingly smaller. The temperature increase also depends on the specific heat, which is generally twice as large for water as for land. Furthermore, since water is a fluid, it spreads heat both by convection and by ocean currents, which attenuate changes in its temperature. Lastly, consider the water cycle. It may seem tautologous to say that liquid water is never absent from the surface of the ocean, but the result is that the surface temperature of the ocean is strictly controlled through cooling by evaporation. These factors all ensure that the ocean surface warms less than that of continents.

What lesson can be drawn? These examples from the past show how climate change involving a difference in the average temperature of the Earth of about 5°C affects the various regions of the planet. Continents are more influenced than oceans, and high latitudes more than low latitudes: the stress to living organisms in order to adapt is accordingly greater.

21.2.4 CHANGES IN GREENHOUSE EFFECT GASES

Two important pieces of information relating to changes in the atmospheric composition during the past 800,000 years are contained in air bubbles trapped inside the Antarctic ice. Reconstructions of the atmospheric content of the greenhouse gases CO_2, CH_4 and N_2O show that they fluctuated between two limiting values, low during glacial periods and high in interglacials (Fig. 19.2). The underlying causes were changes in the marine currents, and cooling of the land masses and oceans:

- Although the increase of CO_2 in the atmosphere is driven by the Earth's orbital parameters, its effect on climate is no less important. An increase of CO_2 reinforces warming through its feedback effect. Current simulations estimate that of the 5°C temperature increase between a glacial and an interglacial, between one third and one half is due to the rise in CO_2.
- Past changes of the minor gases CO_2 and CH_4 over the last 800,000 years provide a clear illustration of the enormous perturbation generated by humankind in little more than a century. Man's activities have increased the level of CO_2 by ~40%, and that of CH_4 by ~150% (Part VI). Unless something is done, the CO_2 content of the atmosphere is likely to reach three times its natural level by the end of the 21st century.

What lesson can be drawn? From the beginning of the industrial era (the middle of the 19th century), human beings have modified the atmosphere of the whole planet, and continue to do so. The present composition of the atmosphere lies far outside the limits of the natural fluctuations of the past million years. This is just one example of the immense upheaval that humankind has perpetrated on the planet in such a short time.

21.2.5 EARLIER INTERGLACIALS: WAS THE SEA LEVEL HIGHER THAN NOW?

Fluctuations of the global sea level are mainly the result of changes in the volume of ice stored on continents. Sea level is therefore an ideal proxy for glaciation (a low sea level corresponds to a large volume of ice sheet and a high sea level to a warm prevailing climate). It is instructive to examine the various interglacials that preceded ours. Although the flora and fauna appear fairly similar, interglacials nevertheless differ from each other in their degree of warming and in their sea levels. We already saw that the previous interglacial period, the Eemian MIS 5.5 (about 125,000 years ago) was warmer than the present Holocene by about 2°C (the global mean temperature). The Eemian appears to have been one of the warmest of the Quaternary interglacials. Reconstructions show that the sea level at that time was about 7 m higher than now. Modelling studies, confirmed by different cores from the Greenland Ice Sheet, reveal that the volume of the ice sheet was about one half of its present value. The rise in sea level is currently believed to be due in part to the melting of the present Greenland ice sheet (~2–3 m) and in part to that of the present West Antarctic Ice Sheet (~3–4 m).

Reconstruction of the high sea levels during the warm Quaternary stages is still in progress, in particular for the short-lived stages in which the records of high levels are faint. It appears, for example, that a sea level 11 m higher than now did briefly occur during a previous interglacial, but the scientific community is still undecided as to which interglacial it should be attributed.

What lesson can be drawn? In the interglacial prior to ours, the Eemian, the global mean temperature was 2°C higher than now and was accompanied by a long-lasting rise in sea level of several metres. That increase in global mean temperature is the lowest estimate for warming from among the 21st-century scenarios.

21.3 When will the next glaciation come?

Another interglacial that is rich in information is MIS11, which began about 425,000 years ago. *Why is this stage interesting?* Because of its similarity to the current Holocene. In Chapter 14, we saw that every 400,000 years the eccentricity of the Earth's orbit goes through a minimum (Fig. 14.1). This means that the orbit remains practically circular for several precession cycles; that is, several tens of thousands of years. In these conditions, during an interglacial, the decrease in NH summer irradiance becomes too small to tip the climate system into a new glacial stage. This is what happened during interglacial stage MIS11, which lasted about 30,000 years (from ~425,000 to 395,000 years ago). After the normal period of an interglacial (~10,000 years), a further full precession cycle was needed to bring glaciation back.

Our present situation is analogous to MIS 11, but our eccentricity is even closer to zero: the trajectory of the Earth is again becoming almost circular, and the current interglacial will last not merely for one more precession period, but more likely two. A favourable situation for a return to glaciation may be reached in 50,000 years, when the Earth–Sun distance in June is greater (greater eccentricity combined with minimum obliquity, designated by the blue dotted lines in Fig. 20.3). This analysis is in agreement with the results of various simulations of the next glacial stage (Berger & Loutre 2002). It should be noted, however, that on the basis of complex mechanisms, some workers have suggested that if atmospheric CO_2 concentrations did not exceed 240 ± 5 ppm, the end of the current interglacial would occur within the next 1500 years (Tzedakis et al. 2012).

In conclusion, a return to a glacial climate will still require several tens of thousands of years. So we may say goodbye to the apocalyptic scenario of the movie *The Day After Tomorrow*! (Box 21.1).

What lesson can be drawn? We cannot count on the coming of a new glaciation to compensate for the projected future warming due to human activities. The records of past glaciations tell us that even without human activity, the current interglacial will be exceptionally long, lasting several tens of thousands of years, as was the case about 400,000 years ago.

BOX 21.1 THE APOCALYPTIC FILM *THE DAY AFTER TOMORROW*

The film *The Day After Tomorrow*, directed by R. Emmerich (2004), is set in the present. It describes how a sudden onset of glaciation, with its train of catastrophes, brings about complete disruption of society. An exciting science fiction novel, *The Sixth Winter*, by D. Orgil and J. Gribbin (1982) was based on a similar scenario. In this book, after six consecutive cold winters, a persistent snow cover develops, which opens the way to a return to an ice age. The impact of this violent change of climate on different societies is recounted, from the nomad peoples of the Sahara to the Eskimos: few manage to adapt.

The book is a landmark that reveals the paradigm shift that has taken place in palaeoclimatology since the 1980s, and it highlights the extraordinary progress in scientific knowledge that has been made in this field in the past two decades. At that time, a general record of the successions of glacial and interglacial periods obtained from ocean core drilling had only recently become available. The astronomical theory of the palaeoclimate had just been brilliantly confirmed by spectral analysis of the isotopic signal of oxygen in the ocean sediments. Ice records were only partially available at that time. The data indicated that, like the previous interglacial 125,000 years before (the Eemian), the lifespan of an interglacial period was of the order of 10,000 years. Since the present warm period has lasted for about 10,000 years, it was plausible that a rapid descent into an ice age was likely in the near future. Hence the scenario of the book.

Since then, the ice cores, with an abundance of detail, have revealed the changes that have occurred over the past hundreds of thousands of years. The chronological record from the sediments of the oceans, of the ice and of the continents has been refined, and the effect of irradiance has been much more rigorously tested. The results of the EPICA drilling in the Antarctic ice cap, the analysis of which was published in 2008, have produced for the first time a detailed record of the temperature in the Antarctic and the composition of the atmosphere covering the past 800,000 years.

In particular, the record provides detailed information on the MIS11 interglacial stage, which lasted from approximately 425,000 to 395,000 years BP; that is, about 30,000 years. This has confirmed that the current planetary configuration, in which the Earth's orbit is becoming increasingly circular, will not allow the high latitudes in the NH to cool sufficiently for glaciation to take place, just as happened about 400,000 years ago. The next ice age (without human GHG emission) would need at least 50,000 years before its return: see the text and Berger & Loutre 2002.

It follows that the scenario of the film, like that of the book, is a piece of pure fiction that has nothing to do with reality.

Chapter 22

The past 12,000 years: the warm Holocene

22.1 The Holocene

For 12,000 years, a warm interglacial climate has prevailed over the Earth without any major disturbance. This period is called the Holocene. In comparison with the fluctuations that occurred during the preceding glacial period, this climate appears particularly stable.

What are its main features? Variations in the global mean temperature are small and, at the beginning of the Holocene, the sea level was some 40 m lower than at present. The Earth was then in the middle of its great mutation, a process that would take several thousand years more to complete, with the melting of the ice sheets in the Northern Hemisphere (NH). Melting started about 18,000 years ago, and stopped about 6000 years ago, leaving Greenland still under ice. Since then the ice sheets have remained stationary. Melting has stopped but the ice sheets have not started to grow again, and the sea level is almost constant (Fig. 19.7). The Holocene was disturbed only by a slight and brief cooling event, 8200 years ago, caused by the rupture of a reservoir of meltwater from the Laurentide Ice Sheet in North America. This reservoir was much smaller than that which caused the Younger Dryas. Its effect, which can be detected at mid-latitudes in the NH, lasted no longer than 200 years. Apart from that, no other major event has disturbed the Holocene climate, although there have been a few fluctuations.

Climate Change: Past, Present and Future, First Edition. Marie-Antoinette Mélières and Chloé Maréchal.
© 2015 John Wiley & Sons, Ltd. Published 2015 by John Wiley & Sons, Ltd.
Companion website: www.wiley.com\go\melieres\climatechange

We should not deceive ourselves, however. Beneath this appearance of stability, a slow change can be perceived over these 12,000 years. Insidiously, since the beginning of the Holocene, the climate pendulum had already started to swing as explained below in Timescale 1. In response to the precession cycle, a slow cooling process was gradually set in motion at high NH latitudes. Under certain circumstances, this cooling could set the scene for the next glaciation. *Is this cooling sufficiently strong for glaciation to build up as it did in the previous interglacial?* We have just seen that this is not the case, and that this time several tens of thousands of years will elapse until the next glaciation sets in.

22.2 Deciphering climate changes during the Holocene

The history of the climate during this period can be deciphered using the analogy of an imaginary 'telescope' that is able to look at past periods of time with different magnifications. We employ three different timescales that correspond to three different physical phenomena, which are related in turn to mechanisms described in Part III:

- *Timescale 1.* At this timescale magnification, we see the slow continuous change of seasonal irradiance due to the ~22 kyr precession cycle (Chapter 14). At the beginning of the Holocene, 12,000 years ago, during the NH summer solstice the Earth was at perihelion (see Fig. 14.6). Since then, the Earth at the NH summer solstice has gradually moved further from the Sun, reaching aphelion one half-cycle later; that is, about 1000 years ago. Over the millennia, this continuous decrease of NH summer irradiance during the Holocene brings slow changes to the climate.
- *Timescale 2.* This magnification focuses on a much shorter timescale, a few centuries or at most a few millennia. The fluctuations that it detects are those of changes in the Sun's activity. Our star is a flickering light. In the course of time, the energy flux that it delivers fluctuates around an average value. Episodes in which the intensity decreases bring on periods of overall cooling that appear roughly synchronously in different regions of the planet. Such fluctuations favour the appearance, or the advance, of glaciers. Glaciers advance over periods of several centuries, and then retreat.
- *Timescale 3.* Lastly, with the third magnification we observe climate fluctuations on much shorter timescales, from one decade to the next, or even from one year to the next. These fluctuations generally have two different causes. On the one hand, they stem from the coupling between the atmosphere and the ocean and, on the other, from volcanic eruptions, which disturb the energy balance of the surface of the planet.

Viewing the climate of the past at these different magnifications enables us to distinguish the underlying causes of the climate changes that have punctuated the great warm period of the Holocene.

22.3 Slow changes in irradiance (Timescale 1: millennia)

The gradual decrease of the NH irradiance in June, with its cooler summers, marks the change in climate during the Holocene in two independent ways: (i) progressive cooling at high and mid-latitudes in the NH; and (ii) decreasing intensity of the monsoon in the tropics. To obtain a better description of these changes, let us look again at the changes in the solar energy flux received during the Holocene.

During the Holocene, the amount of solar energy received by the Earth in June gradually decreases according to the precession cycle: starting from its maximum value 12,000 years ago (when the June solstice occurred at perihelion), it continued to decrease until the year AD 1000 (when the solstice took place at aphelion; see Fig. 14.6). During this half-cycle of precession, the irradiance in June decreased by 7.1% (Chapter 14). At middle and high latitudes, this decrease is slightly weighted by the obliquity of the polar axis, which has been decreasing for 10,000 years, and which in consequence delayed the irradiance minimum by 1000 years. That date coincides with our present time (Part IV, Note 7).

The decrease in summer irradiance during the Holocene gradually brought cooler summers to the NH. *What of the solar energy received over a whole year at a given latitude?* The orbit of the Earth around the Sun is described by Kepler's laws (Part IV, Note 8), which state that this annual energy remains constant, because the velocity of the Earth in its orbit varies (the Earth moves faster when it is closer to the Sun).

It follows, therefore, that the large changes that have occurred at each latitude during the Holocene are unrelated to the annual energy received in the NH, because this has been constant. By contrast, the solar energy absorbed in a given season is not constant. *So why are the large changes in the NH related to summers that gradually become cooler?* It is because the decrease in summer irradiance creates mechanisms that affect the global mean temperature. One of the easiest effects to picture is the annual increase in snow cover and ice floes, which, owing to their high albedo, reduce the absorption of solar radiation and modify the annual amount of energy absorbed. Another effect is the change in photosynthetic activity, which modifies vegetation cover, thus bringing about changes in the albedo.

22.4 Slow cooling at middle and high latitudes in the Northern Hemisphere

The consequence of this reduced heating is that since the beginning of the Holocene, the land masses in the NH have gradually cooled, notably in the high latitudes. Some examples are presented below.

Figure 22.1 shows the changes in three proxies situated in different locations (Bradley 2000). First, it shows the changes that occurred in a small ice sheet, the Agassiz Ice Cap, situated to the north of Ellesmere Island in Canada, where the number of ice-melt episodes

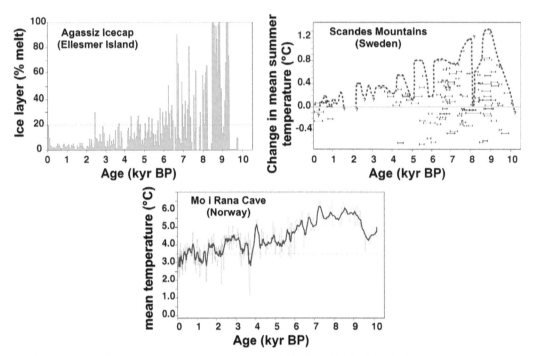

Fig. 22.1 Evidence of overall cooling at high latitudes in the Northern Hemisphere during the Holocene: (top left) reduction in the number of summer melting episodes of the small ice sheet in Ellesmere Island, Canada (Koerner and Fisher 1990); (top right) reduction in summer temperatures in the Scandes (Sweden), reconstructed from changes in altitude of the pine forest (data from Dahl and Nesje 1996); (bottom) the reduction in global mean temperature deduced from isotopic measurements in a stalagmite in the Mo i Rana cave, Norway (Lauritzen and Lundberg 1998).

Source: Bradley et al. (2003). Reproduced with permission from Springer Science+Business Media and R.S. Bradley.

(revealed by ice-core analysis) strongly decreased shortly after the beginning of the Holocene and practically disappeared in recent millennia. Reconstruction of the altitude of the tree line of pines in the centre of Sweden also provides evidence of this cooling; the data suggest a drop in the summer temperature by more than 1°C in this region (on the assumption of a lapse rate of 0.65°C per 100 m). Finally, the record of annual temperatures in the Mo i Rana cave in northern Norway indicates a drop of 3°C. Other evidence, from marine sediments sampled near the east coast of Greenland, show that sea ice was rare or absent in the region at the beginning of the Holocene. In the past 5000 years or so, this situation has undergone a major change. In Asia and in North America, the limit between tundra in the north and the northern forest further south is also a measure of this cooling. In the north of Canada, the forest has retreated by 300 km over the past 6000 years. A final example is permafrost; that is, permanently frozen soil of thickness ranging between 20 and 600 m (in northern Siberia, it can attain depths of more than 1000 m). When the polar ice sheets melted, the permafrost retreated to the highest latitudes; since then, it has been gradually moving back down to ever lower latitudes (Fig. 22.2) (Van Vliet Lanoë & Lisitsyna 2001).

Although the NH high latitudes were gradually cooling in the Holocene, ice sheets nevertheless did not start to grow again. The ground was not yet sufficiently cold throughout the

CLIMEX - CGWM

BVVL1999

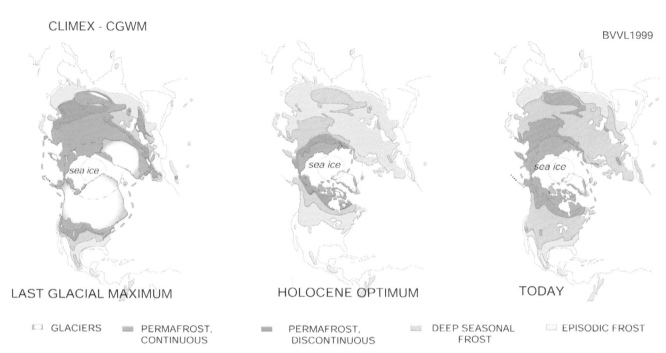

LAST GLACIAL MAXIMUM   HOLOCENE OPTIMUM   TODAY

| ▢ GLACIERS | ▨ PERMAFROST, CONTINUOUS | ▨ PERMAFROST, DISCONTINUOUS | ▨ DEEP SEASONAL FROST | ▢ EPISODIC FROST |

Fig. 22.2 Changes in permafrost between glacial and interglacial stages: (left) the last glacial maximum, about 20,000 years ago; (middle) the Holocene Climatic Optimum (9000–5000 years ago); and (right) the 20th century.

Source: Van Vliet-Lanoë and Lisitsyna (2001). Reproduced with permission from Springer Science+Business Media and Brigitte Van Vliet-Lanoë.

year for snow to accumulate, and, for the past 6000 years the sea level has remained stable, a sure sign that the ice sheets are stationary. The ground, however, became progressively cooler, and the cold made its presence felt, timidly at first, at the highest latitudes, and then slowly extended to lower latitudes. We have seen that the decrease in NH summer irradiance during the precession half-cycle (11,000 years) varies from one cycle to another. It is therefore easy to imagine that, *depending on the cycle, the cooling period can be more or less pronounced, and may or may not reintroduce ice sheets.*

The corollary of this slow change in irradiance is that the climate in the first part of the Holocene was warmer than now, notably at middle and high NH latitudes. This period is sometimes called the Climatic Optimum. It is estimated that the average temperature of the Earth was about 1°C higher than at present. The second part of the Holocene is marked by a cooler climate. Throughout Western Europe, this change appears clearly in the history of vegetation. The cold steppe of the glacial period first gave way to the deciduous forest of the interglacial, which became entrenched after the Younger Dryas cold event, about 12,000 years ago. A first phase of expansion of thermophilic (warmth-loving) species was then established and reached its maximum during the Climatic Optimum. These were followed by species that were the signature of a cooler, more humid, climate. The work of de Beaulieu et al. (1994) illustrates this change in the Jura region (Western Europe). At the beginning of the Bølling interstadial (14,700–14,000 years BP), the first to arrive was juniper, marking the reconquest of herbaceous habitats by bushes. Soon after follow birch and pine, during the Allerød period (14,000–12,800 years BP). Shortly after the end of the cold Younger Dryas around

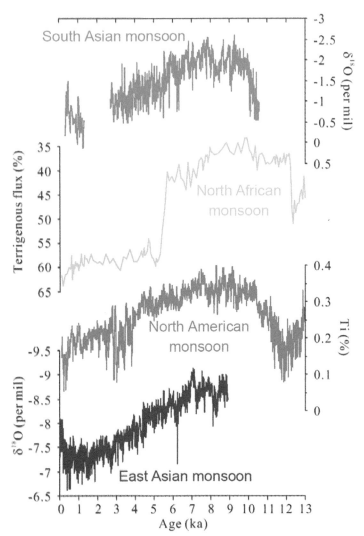

Fig. 22.3 The changes in monsoon intensity on the three continents during the Holocene, as recorded by different proxies. The intensity of the monsoon is at a maximum in the first half of the Holocene. The South Asian monsoon trace shows the $\delta^{18}O$ signal from a stalagmite in the Ounf cave in south Oman (Fleitman et al. 2003). The West African monsoon trace shows terrigenous deposits (in %) in the tropical Atlantic Ocean sediment off the West African coast (deMenocal et al. 2000). The North American monsoon trace shows titanium, Ti (%) in the layered deposits of the Cariaco basin off the coast of Venezuela, revealing input from rivers due to the hydrological cycle (Haug et al. 2001). The East Asian monsoon trace shows the $\delta^{18}O$ signal from a stalagmite in the Dongge cave in South China (Wang et al. 2005).

Source: Wang et al. (2009). Reproduced by permission of PAGES News.

11,000 years BP, thermophilic species started to spread, especially those of mixed oak groves (oak, elm and lime). Finally, at the end of the Climatic Optimum, about 6,000 years ago, another period started with tree taxa that preferred a cooler and wetter climate: beech and fir flourished, and spruce now prevails. As we saw in Figure 21.1, this course of events has been repeated many times over. Each interglacial involves practically the same sequence of species, with only minor variations.

Changes in climate also modify the landscape. The Climatic Optimum, during which the climate at mid-latitudes in the NH was warm and wet, shaped some of the reliefs through intense erosion, especially on soils such as Jurassic clays, which are easy to erode. This process laid down vast fertile deposits and created valuable agricultural lands. But now we turn our attention to the tropics, where, for a few millennia, the change in summer irradiance was illustrated in a spectacular fashion.

22.5 Strong monsoon in the Early Holocene: the 'Green Sahara' episode

Using our fictitious 'telescope' set at Timescale 1, we focus on the beginning of the Holocene, when the monsoon was reinforced on the three continents in the NH.

22.5.1 THE THREE CONTINENTS

Figure 22.3 shows examples of the changes in monsoon activity on each of these continents (Wang et al. 2009). In East Asia, South Asia and North America, and in the northern half of Africa, all of which responded in the same way to the increased summer irradiance at the beginning of the Holocene, the monsoons at that time were more intense. In the African tropics, the hydrological changes since the Last Glacial Maximum have been reviewed by Gasse (2000). The West African monsoon gave rise to the spectacular 'Green Sahara' episode. These rains, which at the beginning were abundant, have decreased more or less regularly, in such a way that the climate in Africa has become far more arid in the past 5000 years. That was the time at which the climate in many regions in the middle and high latitudes of the NH reverted to conditions similar to those of the present day.

22.5.2 THE 'GREEN SAHARA' EPISODE

At present, the Sahara is a desert, but for several thousand years in the first part of the Holocene, the rainfall on part of its surface was sufficient for lakes to subsist and to harbour life. It was wetter in both the north and the south, but it remained arid in its centre. These regions are now scattered with basins where strange whitish formations emerge (Fig. 22.4). They are the remnants of sediments that accumulated at the bottom of the then-existing lakes. Since then, aridity has returned. The lakes have dried up and the wind erodes the ancient lake depressions. The sediment layer survives only in certain parts, giving the impression of mushroom-shaped forms several metres high. In these former times, humankind could settle and raise herds of cattle. A 'rupestrian civilization' developed that has left us a heritage of splendid engravings and rock wall paintings, notably on the cliffs of Tassili n'Ajjer in the south-east of Algeria, and also in the south-east of Libya (Fig. 22.4). Engravings and bone remains bear witness to the fauna of that time: giraffes, fish, crocodiles, hippopotamus and so on, where now only desert survives. The rainfall filled the great water table that still stretches beneath Algeria, Tunisia and Libya. Such aquifers are fossil water, because they are no longer replenished. It was also at that time that the lakes in Africa between 15°N and 30°N reached their highest levels (Liu et al. 2007) (Fig. 22.5). The wet phase did not last. Gradually the desert returned and, about 5000 years ago, a far drier climate settled in.

Figure 22.6 shows the different stages of human occupation in the eastern Sahara during the main phases of the Holocene (Kuper & Kroeplin 2006). Also displayed is the change in annual precipitation. During the wet phase between 8500 BC and 5300 BC, permanent settlements were established in areas that had been totally devoid of habitation before 8500 BC (~10,500 years BP). Prior to that date, human settlement was restricted to the Nile Valley. After 5300 BC, a gradual decrease in the number of permanent settlements appears, mirroring the progressive occupation of the Nile Valley. In order to survive, human beings had then to take refuge in more clement regions. The Nile Valley became a natural refuge for migrating populations. This was where, with the permanent settlements and the concentration of population, the ancient Egyptian civilization was born, of which one of the earliest records is the oldest hieroglyphs. These can still be admired in the Saqqara pyramid, and date from about 3000 BC (5000 years BP).

The role of irradiance

As NH summer irradiance increased over thousands of years, the warming of North Africa increased and the low pressure in the Intertropical Convergence Zone became more pronounced. The two branches of the trade winds became more active, which reinforced the

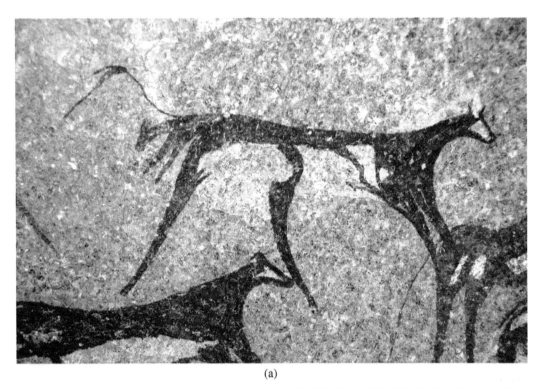

(a)

(b)

Fig. 22.4 (a) Cave paintings in Djebel Aweinat (south-east Libya), now a desert region, indicating Neolithic pastoral activity during the wet phase of the Holocene. (b) Erosion sediments in north Mali, now a desert region, showing lake deposits that existed during the wet phase of the first half of the Holocene.

Sources: (a) Thierry Tillet. Reproduced with permission (b) © CNRS Photolibrary/Nicole Petit-Maire.

water cycle and pushed the rains deeper into the heart of the Sahara. During the Green Sahara episode in the earlier part of the Holocene, when summer irradiance was at its highest, the Sahel zone that borders the south of the Sahara Desert (present position 15°N) was then at 20°N. *Which are the cycles that govern the summer monsoon maxima?*

The irradiance in June at low latitudes depends mainly on the Earth–Sun distance, since, contrary to high latitudes, it is affected only weakly by the obliquity. We therefore need only the curve of the Earth–Sun distance to understand qualitatively when the monsoon increases: a strong monsoon corresponds to a minimum distance. We have already seen in Chapter 14 that this distance depends on the orbital parameters: it is governed by the precession cycle (~22 kyr) and modulated by the eccentricity (~100 kyr) (Figs. 14.5 and 14.2). It was at a minimum 12,000 years ago, when the irradiance in June was at a maximum. This is consistent with the wet phase that lasted from about 10,000 years BP until 7000 years BP, which corresponds in fact to a maximum of sunshine over the *three* summer months (Fig. 22.5 and Part IV, Note 9).

We may therefore conclude that the cycles that drove previous summer monsoon maxima ought also to be those that govern the minima of the Earth–Sun distance in June.

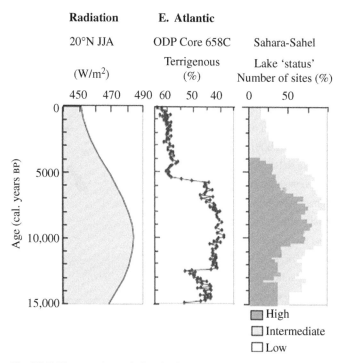

Fig. 22.5 The wet phase during the Holocene in Africa (equatorial zone, Sahel and Sahara). Left: average irradiance over the months of June, July and August at 20°N (Part IV, Note 9). Centre: wind-borne deposits in East Atlantic Ocean sediments, reflecting the onset of an arid phase that promotes erosion on the African continent (deMenocal et al. 2000). Right: the percentage of lakes in the Sahara–Sahel zone for which the level is high, intermediate or low. The wet phase in the Sahel region terminated about 5000 years ago.

Source: Liu et al. (2007). Reproduced by permission of Elsevier and Z. Liu.

22.5.3 INTENSE MONSOONS LONG BEFORE THE HOLOCENE

We now concentrate on the history of the summer monsoon in the northern half of Africa (West Africa, Ethiopia and the Nile). Past episodes of these monsoons are clearly illustrated in the sapropel deposits.

The sapropel deposit in the Mediterranean

This wet phase is recorded in the sediment deposited on the Mediterranean seafloor. The rains that fell in greater abundance than now on the Sahara and over North Africa brought fresh water into the Mediterranean (and the Red Sea) through the rivers that flowed into it, notably the Nile. As the salinity of the surface water decreased, it became less dense, remaining too buoyant in winter to sink and feed the deep water. As a result, deep-water circulation stopped for several thousand years during the wet phase. Earlier, the oxygenated surface water had

Fig. 22.6 Over
150 archaeological
excavations have been
undertaken in the now
hyper-arid eastern
Sahara of Egypt, Sudan,
Libya and Chad (Kuper
& Kroeplin 2006).
They reveal the close
relationship between
climatic variations and
prehistoric occupation
over the past 12,000
years. Permanent
settlements developed
during the wet period
between 8500 and
5300 BC, in a region that
is now hyper-arid. The
southward shift of the
desert edge facilitated
the emergence of
the ancient Egyptian
civilization along the
Nile. Annual precipitation
ranges from light green
(50–150 mm/yr) to dark
green (> 450 mm/yr). Red
dots, major occupation
areas; white dots,
isolated settlements in
ecological refuges and
sporadic transhumance;
pink area, the Pharaonic
state along the Nile
valley.
Source: Reproduced with
permission of Stefan
Kröpelin.

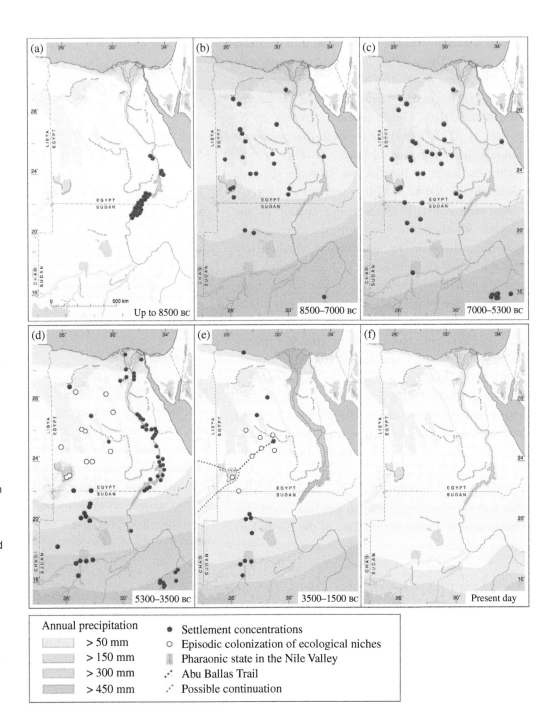

sunk and ventilated the bottom water. But when the circulation stopped, oxygen depletion in
the bottom water prevented the decay of sinking organic matter, and gave rise to organic-rich
sediments. The deposit, called *sapropel*, owes its dark colour to the wealth of organic matter
it contains.

Cores drilled in the Mediterranean seabed show that sediments of this type were deposited
during the different climate cycles (Cita et al. 1977; Emeis et al. 2003). Figure 22.7 shows a

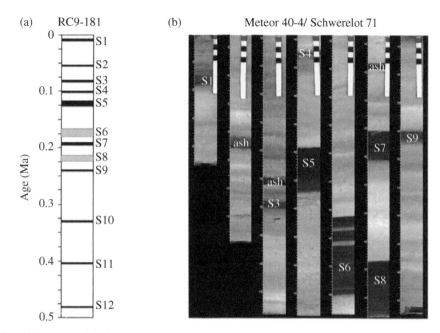

Fig. 22.7 (a) The Eastern Mediterranean sea sediment record (core RC9-181) of sapropel deposits over the past 500,000 years. These deposits are contemporaneous with the wet phases in the Sahara and coincide with the summer irradiance maxima in the Northern Hemisphere tropics. They usually take place during interglacials, except for S6 and S8 (in blue), which occurred during a glacial stage. (b) A section of a 6.14 m long core from the Mediterranean, south of Crete, showing sapropel deposits (S1–S9, deposited about 240,000 years ago). The deposits appear as dark bands because of their high organic content. S2 (the so-called 'ghost sapropel') is absent here. Deposits of volcanic ash are indicated. The core is sectioned into seven parts, with the surface of the sediment at the top of the section on the left.

Sources: (a) adapted from Strasser et al. (2006) (modified after Hilgen 1991; Lourens et al. 1996). (b) K. Emeis, University of Hamburg.

sediment core that records these deposits over the past 500,000 years. The most recent sapropel (S1) was deposited during the Holocene between 10,000 and 7000 BP. According to the different sites in the Mediterranean, the dates of the deposit vary slightly, since they are influenced by depth and geographical position. In summary, the deposits of this blackish layer, rich in organic matter, in the Mediterranean and in the Red Sea are the irrefutable signature of the wet episodes in the Sahara.

The last glacial–interglacial climate cycle

Figure 22.7 illustrates the history of the Sahara wet phases over the past 500,000 years. The sapropel record in the Mediterranean sediments establishes a precise chronology. They occur during June irradiance maxima at low latitudes (Fig. 22.8), which correspond to minima of the Earth–Sun distance in June (Rossignol-Strick 1985). In most cases, they take place during interglacials. In the last climate cycle, four sapropel deposits occurred: S5 in the previous interglacial (MIS 5.5, which was in full swing about 125,000 years ago), S4 and S3,

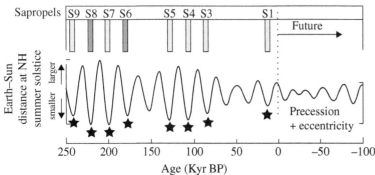

Fig. 22.8 The wet phases in the Sahel and the Sub-Sahara of Africa over the past 250,000 years, as recorded by sapropel deposits in Mediterranean sediments, during interglacial stages (pink) and glacial stages (blue). These phases are associated with maxima of summer irradiance in the NH tropical regions and correspond to minimum Earth–Sun distances in June (star symbols, see text). The position and duration of sapropels are only approximate, with an uncertainty of a few thousand years.

deposited during the two following interstadials (MIS 5.3 and 5.1, centred at ~105 kyr BP and ~85 kyr BP, respectively), and S1, in the Early Holocene. S2 occurred during a period of weaker irradiance, and is only partially recorded. Artefacts discovered in the Sahara, indicating the presence of human occupation well before the Holocene, appear to be contemporary with these wet phases.

22.5.4 MONSOON AND GLACIAL CYCLES

The existence of sapropels in the Mediterranean sediments from well before the Quaternary glaciations (i.e. much earlier than 3 million years ago) has been recorded. These sediment deposits, which go back at least to the Middle Miocene, have been widely documented on continental sites with a high uplift rate (Sicily, Italy); see, for example, the recent study of Mourik et al. (2010) on the La Vedova High Cliff section (Ancona, Italy), which covers sapropel deposits between ~14.2 and 13.5 million years ago. This indicates that these wet phases, originating from intense tropical monsoon activity in the NH, occurred well before the Quaternary glaciations. This is because the strong irradiance in summer, which regularly intensifies the African monsoon, is governed by the precession cycle, and is independent of glaciation. In general, these wet phases arise during interglacials, because the same increase of irradiance in summer accelerates thawing in the middle and high NH latitudes, thus favouring the onset of interglacials.

If, however, the increase in irradiance in summer in the NH is not sufficiently strong at high latitudes during one cycle, deglaciation is aborted. But in the tropics the increase in irradiance always intensifies the monsoon, and can be more or less pronounced. This is revealed in the sapropel record, since proxies of the sea surface temperature (alkenones, planktonic fauna etc.) are also found in the same sediment deposit, and these determine whether the deposit was laid down during a glacial or interglacial period (Emeis et al. 2003).

Of the nine sapropels deposited in the past 250,000 years, seven occurred in interglacials and two in glacials (Figs. 22.7 and 22.8). The latter was the case for sapropel S6, which occurred 175,000 years ago in the cold interstadial MIS 6.5, during the Riss glaciation. The strong irradiance in summer at low latitudes, corresponding to the minimum in Earth–Sun distance, gave rise to the wet phase attested by the S6 deposit. By contrast, at high latitudes, this summer irradiance was not reinforced by obliquity, and so could not initiate an interglacial. The sea level rose only moderately, but meanwhile, in the Antarctic, glaciation was in full control. The CO_2 content of the atmosphere remained unchanged, indicating that the wet phase in the tropics was unable to modify the CO_2 cycle. This also reinforces the idea that variations of CO_2 in the atmosphere are controlled mainly by the global ocean circulation (which hardly changed 175,000 years ago), and are practically unaffected by the tropical monsoon cycles.

This was also true for sapropel S8, which was deposited about 220,000 years ago, towards the end of the cold period of stage MIS 7.4. At that time, the obliquity maximum occurred slightly later than the minimum in the Earth–Sun distance in June (Fig. 20.3). The result was to delay the onset of the warm MIS 7.3 stage with respect to the intense monsoon phase; that phase arrived when deglaciation had not yet fully started. As we have previously seen, this is also illustrated by the fact that the warm stages MIS 7.3 and MIS 7.1 are very closely spaced in time, since MIS 7.3 had been delayed. In sediment cores, when the resolution is poor, these two stages appear as a single warm stage (Fig. 18.4).

22.5.5 $\delta^{18}O_{ATM}$, A PROXY FOR THE WATER CYCLE IN THE TROPICS

Photosynthesis creates molecular oxygen (O_2) from the oxygen atoms of water in plants (see Section 29.1.3), and the isotopic composition of that water is transferred to the released O_2, without fractionation. In principle, therefore, the isotopic composition of O_2 (known as $\delta^{18}O_{atm}$) should contain information about the water cycle in areas where photosynthesis is vigorous. From air trapped in ice cores, it was found in the 1980s that the isotopic composition of atmospheric O_2 changed over glacial–interglacial cycles (Bender et al. 1985). Towards the end of the 1980s, it was believed that the variations in $\delta^{18}O$ of O_2 in the atmosphere ($\delta^{18}O_{atm}$) were largely controlled by the $\delta^{18}O$ of ocean water ($\delta^{18}O$ of H_2O), and could thus be used as a measure of the volume of ice stored and hence of the sea level. The Dole effect is the difference between $\delta^{18}O_{atm}$ and the seawater $\delta^{18}O$ (H_2O); these studies led to the belief that the Dole effect had been constant over time.

This isotopic signal is recorded in the ice of the Antarctic over several climate cycles. The large variation in the signal of the atmospheric oxygen $\delta^{18}O_{atm}$ at 175 kyr (stage MIS 6.5) raised doubts about this interpretation, and the occurrence of an intense monsoon during the same period that was revealed by the sapropel S6 during the Riss glaciation (MIS 6) suggested that *the isotopic signal $\delta^{18}O_{atm}$ of O_2 in the atmosphere is governed by the water cycle in the tropics. Moreover, this cycle is independent of the glacial–interglacial alternations* (Mélières et al. 1997). This hypothesis has since been confirmed and developed by J. Severinghaus (Severinghaus et al. 2009). What is recorded in the isotopic signal of atmospheric O_2 is therefore the water cycle at low latitudes in regions of intense photosynthesis. This signal is the response to the precession cycle, and is at the same time a measure of monsoon activity.

22.5.6 THE IMPACT OF FUTURE WARMING ON THE SAHARA

May we conclude that future climate warming due to the increase of greenhouse gases will make the Sahara Desert more hospitable? It was not the *warming* that caused rainfall to increase over certain parts of the Sahara, but the *source* of the warming, namely stronger NH summer irradiance. This irradiance is what modifies the water cycle and the path of rains over the Sahara region. Its mechanism is fundamentally different from that of future warming due to the increase in greenhouse gases. The former acts through the *difference* in irradiance between low and high latitudes: the difference between the Equator and the North Pole, which drives the global weather patterns, is enhanced when the NH receives more solar energy in summer. The second mechanism, by contrast, acts through an increase in greenhouse gases, and is uniform at all latitudes. The gases are dispersed in the atmosphere over the whole of the surface of the Earth, and their increase is therefore comparable at all latitudes. There is no *a priori* similarity between these two types of warming, and their subsequent effects on the circulation and the rainfall could be different. As we shall see from the simulations (Part VI), increasing greenhouse gases does not reinforce rainfall in the Sahara zone. Quite the contrary.

22.6 Solar fluctuations (Timescale 2: centuries)

The above trend, which has lasted for thousands of years, is punctuated by smaller climate fluctuations with cooling episodes that lasted for several centuries, or even longer, and were most probably global. G. Bond and his group (see Bond & Lotti 1995) clearly identified these events through the layers of ice-rafted debris from iceberg discharges detected in North Atlantic deep-sea cores. Such discharges are associated with *cold events* that occurred throughout the Holocene. Nine cold events, called Bond events, have been identified. These suggested a 1500-year cycle during the postglacial era (Bond et al. 1997), and are attributed to solar influence (Bond et al. 2001). Bond events were believed to be the interglacial counterparts of glacial Dansgaard–Oeschger events.

Wanner et al. (2008) addressed the following questions: *'Where in the palaeoclimate records are Bond cycles postulated? … Do the proxy time series exhibit similar cyclic behaviour? … What are the possible processes that drive the Bond cycles?'* Their collation shows that *'the existence of Bond Cycles has been documented in many areas of the globe from a great variety of data sources (isotopes or organic and inorganic compounds detected in ice cores, lake or ocean sediments cores, stalagmites, and records from peat bogs)'.* Wanner and his colleagues conclude that *'the 1500 year cycle existed at least in areas of the North Atlantic and its surroundings'.* Possible causes of the Bond cycles were reviewed. This is where the key role of solar activity is revealed (our Timescale 2), combined occasionally with volcanic activity (Timescale 3), with an overall trend due to the slow decrease of irradiance in the NH (Timescale 1). Let us now return to the evidence of these climate fluctuations, as recorded by glaciers.

22.6.1 THE LITTLE ICE AGE OF THE LAST MILLENNIUM

The most recent of these cooling episodes was the Little Ice Age (LIA). This period has been described comprehensively by many authors (e.g. Grove 2004; Le Roy Ladurie 2011).

Well documented in Europe, it is just one episode in the history of glacier fluctuations that have been occurring for several thousand years. Reconstruction of the position of the Aletsch glacier front in the Swiss Alps over the past 3000 years (Figs. 22.9 and 22.10) tells the magnitude of this climate episode (Holzhauser et al. 2005; Holzhauser 2009). The peak of the LIA in Europe occurred between the 16th and 19th centuries, but its beginnings were strongly felt from the 14th century on. That epoch, which was cooler than either the medieval period or the 20th century, saw a general advance of glaciers worldwide. Harvests were damaged (wheat disappeared in Norway, for example, and vineyards in England), with severe economic consequences. As usual, the cooling was more pronounced at

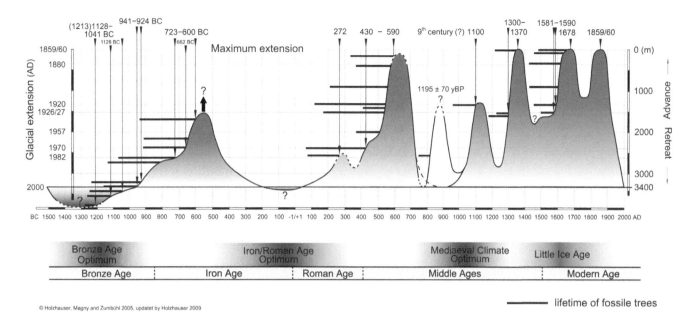

© Holzhauser, Magny and Zumbühl 2005, updatet by Holzhauser 2009

———— lifetime of fossile trees

Fig. 22.9 The variations in the length of the Aletsch glacier (Swiss Alps) over the past 3000 years (reconstructed). The maximum length occurred during the Little Ice Age, between the 14th and 19th centuries. Since 1850, the glacier front has retreated by more than 3 km. The left-hand scale shows the length of the glacier at different periods. Horizontal lines indicate the lifespan of tree trunks that have been discovered.

Source: Holzhauser (2009). Reproduced with permission of Hanspeter Holzhauser.

Fig. 22.10 The Aletsch glacier (Switzerland) in 1856 and in 2011: views from Belalp. Left: an original photograph from 1856 (F. Martens, London Mountaineering Club library; archives of H.J. Zumbhül); right: a photograph from 2011 (by H. Holzhauser).
Sources: Left, courtesy of H.J. Zumbhül. Right, Hanspeter Holzhauser. Reproduced with permission.

Fig. 22.11 The changes in the Charquini glacier (16°S, 5392 m) in the Royal Mountain Range in Bolivia (tropical Andes), illustrating the length of this glacier in the Little Ice Age (Rabatel et al. 2006). The sequence of moraines, dated by lichenometry (dates in italics, precision about 20 years), reveals the succession of glacier retreats after its advance in the Little Ice Age. Adapted from Rabatel et al. (2005).

Source: Adapted from Rabatel et al. (2005). Photograph by V. Jomelli, reproduced with permission.

higher latitudes, notably in Greenland, where two European communities that had settled there suffered dramatic consequences. Viking colonies had started to settle on this island from the year AD 985, when the climate was milder. The territory became witness to a drama in which the actors were not 'climate refugees', but 'climate victims'. The harsher climate conditions and the isolation of these communities were fatal for them, causing their disappearance (Part IV, Note 10).

Described at length by many authors, the effects of the LIA are portrayed in many paintings depicting the advance of numerous alpine glaciers into the valleys. Palaeoclimate reconstructions show that these glacier advances indeed affected the whole of the planet, both in the Southern Hemisphere (SH) (Rabatel et al. 2006; Fig. 22.11) and in the NH (Fig. 22.12). The LIA finally came to an end in Europe (Fig. 22.13), in the middle of the 19th century. An investigation by Oerlemans (2005) of 169 glaciers, distributed between the

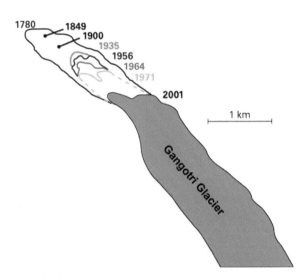

Fig. 22.12 A composite satellite image showing the successive positions of the Gangotri glacier front (source of the Ganges in Uttarakhand, India) since 1780.

Source: IPCC 2007. *Climate Change 2007: Impacts, Adaptation and Vulnerability*. Working Group II Contribution to the Fourth Assessment Report of the Intergovernmental Panel on Climate Change, Figure 10.6. Cambridge University Press.

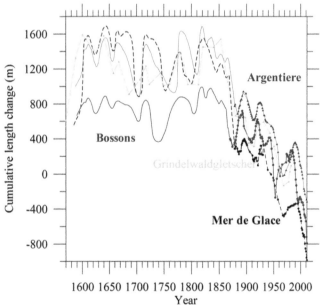

Figure 22.13 The variation in length of four large alpine glaciers since 1600 (Mer de Glace, Argentière, Bossons, Grindelwald). The origin of the vertical axis is arbitrary. The end of the Little Ice Age brought on the retreat of the glaciers in the middle of the 19th century.

Adapted from Vincent et al. (2004). Reproduced with permission of C. Vincent.

tropics and high latitudes over both hemispheres, reveals a similar regression in glacier length starting from the first part of the 19th century. Notwithstanding some variation in timing due to regional differences in climate, the LIA ended in the middle of the 19th century almost all over the globe.

It is now generally accepted that the main cause of the worldwide glacier advances was a temporary 'weakness' of the Sun (less energy radiated). Fluctuations in solar activity are detected by different proxies (Chapter 13). The oldest and most direct observation is the number of sunspots visible on the surface of the Sun, which is a measure of its activity. Sunspots and solar activity adhere to the same 11-year cycle, and their minima and maxima are locked in phase. Observations made since Galileo's invention of the astronomical telescope in 1609 reveal periods in which the maximum number of sunspots during a cycle decreases, sometimes even falling to zero. These are periods of lower solar activity (Fig. 22.14). In the second half of the 17th century, in particular, no sunspots were observed. This was the 'Maunder minimum'. These observations are confirmed by measurements of cosmogenic radionucleotides (i.e. produced by cosmic rays), from which the irradiance can be deduced over a more extended period (Fig. 22.15). Four minima were recorded during the LIA. The accompanying decrease in solar activity is now held to be the cause of the global cooling. Volcanic eruptions are also a secondary factor that can cause global cooling. The impact of such eruptions lasts only a few years, however, and their cooling effect due to reduced solar flux at the Earth's surface is only temporary. Figure 22.15 shows this volcanic activity over the last millennium.

How much colder was it during this period?

The reconstruction of the annual temperature in both the NH and the SH during the last millennium shows that the temperature decreased by about half a degree between the LIA and each of the warm periods, the Medieval Climatic Optimum and the beginning of 20th century (Fig. 22.16).

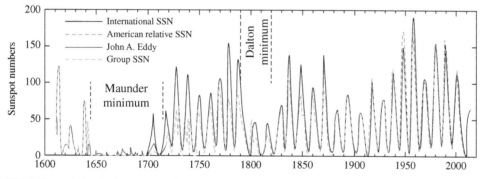

Fig. 22.14 The variation in the number of sunspots visible at the Sun's surface since 1610. The period of total absence of sunspots during the 17th century coincides with the minimum of irradiance, called the *Maunder minimum*. The number of sunspots is strongly correlated with the solar radiation received by the Earth (see Fig. 13.1).

Source: Reproduced with permission of Makiko Sato.

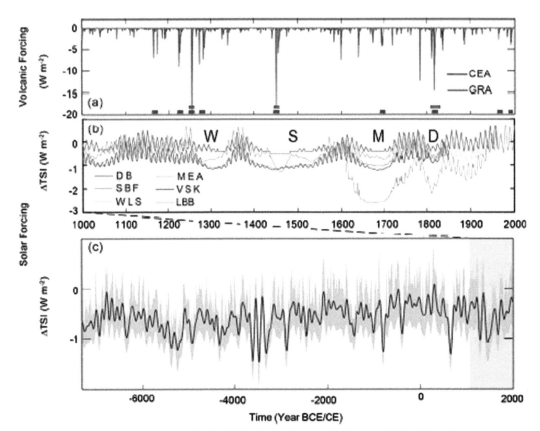

Fig. 22.15 (a) Two reconstructions of volcanic forcing in the past 1000 years, derived from ice core sulphates (green squares: Greenland; brown squares: Antarctica). (b) Reconstruction of fluctuations with respect to the year 2000 in the total solar irradiance (TSI) for the last millennium. The 11-year cycle prior to 1600 has been added artificially. Four irradiance minima occurred in the 14th, 15th, 17th and 19th centuries, respectively named the Wolf (W), Spörer (S), Maunder (M) and Dalton (D) minima. (c) Reconstruction of fluctuations with respect to the present of the total solar irradiance in the past 9300 years based on [10]Be.

Source: IPCC 2013. *Climate Change 2013: The Physical Science Basis*. Working Group I Contribution to the Fifth Assessment Report of the Intergovernmental Panel on Climate Change, modified from Figure 5.1. Cambridge University Press.

What was the climate in Europe like during the Little Ice Age?

Glaciers advance as a result of a positive annual snow balance that subsequently transforms into ice. A positive snow balance can be achieved in two ways: either the weather is colder and less snow melts in the summer, or more snow falls. Recent reconstructions of changes in the average temperature in the NH based on different proxies confirm the cooling during the LIA. In parallel, historical records note an increase in river flooding during this period, suggestive of greater rainfall. Historical archives and the various palaeoclimate reconstructions tend to show that the LIA was both colder and wetter in Europe, thus contributing in double measure to the advance of the glaciers. The same tendency appears fairly systematically during the various

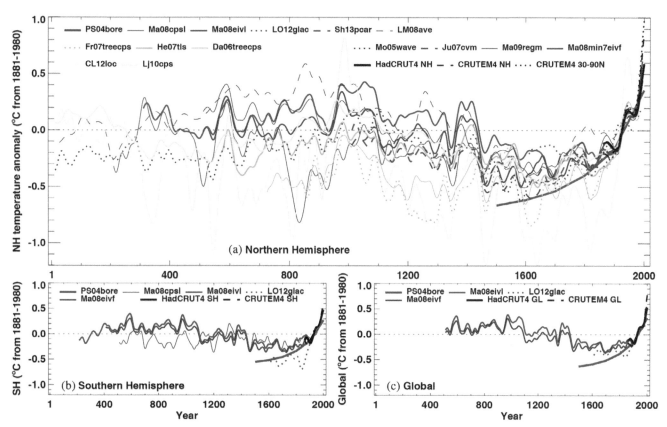

Fig. 22.16 Reconstructed annual temperatures in the last 2000 years: (a) Northern Hemisphere, (b) Southern Hemisphere, (c) global. This period includes the Medieval Warm Period (950–1250) and the Little Ice Age (1400–1700). Individual reconstructions are grouped by colour according to their spatial representation (red: land-only, all latitudes; orange: land-only, extra-tropical latitudes; light blue: land and sea extra-tropical latitudes; dark blue: land and sea all latitudes). Instrumental temperatures are shown in black. All series represent the deviation in °C (anomaly) from the average temperature between 1881 and 1980 (horizontal dashed line).

Source: IPCC 2013. *Climate Change 2013: The Physical Science Basis*. Working Group I Contribution to the Fifth Assessment Report of the Intergovernmental Panel on Climate Change, Figure 5.7. Cambridge University Press.

cooling events that occurred in the Holocene period. A similar conclusion is also arrived at from the reconstruction of the lake levels in Western and Central Europe, demonstrating that lake water levels rose during these cold periods (Magny 2013).

The natural end of the Little Ice Age

The period of advancing glaciers came to an end in Europe in about the middle of the 19th century, as it also did in most of the regions of the globe. This epoch happens to coincide with the beginning of the Industrial Revolution, which marks the start of anthropogenic emissions of CO_2. Confusion sometimes arises in attributing the retreat of the glaciers at that period to the impact of human activity. This is not the case. The increase in anthropogenic greenhouse gases in the atmosphere at that time was too small to produce the warming required, and the effects of the increase in CO_2 remained minor until about the middle of the 20th century. Prior to that, the advance and retreat of glaciers (movements that extend over several decades) were governed mainly by natural climate fluctuations.

22.6.2 NEO-GLACIATIONS IN THE HOLOCENE

To record the history of glaciers throughout the world over the past 12,000 years of the Holocene is no light task. Reconstructions of the advance of glaciers in a large number of countries, however, enable us to determine their dynamics. In the NH, the glaciers regressed sharply in the first half of the Holocene, and over several thousand years they were reduced or even eliminated. Then, starting from the middle of the Holocene, between 6000 and 5000 years BP, a phase of *neo-glaciation* took place almost everywhere, in which vanished glaciers reappeared and existing glaciers advanced (Denton & Karlen 1973; Wanner et al. 2008). Changes in the glaciers are only one facet of the general cooling that marked the end of the Climatic Optimum. This glacier expansion is seen in regions as diverse as Franz Joseph Island, Svalbard, Scandinavia, Alaska, Canada, the European Alps and the Himalayas. In the past 3500 years, there have been several phases of advance. The glacier progressions were interspersed with three main warm phases, in the Bronze Age, in Roman times and in the Middle Ages. These are illustrated by the glaciers in Switzerland (Fig. 22.9).

Current interpretations attribute the glacier advances to periods of lower solar activity. The solar activity, reconstructed qualitatively from the flux of cosmogenic radionucleotides, is retraced in Fig. 22.15, and shows that the above-mentioned warm phases are not associated with minima of solar activity. Among all the phases in which glacier activity resumed, the LIA appears to be the most pronounced glacier advance since the beginning of the Holocene. The findings are consistent with the following general picture.

In this picture, we observe the past through our time telescope, and combine the first two timescales. Throughout the Holocene, fluctuations in solar activity, similar to that in the LIA, occurred continually. When conditions are favourable, they can successively force the advance of glaciers during a decrease in solar activity, and then their retreat when full solar activity resumes. That is Timescale 2. But the solar fluctuations add to the slow cooling due to the decrease in NH June irradiance that has been taking place throughout the Holocene (Timescale 1). Temporary reductions in solar activity are thus more likely to cause glacier advance in the cooler second half of the Holocene. It is therefore no surprise that the most recent reduction in solar activity should produce the most pronounced advance of the last millennia, the LIA.

In the SH, where on the contrary the summer (December) irradiance increases during the Holocene, this process follows a different course of events. By the same argument, the sequences of glacier advances should have been more pronounced at the beginning of the Holocene than in the most recent millennia. However, information on the past history of glaciers in the SH is still too patchy, which makes it difficult to get an overall picture of the climate changes in the Holocene. Nevertheless, certain reconstructions, for which supplementary material is still lacking, could indeed corroborate this picture: some authors report a stronger advance of glaciers at the beginning of the Holocene than in the later millennia.

22.6.3 THE WARM PHASES OF THE LAST MILLENNIA

The warm phases of the last millennia took place mainly during the Bronze Age, in Roman times and in the Medieval Climate Optimum. *How do those earlier millennia compare with the particularly warm decades at the end of 20th century and the beginning of the 21st century?* The recent discovery of *Ötzi the Iceman* casts new light on this question (Baronni & Orombelli 1996).

Ötzi is the name given to the mummified body of a man discovered in 1991, who was found half-emerged from the snow at 3210 m altitude in the Ötzal Alps at the Hauslabjoch saddle, between Italy and Austria. Dating by ^{14}C established that he lived in the Copper Age, 5300 years ago. He had been buried under a layer of snow, and then a layer of ice. His body was found in the summer, frozen and mummified, and exposed to the air when the ice melted (Fig. 22.17). The artefacts found with him, clothes, footwear and a woven bag, were made of perishable materials. The fact that the body was almost intact, as were the artefacts that accompanied him, suggest that it had never been exposed to the open air before, because otherwise it would have been attacked by birds and degraded by ultraviolet light.

This implies that the intensity and duration of the warm phases that followed the death of Ötzi, whether the end of the Bronze Age, in the Roman period or in the Medieval Climatic Optimum, were never comparable to the present. More recent findings in the Swiss Alps (Schnidejoch) after the hot summer of 2003 indicate that the permanent ice cover that lasted from approximately 4900 years BP until AD 2003 has now disappeared (Grosjean et al. 2007).

Temperature compilations covering the past 1000–2000 years have been carried out by the PAGES 2k Consortium (2013) in seven continental regions (Europe, Asia, Australia, the Arctic, North America, South America and Antarctica). The results show that during the three decades of 1971–2000, the area-weighted average reconstructed temperature in the majority of those regions was higher than at any other time in the past 1400 years.

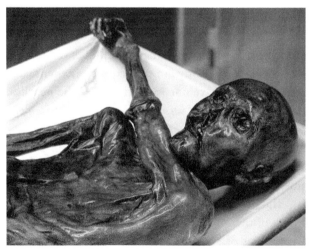 

Fig. 22.17 Left: Ötzi the Iceman is the well-preserved mummy of a man who lived 5300 years ago in the Copper Age, a period of warm climate. His perfectly preserved body was found in the European Alps between Italy and Austria in September 1991, emerging from the snow at 3210 m altitude. Right: Hauslabjoch; the red dot shows the location at which Ötzi was discovered.

Sources: Left, South Tyrol Museum of Archaeology, http://www.iceman.it. Right, photograph by Kogo [GFDL (http://www.gnu.org/copyleft/fdl.html)].

22.7 The Holocene and the birth of agriculture and animal husbandry

Among all the interglacials that punctuated the Quaternary, the Holocene occupies a special place. As human beings progressed through earlier interglacials, they gradually improved their hunting techniques. The Holocene, however, witnessed the birth of an entirely new relationship between humankind and the natural world: domestication of plants and animals made its appearance. People were gradually transformed from hunter–gatherers into farmers. An excellent description of this aspect, which we outline briefly, is to be found in *The Holocene*, by Roberts (2014).

The transition took place within a time span of only a few thousand years. Compared to the previous hundreds of thousands of years during which *Homo sapiens* had existed, the change was practically simultaneous. That it took place on different continents, with different starting materials, different local plants and different wild animals, is proof that this tremendous revolution originated independently in different places, and was not the outcome of a discovery that spread from a single centre. It was, for example, in South-West Asia that cereal crops such as wheat, barley, peas and lentils first made their appearance ~10 kyr BP. From East Asia to South Asia, wild rice was domesticated ~8.5 kyr BP, while in Mesoamerica the original staple crop was maize, ~9 kyr BP. These innovations give rise to societies that gradually became sedentary, organized around agriculture and rearing animals, societies in which the social structure became increasingly complex. The face of the planet was modified, albeit very slightly at first, when the total human population consisted of no more than a few million individuals. In the last millennium the change became more pronounced, with the population reaching a few hundred million individuals in the Middle Ages, 1 billion in 1800 and 7 billion in 2012.

Mankind–climate interaction

Why did man's discovery of agriculture emerge only in the current interglacial and not in those that went before? What was the synergy that gave rise to this prodigious discovery? Was it that the population was sufficiently advanced and structured for new techniques to be developed? Was human life expectancy in previous ages sufficiently long to acquire experience and knowledge of the natural cycles? Those questions we leave unanswered … But the next two questions do have answers: *Has the climate of our interglacial been much more clement than previous interglacials? Did humans benefit from extraordinary conditions and resources?* In fact, the Holocene climate is comparable to that of previous interglacials, displaying similar successions of vegetation, as illustrated by the succession of pollen spectra (Fig. 21.1). The conditions of the present interglacial provided the same climate advantages that our ancestors enjoyed in earlier interglacials, no more, and no less. The innovation may therefore be attributed to the evolution and the sociocultural development of the species itself.

Human beings, in their gradual development of agriculture and animal farming, modified their environment, mainly through clearing new land for crops. Deforestation, for example, is accompanied by release of CO_2 into the atmosphere. Prior to the 19th century, most of Europe's original forests had in this way been replaced by fields and farms. In Asia, the development of paddy fields generated CH_4 emissions. What effect did this have on the environment and on the climate? Could it be that human activity over the past millennia has modified the world's climate and changed the climate history of the present interglacial? This notion, suggested by W. Ruddiman, is addressed in Box 22.1.

BOX 22.1 RUDDIMAN'S EARLY ANTHROPOGENIC GHG THEORY

This theory seeks to explain the small increase in the greenhouse gases CO_2 and CH_4 that took place during the Holocene. Ruddiman (2003, 2005, 2007) attributes the change to human activity and suggests that this was what saved the Earth from returning to glaciation:

a. The changes in CO_2 and CH_4 in the Holocene differ from those of the previous interglacial (MIS 5.5, 130–115 kyr BP): by the 18th century, CO_2 had increased by about 20 ppm since 8000 years ago, and CH_4 by about 100 ppb since 5000 years ago (Fig. B22.1; see also Fig 19.2). In the Ruddiman interpretation, this change is *anomalous*: if the effects of human activity are removed, the conditions should be similar to those of the preceding interglacial, where there was no such increase.
b. The theory assumes that these increases are the result of human activity, through changes in early land use and forest clearing (emission of CO_2) and early rice cultivation (emission of CH_4).
c. By analogy with the previous interglacial, the climate during the Holocene ought, in this scenario, to have tended 'naturally' towards glaciation, which is not the case. Ruddiman is thus led to conclude that return to glaciation was avoided by virtue of the increase in anthropogenic greenhouse gases over several millennia.

This attractive hypothesis has encountered much criticism. We consider the different arguments in turn:

a. As is emphasized with great insistence in this book, interglacials are never identical. The different orbital configurations make each glacial and interglacial period unique, since they stem from a conjunction of the three orbital parameters – eccentricity, precession and obliquity – each of which varies independently of the others (see Chapters 14 and 20). The preceding interglacial, MIS 5.5, took place when the eccentricity

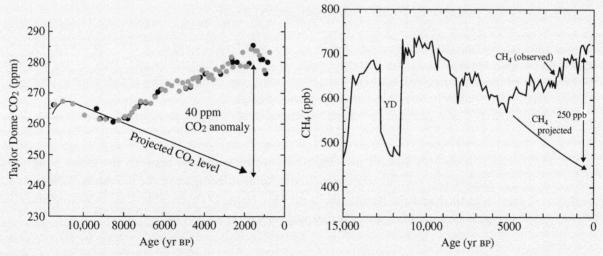

Fig. B22.1 Observed trajectories of atmospheric carbon dioxide and methane concentrations during the Holocene, together with trajectories expected in Ruddiman's hypothesis.

Source: Modified from Ruddiman (2003). Reproduced with permission of Springer Science+Business Media.

was much greater than that of the current Holocene interglacial (MIS 1, Fig. 14.2). A difference of this magnitude brings to the NH a summer irradiance that, during the first part of the interglacial MIS 5.5, was far stronger than during the present interglacial (see Section 20.5).

b. An investigation by Singarayer et al. (2011) addresses this difference and offers an explanation for the apparently anomalous increase of atmospheric methane concentrations since the mid-Holocene with respect to previous interglacials. Singarayer et al. note that '… *the climate and wetland simulations of the past global methane cycle over the last 130,000 years recreate and capture the Late Holocene increase in methane concentrations. It results from natural changes in the Earth's orbital configurations, with enhanced emission in the Southern Hemisphere tropics, linked to precession-induced modification of seasonal precipitation … [These] findings suggest that no early agricultural sources are required to account for the increase in methane concentrations in the 5,000 years before the industrial era.'* As for the change in CO_2 content, it is notable that although there was no increase during the last three interglacials, there was one in the MIS 11 interglacial, about 400,000 years ago (Siegenthaler et al. 2005; see also Fig. 19.2), when the orbital parameters were very close to those of the Holocene (see Section 21.3). The causes of this slight increase may be related to oceanic processes (changes in carbonate chemistry) and/or terrestrial processes (see IPCC 2013, Chapter 6). Terrestrial changes in the carbon storage of the biosphere would alter the ratio between ^{12}C and ^{13}C in the atmospheric CO_2, since plants preferentially take up ^{12}C. Measurements of $\delta^{13}C$ in atmospheric CO_2 trapped in ice cores suggest that the terrestrial carbon storage increased between 11,000 and 5000 years ago (due to peatland formation), and that only small overall terrestrial changes have occurred thereafter (Elsig et al. 2009). This finding suggests that the impact of vegetation clearance was small. Nonetheless, it is possible that this investigation underestimates the amount of ^{12}C taken up by peatlands during the second half of the Holocene. Finally, and still with reference to point (b), it has been pointed out that the human population was probably too small to generate the amount of early anthropogenic carbon and methane emissions required by the Ruddiman theory. These questions, however, are still under discussion.

c. A return to a glacial period is governed by changes in the irradiance and hence by the conjunction of orbital parameters. The present interglacial differs from those that preceded it in the very small eccentricity of the Earth's orbit round the Sun, but is closely similar to what happened about 400,000 years ago, when the interglacial was extended by 20,000 years (see Section 21.3).

Although the above arguments are unfavourable to the Ruddiman hypothesis, the increasing impact of mankind on his environment during the Holocene remains an unquestionable reality. They simply show that until the arrival of the Industrial Revolution in the 18th century, the main features of the climate changes in the Holocene can be explained without recourse to human agency.

If we leap forward over the many centuries that lie between the dawn of agropastoral civilizations and our recent epoch, the worldwide impact of the human activities that started with the Industrial Revolution in the 18th century is irrefutable. One of its consequences, the profound modification of the atmosphere, is the subject of Part V.

Chapter 23
Global and regional fluctuations (Timescale 3: decades)

W e now return to our time telescope, set at Timescale 3. *What can be said of climate fluctuations from one decade to the next, or even from one year to the next?* For almost two centuries, direct observations in the atmosphere, on land and in the oceans have amassed evidence of rapid climate fluctuations that last for one or more years. These involve large areas, sometimes encompassing the whole planet, and have repercussions on the mean global temperature. They have several causes, and different mechanisms. Among these, two main mechanisms produce fluctuations that last from one year to the next. *The first are internal instabilities that develop in the climate system between the ocean and the atmosphere*, and redistribute the stored energy among the different regions and latitudes. The El Niño and La Niña events are illustrations. *The second comes from changes in the composition of the atmosphere due to volcanic eruptions, or from changes in solar activity*. These introduce *external* forcing: the alteration to the climate system comes from outside (injection of aerosols into the atmosphere, change in irradiance), modifying the energy supplied to the system. The perturbations from these two sources and their impact on the climate were described in Part III. Here, we focus on the past century, which is a sufficiently long period to track such fluctuations, and where the climate archives are exceptionally rich.

Climate Change: Past, Present and Future, First Edition. Marie-Antoinette Mélières and Chloé Maréchal.
© 2015 John Wiley & Sons, Ltd. Published 2015 by John Wiley & Sons, Ltd.
Companion website: www.wiley.com\go\melieres\climatechange

23.1 From global …

23.1.1 ENSO EPISODES (EL NIÑO AND LA NIÑA)

The El Niño Southern Oscillation (ENSO) is due to coupling between the atmosphere and the ocean. It affects in particular the system of trade winds and ocean surface currents in the tropical region of the Pacific. The mechanisms involved and the different effects that it produces on the climate are described in Chapter 17. Here, we simply recall that an El Niño episode is accompanied by weakening of the trade winds. Among its effects, it gives rise to warming in the tropical zone that is intense enough to affect the global mean temperature. It increases rainfall, notably in the central region and on the eastern border of the Pacific, and it also creates zones of drought in South Asia. The La Niña episodes have the opposite effects. El Niño and La Niña episodes occur at irregular intervals of between 2 and 7 years.

Extensive studies of the historical archives have been carried out into the intensity of these events over recent centuries (Quinn et al. 1987; Ortlieb 2000). For the most recent periods, a range of indices has been defined to quantify this intensity. Figure 23.1 shows the fluctuations in a multivariate index since 1950. The index contains six variables from the tropical Pacific Ocean that are affected by the ENSO. The strongest El Niño in the 20th century was that of 1997–8, which gave rise to a series of fires and drought, along with its legacy of exceptional famines, in regions of South-East Asia that are generally very wet. To find an El Niño of comparable intensity, we have to go back to that of 1878, which was exceptional in the historical records. The perturbation of an El Niño generates global warming that can attain several tenths of a degree Celsius.

The intensity of the 1997–8 event, where the regional perturbations affected the global average, is often used to illustrate the impact of an El Niño episode on the planet. Some examples of the effects include the following:

- a higher average sea level due to lower precipitation on continents (Chapter 27);
- an increased water vapour content in the atmosphere (by 3%, averaged over all the oceans) from the stronger evaporation in this zone; and

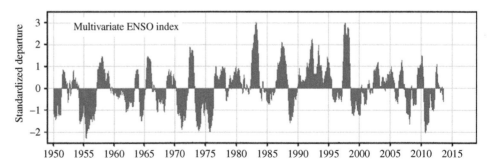

Fig. 23.1 The variation of the El Niño Southern Oscillation (ENSO) multivariate index since 1950. This index incorporates information from six variables from the tropical Pacific Ocean affected by the ENSO (pressure, the zonal and meridional wind components, the surface temperature of the sea and that of the air, and the cloud cover). A positive index corresponds to a warm ENSO period; that is, an El Niño event. The magnitude of the El Niño episodes of 1983 and 1997–8 is remarkable.

Source: NOAA/ESRL; http://www.esrl.noaa.gov/psd/enso/mei/

- a perturbed carbon cycle and increased CO_2 content in the atmosphere – the source of the perturbation is located in the tropical belt (Chapter 29).

The impact of an El Niño on the temperature of the different layers of the atmosphere is a mine of information. The effect is illustrated in Fig. 23.2, which shows the changes in the average temperature of the atmosphere at four different altitudes, surface level, the lower troposphere, the middle to upper troposphere and the lower stratosphere. The first three layers belong to the troposphere, where convection is the rule. The last layer belongs to the stratosphere, where the atmosphere is stable and stratified. These two reservoirs, which do not mix, are separated by the tropopause. In the exceptionally intense El Niño event of 1997–8, the

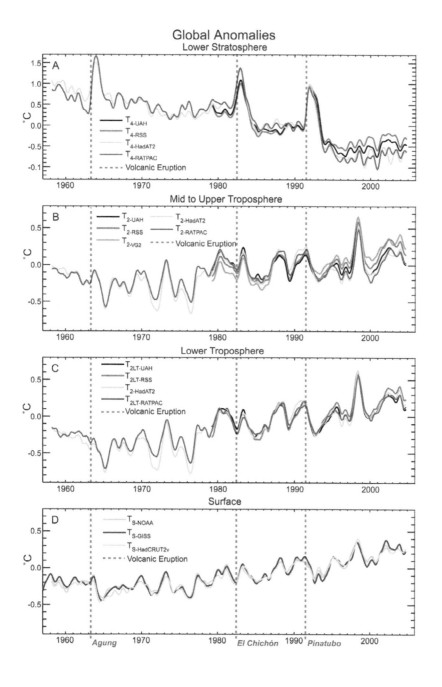

Fig. 23.2 Variation in global mean temperature at the surface of the Earth and in the atmosphere. (a) Lower stratosphere at about 15 km altitude. (b) Middle and high troposphere. (c) Lower troposphere. (d) Surface. Observations made by different groups (coloured curves). The vertical axis is the temperature difference with respect to the 1979–1997 average. Major volcanic eruptions are denoted by dashed vertical lines. **Source:** IPCC 2007. *Climate Change 2007: The Physical Science Basis.* Working Group I Contribution to the Fourth Assessment Report of the Intergovernmental Panel on Climate Change, Figure 3.17. Cambridge University Press.

increase in temperature at the surface was about 0.3°C, while in the mid-troposphere it was about 0.6°C. The El Niño of 1983, the second most intense of the century, is also perceptible, with a temperature increase of about 0.3°C in the mid-troposphere. These signals clearly reveal the mark of the various known El Niño events. Such perturbations are always confined to the troposphere and never penetrate the tropopause barrier. Volcanic eruptions provide another illustration of the behaviour of these two regions of the atmosphere.

23.1.2 VOLCANIC ERUPTIONS

Explosive volcanic eruptions inject a layer of aerosols into the atmosphere that spreads around the planet and reduces the solar radiation arriving at the surface (Box 15.1). The size of the effect, and hence the magnitude of the eruption, is defined by the radiative forcing factor (Chapter 9). The record of this factor in the volcanic eruptions over the past 1000 years is displayed in Fig. 22.15.

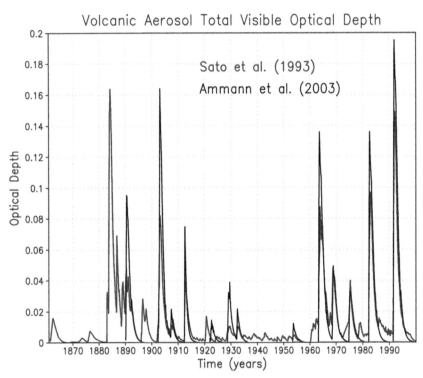

Fig. 23.3 This proxy (optical depth in visible light) indicates the thickness of stratospheric sulphate aerosol layers formed in the aftermath of explosive volcanic eruptions. Results are shown from two different data sets. The eruption of Pinatubo, Philippines (1991) was the largest of the century and was comparable to that of Krakatoa, Indonesia (1883). Recent decades are marked in addition by the eruptions of El Chichón, Mexico (1982) and of Mount Agung, Indonesia (1963). All these eruptions occurred at low latitudes.

Source: IPCC 2007. *Climate Change 2007: The Physical Science Basis.* Working Group I Contribution to the Fourth Assessment Report of the Intergovernmental Panel on Climate Change, Figure 2.18. Cambridge University Press.

Since 1500, two powerful eruptions have appeared in the historical records, Huaynaputina in Peru in 1600, and Mount Tambora in Indonesia in 1815. The ensuing cooling events were very severe, with their effects being felt all over the planet, and in certain regions causing major economic difficulties. The Huaynaputina eruption in Peru generated a pronounced cold period in countries as diverse as Japan, China, Russia, Estonia, Germany and Switzerland, where strong perturbations were recorded. Russia was confronted with one of the most severe food shortages in its history. In Europe, the year 1815 was dubbed 'the year without a summer'; it caused appalling famine (a tangible example of the notion of 'intense perturbation'!). More recently, the explosive eruptions of Krakatoa in 1883 and Mount Pinatubo in 1991 left their mark on the history of the 19th and 20th centuries. The various volcanic eruptions can be graded according to the thickness of the aerosol layer they inject into the stratosphere. The record of the aerosol thickness since 1860 is displayed in Fig. 23.3.

The most powerful eruption of the past few decades is incontestably that of Mount Pinatubo in the Philippines, in 1991. To see its impact on the global temperature, we take another look at the temperature of the atmosphere in Fig. 23.2. Absorption of solar radiation by aerosols causes heating in the surrounding layer; that is, in the stratosphere at an altitude of about 15 km. At that altitude, the warming was intense, exceeding 1°C in the year of the eruption, and decreased progressively 1–2 years as the aerosol layer gradually diminished. In the troposphere and at the surface, the opposite phenomenon occurred: the surface cooled by about 0.3°C. Less energy is available on the ground owing to reflection (and, to a lesser degree, absorption) of solar radiation by the aerosol layer. The separation of the atmosphere into two reservoirs that do not mix (the troposphere and stratosphere) is illustrated by the fact that cooling of the surface is shared with the troposphere alone, and warming of the stratosphere is confined to the stratosphere.

Each powerful explosive-type volcanic eruption is followed by the same sequence of events. Although the Mount Agung eruption (1963) clearly reduced the temperature of the troposphere and the surface, for the El Chichón eruption this reduction is not nearly as clear, even though it was of comparable intensity. Let us study this case more closely.

23.1.3 EL NIÑO AND VOLCANIC ACTIVITY, TWO KEY SOURCES OF INTER-ANNUAL VARIABILITY

Why does the El Chichón eruption of 1982 not show up clearly in the record of surface temperature, where it seems to be submerged in the random noise? The reason is that El Niño episodes must be taken into account. When Mount Pinatubo erupted, there was no El Niño effect. The surface of the Earth cooled visibly and, over a few years, the cooling decreased gradually until the volcanic aerosols disappeared from the stratosphere. By contrast, during the El Chichón eruption, the powerful El Niño of 1982–3 offset the surface cooling from the volcanic aerosols, masking it for about a year (the normal duration of an El Niño episode). In the record of the global mean temperature, therefore, what appears as random noise is often the response of the climate system to El Niño episodes and to volcanic eruptions. *How did these sources of fluctuation contribute to the global temperature changes during the 20th century?* This question is addressed in Part V.

23.2 … to regional: the North Atlantic Oscillation

Numerous other oscillations exist that link neighbouring permanent centres of high and low pressure. Although they do not have the same power as the ENSO in their range and impact, they exert an influence on the climate over large regions of the globe. Their influence is all the stronger the further they are from the central Pacific zone, where the ENSO dominates. This applies to the North Atlantic Oscillation (NAO) (Section 17.3), with its consequences for the climate (Fig. 17.5). Its measure is the NAO index. We recall that a positive index in winter brings more frequent storms and an increase in rainfall over the northern part of Europe. By contrast, rainfall decreases over the southern part of Europe and the Near East. A negative index, however, increases rainfall in the Mediterranean zone. The fluctuations of this index since 1860 are displayed in Fig. 23.4. A distinguishing trait of the index is its strong variability

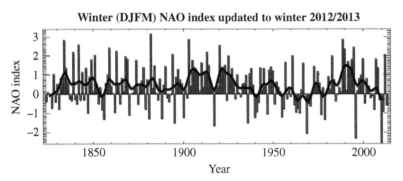

Fig. 23.4 Variations in the North Atlantic Oscillation (NAO) index since 1830. This index strongly affects the climate in the North Atlantic region (the United States, Canada and Europe), particularly in winter (see Fig. 17.5). The positive (/negative) index shown here corresponds to a strong (/small) difference between the atmospheric pressure at sea level at Gibraltar and that over the south-west of Iceland.

Source: Osborn (2011). Reproduced with permission of Springer Science+Business Media and Timothy J. Osborn.

from one year to the next, or even from one decade to another. Over the recent decades, no net tendency emerges. During the decade between 1990 and 2000, the index was generally positive, but its amplitude started to decrease from the beginning of the 2000s.

23.3 The Sun, the other source of change

We have mentioned two sources of inter-annual change on a global scale, the ENSO and volcanic eruptions. A further cause of climate variation, which may last for one or more decades, is our sole energy supply, the solar flux. As already seen, the Sun's radiation varies with a period of 11 years (Fig. 13.1). During this cycle, the solar flux arriving on the Earth fluctuates around $1361\,\text{W/m}^2$, with a maximum amplitude of $1.5\,\text{W/m}^2$ that generates a radiative forcing of $0.25\,\text{W/m}^2$ at the Earth's surface. The resulting variation in global mean temperature is estimated to be about 0.1°C.

Independently of the 11-year cycle, the Sun's activity fluctuates over decades and centuries – which brings us back to Timescale 2 (Chapter 22). Although periods such as the Little Ice Age lasted for several centuries, such fluctuations can also affect the climate for shorter periods – for example, as in the Maunders minimum between 1650 and 1710 (Fig. 22.14), when sunspots were absent. *What contribution has the Sun's activity made to the global mean temperature since the beginning of the 20th century?* This question is addressed in Part V.

Chapter 24
Future warming and past climates

We shall see in Part V that, depending on the amount of greenhouse gases released by humankind, the global mean temperature could rise during the 21st century by between 2°C and 6°C. *Where will this degree of warming lead us? What does the experience of the past few tens of millions of years tell us?* Three different time windows in the past are of particular interest.

24.1 The global 'hot flush' of 55 million years ago

Fifty-five million years ago, the environment at the surface of the planet was very different from that of the present day, but the solar flux was comparable. The environment was different in regard to the position of the continents, the marine and atmospheric circulation and the flora and fauna; moreover, the temperature was much higher (Figs. 18.1 and 18.3). The global mean temperature at the surface and in the deep ocean was roughly 10°C higher than now. Although there is uncertainty in the estimates, the CO_2 content of the atmosphere was most probably greater than now, in the range of 1000 ppm.

An abrupt warming episode, the *Palaeocene–Eocene Thermal Maximum* (PETM), then intervened, which profoundly modified the environment. It was then that the highest temperatures within the past 65 million years were recorded, with a warming of about 5°C at the Earth's surface and in the deep ocean water (Zachos et al. 2003). The extreme

Climate Change: Past, Present and Future, First Edition. Marie-Antoinette Mélières and Chloé Maréchal.
© 2015 John Wiley & Sons, Ltd. Published 2015 by John Wiley & Sons, Ltd.
Companion website: www.wiley.com\go\melieres\climatechange

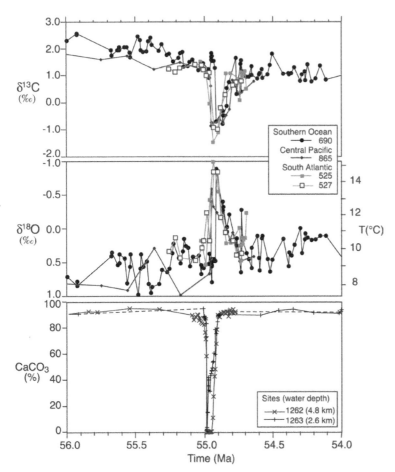

Fig. 24.1 The Paleocene–Eocene Thermal Maximum as recorded in the benthic (bottom-dwelling) foraminifer isotopic records, from sites in the Antarctic, South Atlantic and Pacific. The highest temperatures of the past 65 million years occurred during this period. (top) The rapid decrease in $\delta^{13}C$ indicates a strong increase in greenhouse gases (CO_2 and/or CH_4). (middle) This decrease coincides with an average warming by about 5°C of the deep ocean water, revealed in the $d^{18}O$ signal. (bottom) At the same time a decrease in the carbonate content of the sediments is observed, indicating a decrease in pH of the sea water. In theory much of the additional greenhouse carbon would have been absorbed by the ocean, thereby lowering seawater pH and causing widespread dissolution of seafloor carbonates. The observed patterns indicate that the ocean's carbonate saturation horizon rapidly shoaled by more than 2 km, and then gradually recovered as buffering processes slowly restored the chemical balance of the ocean. The temperature anomaly lasted for about 100,000 years. Note that some datings situate this even closer to 56 Myr ago.

Source: IPCC 2007. *Climate Change 2007: The Physical Science Basis.* Working Group I Contribution to the Fourth Assessment Report of the Intergovernmental Panel on Climate Change, Figure 6.2. Cambridge University Press.

warming set in very rapidly (within 1000–10,000 years, or even less) and lasted about 100,000 years. The whole of the Earth was affected, with disruption in terrestrial eco-systems and net decrease in biodiversity of the oceanic benthic community. This marked the termination of the Palaeocene, a geologi-cal period that extended between 66 and 56 million years ago. The event is recorded in the sea sediments; in particular, in the iso-topes of carbon and of oxygen in the tests of benthic foraminifera (zooplankton from the ocean floor) (Fig. 24.1). The deep ocean waters warmed by about 5°C, as witnessed by the increase in $\delta^{18}O$. The change in $\delta^{13}C$, both in the marine and continental records, indicates that a large amount of ^{13}C-depleted carbon was injected into the atmosphere and into the ocean. Possible sources of this car-bon are methane (from the decomposition of clathrates on the seafloor) or CO_2 (from vol-canic activity), or oxidation of sediments that are rich in organic matter (Part IV, Note 11). Based on these data, the numerical models estimate that between 1000 and 2000 Gt of carbon were injected into the atmosphere. This happens to be the order of magnitude of what is likely to be emitted in the near future (Part IV, Note 12). To reach equilibrium of the CO_2 partial pressure between atmosphere and ocean, the amount of CO_2 dissolved in the ocean water rose. This increased the acid-ity of the water and dissolved the carbonates deposited on the ocean floor (Zachos et al. 2005). The change in the composition of the ocean sediments appears as a sudden drop in the amount of calcium carbonate deposited at that period (Fig 24.1). The subsequent return to the previous conditions was gradual, and took about 100,000 years.

That event in the distant past is of particu-lar interest now for two reasons. First, it shows that an abrupt increase in the average temper-ature by about 5°C can really occur within a geologically short timespan (less than 1000–10,000 years, according to present resolution), and that it disrupts the environment profoundly, exacting its toll on biodiversity. Biodiversity, as usual, requires a long period to regenerate. Second, it offers the possibility of testing current models. Simulations show that a release of CO_2 of this order does indeed produce warming of several degrees, and that the time required for the previous conditions to return (excluding the time needed to sequester the excess CO_2) is of the order of 100,000 years. One discrepancy must, however, be highlighted. The simulations fail to predict warming of such magnitude at high latitudes. *Does that mean that the models underestimate warming at high latitudes?*

24.2 Three million years ago

A second time window is that of the Middle Pliocene, which took place between 3.3 million and 3 million years BP. In that epoch, large ice sheets had not yet colonized the high Northern Hemisphere (NH) latitudes. On the basis of the different proxies, Global Climate Models (GCMs) estimate that the mean global temperature then was 2–3°C higher than in pre-industrial times. In the Middle Pliocene, the position of the continents was practically the same as now. That epoch can therefore serve as an illustration of a world that is globally warmer than ours. The amount of CO_2 in the atmosphere was in the range 360–400 ppm. Data collected both on the continents and in the oceans show that the temperature in the tropics was hardly different from now, but at high latitudes the temperatures were substantially higher. The data also show that conditions on the continents were wetter. Reconstructions of the temperature using models based on fossil flora and fauna suggest that at high NH latitudes in winter it was from 10 to 20°C warmer on the continents, and from 5 to 10°C warmer at 60°N in the North Atlantic. The sea level was at least 15–25 m higher than the present sea level, which is typical of interglacial sea levels since 500 kyr ago. It follows, for example, that not only was there no Greenland ice sheet (thereby increasing the sea level by ~7 m), but also the Antarctic ice sheet was smaller than now (contributing between 8 and 18 m to the rise in sea level).

Why are these reconstructions so important? First, it is because they are based on observations, and they tell us of the environment in a world that was closely similar to our own, but warmer by about 3°C. This relatively modest difference in mean temperature corresponds to an environment that was very different from the present (sea level, distribution of the temperature from the tropics to high latitudes, rainfall distribution etc.). The reconstructions are also important because they test the ability of the models to replicate the observations: at present, models underestimate the warming at high latitudes that occurred in that past epoch. It is not unreasonable to conclude that they could similarly be underestimating the warming at high latitudes that may well happen in the near future.

24.3 Warmer periods in the past 2 million years?

In this time window, we turn to the more recent Quaternary epoch, when glaciations appeared in the NH.

During the warm phases (interglacials), was the climate warmer than in the Holocene? And if so, by how much? Current reconstructions based on both marine and ice cores show that the warmer interglacials were the most recent (high sea levels and high temperatures; see Fig. 18.4). *How do these compare with the Holocene?* The previous interglacial (Eemian MIS 5.5, 125,000 years ago) is representative of the warmest stages. Its global mean temperature was 1–2°C higher than in the Holocene, and the sea level was about 7 m higher. In Antarctica, the increase in temperature reached 4°C (Fig. 19.2). One other warmer interglacial is of outstanding length: stage MIS 11 lasted from approximately 425,000 to 395,000 years ago. Its climate, albeit warmer than that of the Holocene, does not appear to have been warmer than the Eemian, since the temperature rise in the centre of the Antarctic was limited to 2°C.

Finally, within the present Holocene warm period, were there any especially warm episodes? We already know the answer: the first few millennia of the Holocene enjoyed a milder climate than now, with an average temperature about 1°C or even 2°C higher.

In this rapid survey, one point stands out clearly: among the warm climates that have occurred periodically in the glacial–interglacial alternations over the past 2.6 million years, the increase in global mean temperature with respect to our present (pre-industrial) climate was never greater than about 2°C.

PART IV SUMMARY

Past records show that glaciations, with great extensions of ice sheets, developed intermittently during the last three billion years of the Earth's existence. Their appearance was episodic. Since about 35 million years ago, we have entered into one of these epochs. This has been followed in the past 3 million years by intermittent ice sheets at high latitudes in the Northern Hemisphere. The onset of glaciation is the consequence of gradual cooling that has been taking place over the past 50 million years. The cooling is illustrated by the decrease by about ten degrees in the temperature of the water in the oceans. Since about 3 million years ago, glacial–interglacial oscillations have been leaving their mark on the climate of the whole planet. The difference in temperature between the two types of climate, averaged over the surface of the planet, is about 5°C.

This alternation of climates is triggered by astronomical factors and is reinforced by different feedback mechanisms (among which albedo and greenhouse gases play an important role). The variations in irradiance at the different latitudes and at the different seasons are cyclical, with a fundamental period of the precession cycle of the order of 22,000 years. Every 11,000 years, at the middle and high latitudes of the Northern Hemisphere, this cycle creates situations that first favour the development of ice sheets (cool summers) or their melting (warmer summers). The swing of this pendulum is all the more pronounced when the elliptical orbit of the Earth around the Sun deviates from a circle. This happens with a period of about 100,000 years, which is the periodicity of the last five strongest interglacials. They lasted on average for about 10% of the total time.

The present interglacial period, the Holocene, began 12,000 years ago. At that time in June (i.e. during summer in the Northern Hemisphere), the Earth was close to the Sun. During the 11,000 years of the succeeding half-cycle of the precession, the Earth–Sun distance has slowly increased. In the warm period of the Holocene, the climate appears to be stable, and climate fluctuations have been much smaller than those that punctuated the preceding glacial period. The sea level, which rose by about 120 m when the ice sheets melted, has remained almost constant for the past 6000 years, which means that melting has stopped. *What are these climate fluctuations?* They can be stated roughly as follows. Over 10,000 years, gradual cooling has prevailed in the Northern Hemisphere as a result of the decreasing irradiance in summer. Over decades or centuries, the effects of solar fluctuations prevail, and are revealed, for example, by the advance or retreat of glaciers. On the scale of a year to a decade, it is the fluctuations in the atmosphere–ocean interaction and volcanic eruptions that are mainly felt. *How, in the course of the Holocene, was the overall temperature of the Earth's surface affected by these variations?* Reconstructions based on different proxies indicate that the global mean temperature did not exceed that of the 20th century by more than one (or two) degrees. In the slow cooling through the Holocene, the temperature of the planet has decreased by about one degree, the fluctuations caused by the flickering of the Sun have been of the order of half a degree, and the climate variability over periods of a century has been less than one degree.

This rapid overview of the climate changes over millions of years brings out the following points, which will later help us to appreciate the impact of present human activity and the magnitude of the projected future warming:

- Since about 3 million years ago, the glacial–interglacial oscillations on Earth have given rise to alternations between two very different worlds. Flora and fauna, particularly at middle and high latitudes, have been selected by their ability to adapt to these huge climate swings. The transition between the two worlds takes place over several thousand years, in which the mean temperature of the Earth changes by about 5°C. At middle latitudes on the continents in the Northern Hemisphere, this temperature difference is twice as large, and at high latitudes in the Northern Hemisphere more than three times as large.
- The present warm period is the most recent interglacial. Its average temperature is comparable with those of previous interglacials. Very few interglacials have been exceptionally warm, but their temperature has never exceeded that of the present interglacial by more than 2°C. This fact provides a common reference for the maxima of global mean temperatures attained over the past couple of million years.
- Ice records, which archive the changes in atmospheric composition over the past 800,000 years, reveal the variations in greenhouse gases. Other indirect proxies extend this range to the past couple of million years. Comparison of the natural variations with those of the past two centuries makes clear the magnitude of the disruption caused by humankind since the Industrial Revolution, and particularly in the most recent decades. Within the space of one century, a new atmosphere is now in place, one that has no parallel since at least 800,000 years ago.

PART IV NOTES

1. $\delta^{18}O$ is the relative difference in the ratio of the abundance of the two isotopes of oxygen, ^{18}O and ^{16}O, in a given sample ('sam') with respect to that of the same two isotopes in a known reference standard ('ref'). Since these two ratios are close to each other in natural specimens (measurements must therefore be very precise to discriminate between them), it is convenient to multiply this relative difference by 1000. The result is expressed in parts per thousand (‰). Thus

$$\delta^{18}O = \{[(^{18}O/^{16}O)_{sam}/(^{18}O/^{16}O)_{ref}]-1\} \times 1000 \qquad (IV.I)$$

Example. In a sample consisting of four specimens of benthic foraminifera of age 60.0 Myr (million years), extracted from a zone in the Indian Ocean corresponding to the palaeolatitude 16°S, it is found that $(^{18}O/^{16}O)_{sam} = 0.00206776$. The isotopic ratio in the standard is $(^{18}O/^{16}O)_{ref} = 0.00206720$ (standard international 'PDB', carbonates). The ratio of these two numbers is thus $(^{18}O/^{16}O)_{sam}/(^{18}O/^{16}O)_{ref} = 1.00027$.

Since the *relative difference* in eqn. IV.1 involves only the figures after the decimal point, the value of $\delta^{18}O$ is therefore 0.00027=0.27‰. For the isotopes of other elements, δ is expressed in the same way; for example, δD or $\delta^{13}C$, which refer to the isotopes of hydrogen (deuterium) and of carbon. In these cases, the ratios D/H and $^{13}C/^{12}C$ are substituted for $^{18}O/^{16}O$ in eqn. IV.1.

2. C4 plants are so called because the first sugar molecule in the photosynthesis reaction contains four carbon atoms instead of the three produced by C3 plants. These plants have developed a mechanism whereby CO_2 is concentrated at carboxylation sites. This makes the photosynthesis more efficient while the stomata (the orifices through which gases penetrate into a plant) are less open. In so doing, the plants lose the least possible amount of water and hence are better adapted to dry regions. C4 plants appeared between 60 and 30 million years ago. Although their presence in the Tertiary was modest, between 8 and 5 million years ago they spread widely over the four continents from the tropics to the temperate climate zones. This change in vegetation, which took place about 8 million years ago, coincides with an occurrence of turnover among terrestrial mammals.

3. Greenland is situated at lower latitudes than is the Antarctic continent, which is centred over the South Pole. Due to the properties of atmospheric circulation, Greenland is therefore exposed to much larger changes in airflow than Antarctica between glacial and interglacial stages. These changes bias the interpretation of the isotopic signal of the ice in terms of temperature. Another, more complex, technique must therefore be used. This is based on direct measurements of the temperature of the ice as a function of depth in the ice sheet (several thousand metres) and the propagation of heat within the ice sheet. Analysis of the resulting signal shows that the temperature during the previous glaciation was about 20°C lower than during the Holocene, which indicates that the decrease in temperature was twice as great as that found from the isotopic thermometer. At the Vostok site in the Antarctic, by contrast, the value of about ten degrees that was obtained from the isotopic measurements is confirmed.

4. The estimated volume of the present ice sheets of Greenland and the Antarctic correspond to changes in the sea level of the order of 7 m and 58 m, respectively. To find the volume of ice stored on the continents during the last glaciation, the 120 m rise in sea level due to the melting of this vanished ice must be added to the volume of the present ice sheet. This is equivalent to a sea-level change of about 185 m. The present Greenland ice cap therefore constitutes about 4% of the volume stored during the last glaciation.

5. The suffix 'BP' ('before present') indicates that the reference date is taken to be 1950. Several conventions exist to estimate the time that has passed. The date of an event can be estimated from the decrease of ^{14}C. An extra correction must be added to take account of the changes of this isotope in the atmosphere over the course of time. This correction can be as much as several thousand years. For clarity, ^{14}C dates are frequently denoted '^{14}C age'. Two other conventions are used, BC (before Christ) and AD (*anno Domini*), expressing the number of years before or after the birth of Christ.

6. The snow accumulation on an ice sheet, both in Greenland and in the Antarctic, is twice as great during interglacial periods as during glacial periods. However, in spite of the increased supply of snow, ice caps retreat during the warm periods. Thus, although the precipitation brings more ice, warming wins by melting it faster.

7. The tilt of the axis of rotation of the Earth with respect to the normal to the plane of the ecliptic (obliquity) varies by a few degrees, with a period of 41,000 years. Eleven thousand years ago, this tilt was at its maximum. Since that time and for the rest of the half-cycle (~20,000 years), it has been decreasing. At present, we are in the middle of this half-cycle, where the rate of decrease is fastest. The less the axis is tilted, the less energy is received at middle and high latitudes in June, and consequently the more energy is received in the tropical belt. When the decrease in obliquity is taken into account, the date at which the Earth receives the least energy in June at middle and high latitudes in the NH is delayed by about 1000 years. That date coincides with our present time.

8. The velocity of the planet in its orbit is defined by Kepler's second law, the law of areas. It stipulates that the planet sweeps out equal areas during equal intervals of time. The area is the surface enclosed by the path AB of the planet on the ellipse and the two segments OA and OB, where O is the position of the Sun (the occupied focus of the ellipse). This law implies that the closer the planet is to the Sun, the greater is its velocity.

9. At latitude 20°N, the irradiance at the summer solstice (21 June) is at a maximum when the Earth is at perihelion. This took place approximately 12,000 years ago. However, roughly 1800 years later, the irradiance was at a maximum on 21 July (since half a precsssion cycle, or 11 kyr, is required for the date of perihelion to shift by 6 months). As a result, the maximum of irradiance averaged over the period June to August is delayed by about 2 kyr after that of the summer solstice. This occurred at about 10,000 years BP.

10. Starting from the beginning of the 14th century, the Little Ice Age extended the sea ice around the coasts of Greenland. This ice increasingly hindered the rotation of boats sailing to and from the mother country, Norway. Then, in 1368, the connecting boat was destroyed, and was not replaced. Worsening climatic conditions, which were already difficult for cattle raising and agricultural production at such high latitudes, proved fatal for these communities and caused their extinction. To exacerbate this situation, environmental management (land clearance, cutting bushes for pasture and fuel) degraded the soil. In 1721, when the Danes and Norwegians rediscovered the island, the Inuits were its sole inhabitants. Their management of the living resources (fishing, and hunting of seal, walrus, whale and caribou, with fat being used for fuel) enabled them to survive through the cold period. It appears, however, that the Inuit economy did not escape unscathed from the rigours of the Little Ice Age either, and that numerous families died of hunger and cold.

11. Methane clathrate is a complex of water molecules in the form of a cage around a CH_4 molecule. The complex is stable only at high pressures. Accumulations of methane clathrate are found buried deep in the subsoil, notably beneath the oceans.

12. An emission of 1000 (or 2000) gigatonnes of carbon (GtC) in this century, in the optimistic assumption that the atmosphere continues to store half of the man-made CO_2, would increase the atmospheric content of CO_2 by 500 (or 1000) GtC. These figures correspond to an increase of 226 (or 452) ppm. These values are consistent with the levels of stabilization considered in the scenarios of IPCC 2007 SR (Fig. 32.1) and of IPCC 2013 (Fig. 32.8).

PART IV FURTHER READING

Bradley, R.S. (2014) *Paleoclimatology: Reconstructing Climates of the Quaternary*, 3rd edn. Elsevier/Academic Press, Amsterdam.

Climate Change 2013: *The Physical Science Basis*. Working Group I Contribution to the Fifth Assessment Report of the Intergovernmental Panel on Climate Change (IPCC), Cambridge University Press; available at http://www.ipcc.ch/report/ar5/wg1/-; see Chapter 5.

Cronin, T.M. (2009) *Paleoclimates – Understanding Climate Change Past and Present.* Columbia University Press, New York.

Houghton, J. (2009) *Global Change – The Complete Briefing*, 4th edn. Cambridge University Press, Cambridge, UK; see Chapter 4.

Orgil, D. and Gribbin, J. (1979) *The Sixth Winter*. Simon and Schuster, New York.

Roberts, N. (2014) *The Holocene: An Environmental History*, 3rd edn. John Wiley & Sons, Ltd, Chichester, UK.

Ruddiman, W.F. (2008) *Earth's Climate – Past and Future*, 2nd edn. W.H. Freeman, New York; see Parts 2–4.

PART V

CLIMATE CHANGE IN RECENT YEARS

Chapter 25
Recent climate change

Climate change in the 20th century is now well documented, and the scientific community acknowledges that the world's climate has changed over the past several decades. The five IPCC reports (IPCC 1992, 1995, 2001, 2007 and 2013) have gradually strengthened the body of observations and analyses leading to this assertion. In this chapter, we give a brief overview of these observations, based principally on the last two IPCC reports.

The term 'warming' is often used to describe recent climate change, as the world's average annual temperature (global mean temperature) has increased over the past decades. However, because the warming does not necessarily affect all areas of the planet in the same way, some authors prefer to speak of 'climate change'. We use the term 'warming', since it expresses the recent physical reality: the fact that average temperature is rising means that more energy is available in the atmosphere and in the land mass and oceans of the Earth (Chapter 3).

But first, we consider the question of how to define and measure climate change. To answer this question, we examine the observations of changes in temperature and precipitation, as well as the consequences of these changes.

25.1 Changes in temperature

25.1.1 OBSERVED CHANGES

Changes in global mean temperature

The global mean temperature that is most frequently quoted is that of the University of East Anglia's Climatic Research Unit (CRU). On land, this is taken to be the air temperature

Climate Change: Past, Present and Future, First Edition. Marie-Antoinette Mélières and Chloé Maréchal.
© 2015 John Wiley & Sons, Ltd. Published 2015 by John Wiley & Sons, Ltd.
Companion website: www.wiley.com\go\melieres\climatechange

measured in the shade 2 m above ground level; for the oceans, it is taken to be the temperature of the sea surface. Measurements are taken at various points all around the globe. In some regions (such as Europe), the number of observation points is far greater than in others (e.g. the oceans of the Southern Hemisphere (SH)). Data coverage is also much denser towards the end of the measuring period (today) than at the beginning (in the 19th century). It is therefore necessary to reconstruct an average from incomplete data, depending on place and on time. It is also necessary to take into account gaps in the data and changes in measuring conditions. The CRU data is openly available on its website, in graph and grid point format, as well as in various data set forms. At a local level, reliable temperature chronicles have been constructed using homogenization techniques, which detect breakdowns in a series that are unrelated to climate (resulting from a change of measuring apparatus, a shift of the measuring station, a change in the environment etc.) and correct for these changes. The techniques are based on behavioural similarities between adjacent stations (the principle of relative homogeneity). The two methods are essentially statistical. Another approach is by re-analysing atmospheric data. This draws on all available observations, not just the temperatures, and, by numerical simulation, reconstructs all the variables that describe the atmosphere, generally in 6-hour steps. Successive changes to the Global Observing System, particularly regarding satellite data, have made these re-analyses a useful tool for studying climate change, but only from 1979 to the present day.

The temperature may be averaged over a month, a season, a year and so on. It is the relative changes in these averages that matter in understanding how our climate is evolving. This is why the figures in the following chapters do not display absolute temperature values, but rather the *temperature anomaly*; that is, the deviation from the average temperature over a reference period, generally taken as 1961–90. The temperature anomaly has been reconstructed back to 1850 with an uncertainty that has decreased from ±0.2°C to ±0.1°C.

Figure 25.1 (left) plots the change with time of the global mean surface temperature for three sets of surface data. Since the middle of the 19th century, it has risen by approximately one degree. Since the 1970s/1980s, the increase is particularly marked and sustained, and in the past four decades the global mean surface temperature has risen by more than 0.5°C. Over the 20th century, the change took place in two stages, with a break from the 1950s to the 1970s:

- A first stage between the 1920s and the 1940s; this temperature increase was most pronounced in the Northern Hemisphere (NH), particularly over the Atlantic Ocean, and was much weaker in the SH (Fig. 25.1, right).
- A second stage starting in the 1970s and 1980s, with a pronounced increase in both hemispheres.

Figure 25.1 specifies the contribution of each hemisphere to global warming. Both curves look very similar, with one slight difference: in the SH, the warming is less pronounced than in the NH. This reflects the fact that oceans occupy a larger surface area in the SH and, in particular, that the oceans warm more slowly than land.

Another way of describing the change in mean temperature is by distinguishing night-time and daytime temperatures. Figure 25.2 shows the change in the number of cold nights and days and warm nights and days between 1951 and 2003. As expected, the number of cold nights and days has diminished, whereas the number of warm nights and days has increased.

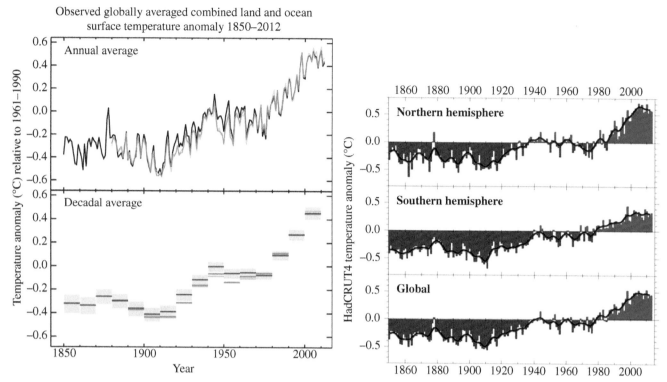

Fig. 25.1 Left: Observed global mean combined land and ocean surface temperature anomalies, from 1850 to 2012 from three data sets. Top panel: annual mean values. Bottom panel: decadal mean values including the estimates of uncertainty for one data set (black). Anomalies are relative to the mean of 1961–90. Right: the observed trend in the mean annual surface temperature with respect to the 1961–90 average: (a) Northern Hemisphere; (b) Southern Hemisphere; and (c) global.

Sources: Left, IPCC 2013. *Climate Change 2013: The Physical Science Basis.* Working Group I Contribution to the Fifth Assessment Report of the Intergovernmental Panel on Climate Change, Fig. SPM 1a. Cambridge University Press. Right, Climatic Research Unit, University of East Anglia.

The remarkable consistency between these four independent curves is a vivid illustration of the climate shift that has occurred, as revealed by the decrease in cold episodes and the increase in warm episodes. *How has the observed global warming affected all the different regions? Is the warming random or is it correlated?*

Land versus ocean, high versus low latitudes

To define the changes of temperature in the various regions of the world over the 20th century, the average temperature change between a chosen starting year and 2011 is assessed for each region. The resulting map indicates whether temperatures in that region have, on average, fallen or risen over that period. IPCC 2013 includes two types of maps, as illustrated by Figure 25.3: one plots changes between 1901 and 2012, and the other changes between 1981 and 2012. Two features stand out. Warming is:

- more pronounced over land than over sea; and
- more pronounced at higher latitudes in the NH than at lower latitudes.

Fig. 25.2 Trends in the frequency of (a) cold nights, (b) cold days, (c) warm nights and (d) warm days between 1951 and 2003. Only regions with at least 40 years of measurement records are included (coloured regions) and with data at least until 1999. Black lines enclose regions where trends are significant at the 5% level. The units are number of days per decade, relative to the 1961–90 average. Warm days (or nights) are defined as belonging to the warmest 10%, and, conversely, cold days (or nights) belong to the coldest 10%.

Source: IPCC 2007. *Climate Change 2007*: The Physical Science Basis. Working Group I Contribution to the Fourth Assessment Report of the Intergovernmental Panel on Climate Change, FAQ3.3 Fig. 1. Cambridge University Press.

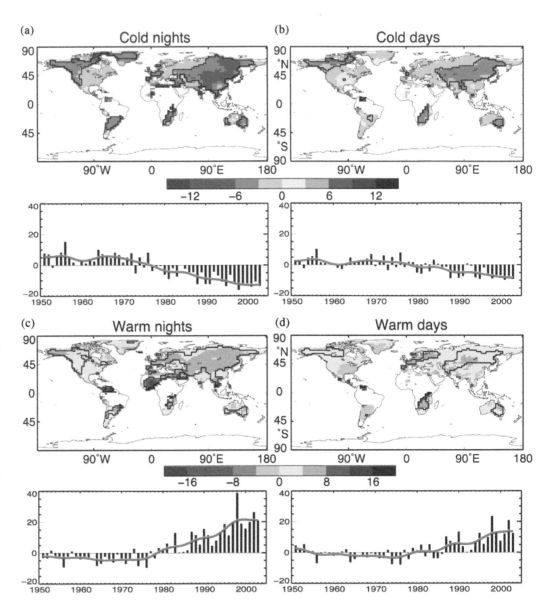

This behaviour is consistent with what is observed with major climate changes in the past (Fig. 18.3): continents warm (or cool) more than the oceans, and in the NH high latitudes are more affected than low latitudes.

The change of the atmospheric temperature with altitude

Independent research groups have estimated how the air temperature has changed at various altitudes between 1979 and 2010. Their estimations are in agreement and reveal two opposing trends: while the troposphere is warming (by about +0.1°C per decade), the lower stratosphere is cooling at a rate five times faster (by about –0.5°C per decade). This

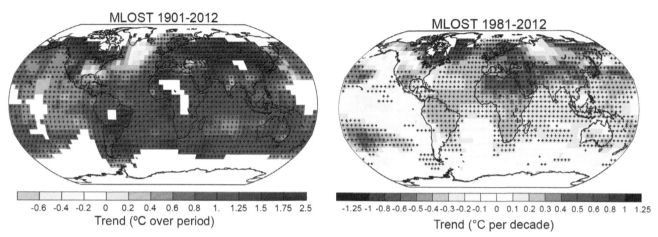

Fig. 25.3 Trends in global mean surface temperature from the NCDC MLOST data set for 1901–2012 (left) and 1981–2012 (right). Trends are calculated only for grid boxes with greater than 70% complete records. Zones with insufficient data displayed in white. Grid boxes where the trend is of higher statistical significance are denoted by +. Interpolation is used in this map.

Source: IPCC 2013. *Climate Change 2013: The Physical Science Basis.* Working Group I Contribution to the Fifth Assessment Report of the Intergovernmental Panel on Climate Change, Figures 2.21 and 2.22. Cambridge University Press.

can be seen in Fig. 23.2, which shows the change in air temperatures at different altitudes over a longer period. The baseline shows a regular downward trend in the stratosphere (average slope −0.5°C per decade) and an upward trend in each layer of the troposphere (average slope +0.1°C per decade). These opposing trends appear to be consistent with an increased greenhouse effect. Since the year 2000, this downward trend has halted, for reasons that are still being debated. It is believed to be related to the pause in temperature rise in the troposphere (see Chapter 29), and is probably the result of coupling between the stratosphere and the troposphere via water vapour, which has a very strong radiative effect in the stratosphere.

25.1.2 MEASURING THE GLOBAL MEAN TEMPERATURE: IMPLAUSIBLE ACCURACY?

The uncertainty in the measurement of the global mean temperature is of the order of ±0.2°C for the 1860s and ±0.1°C at the end of the 20th century. Such a high precision may appear surprising, given that commercial thermometers often exhibit much larger errors. One simple observation, however, reassures us. Let us turn back to Fig. 25.1. We saw in Part III that fluctuations in annual global temperatures are punctuated by strong El Niño events that can warm the planet by a few tenths of a degree. The two warming episodes of 1878 and 1998, during which temperatures increased by about 0.2–0.3°C, are clearly visible in Fig. 25.1. These are not random fluctuations, but the consequence of the two strongest El Niño events since 1850: the fact that they appear so distinctly in the global mean temperature is confirmation that the uncertainty in the yearly measurement is less than a few tenths of a degree.

25.1.3 HOW DO THE DIFFERENT COMPONENTS OF THE EARTH'S CLIMATE SYSTEM WARM?

All the data consistently show that temperatures have, on average, increased over the past few decades. We saw that the atmosphere has warmed, as have the land and the oceans. This means that the 'energy' reservoir available at the Earth's surface (as determined by its average temperature) has increased. We have also seen that the cryosphere, the vast reservoir of ice on the Earth, has partly melted, which also involves an increase of energy. In brief, the increase in energy warms the ocean, the atmosphere and the continents, and melts part of the cryosphere. How exactly is this extra energy distributed?

Figure 25.4 shows the distribution of this energy increase between 1971 and 2010 (IPCC 2013). Ocean warming dominates, accounting for roughly 93% of the total, with the upper

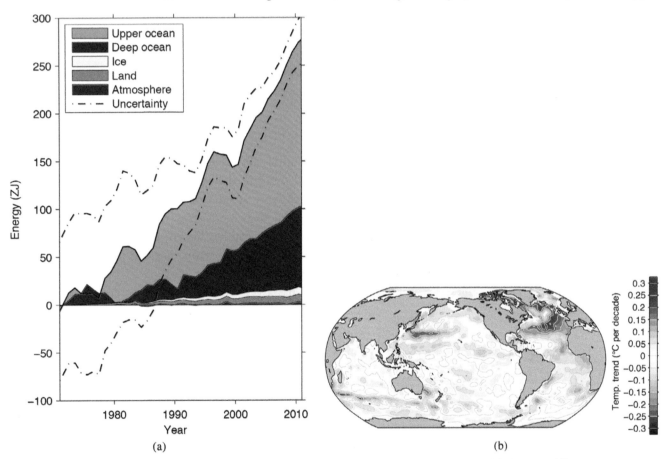

(a)   (b)

Fig. 25.4 (a) A plot of energy accumulation (in ZJ) within distinct components of the Earth from 1971 to 2010 (1 ZJ = 10^{21} Joules). The increase in energy warms the ocean, atmosphere and continents and melts part of the cryosphere. Ocean warming dominates, with the upper ocean (light blue, above 700 m) contributing more than the deep ocean (dark blue, below 700 m). Contributions from ice melt, land and atmospheric warming are much smaller. (b) Trends in ocean temperature for 1971–2010 (colours and grey contours in °C/decade) averaged between the surface and 700 m depth. The greatest warming is in the North Atlantic Ocean, in some places reaching 0.3°C/decade. With ice sheet estimates starting from 1992, arctic sea ice from 1979 to 2008, and atmospheric warming from 1979.

Source: IPCC 2013. *Climate Change 2013: The Physical Science Basis*. Working Group I Contribution to the Fifth Assessment Report of the Intergovernmental Panel on Climate Change, Box 3.1 Figure 1, and Figure 3.1. Cambridge University Press.

ocean (above 700 m depth), contributing more than the deep ocean (below 700 m). Melting ice accounts for 3%, and warming of the continents and atmosphere 3% and 1%, respectively. The distribution of ocean warming, shown in Fig. 25.4b for the upper ocean, is not uniform around the Earth. It depends on ocean circulation. Sinking ocean water was shown diagrammatically in Fig. 5.11, and explains why the greatest heating takes place in the North Atlantic region, where deep-water formation occurs. The ocean component for 1971–2010 is equivalent to a global mean energy flux of $0.55\,W/m^2$ transferred from the atmosphere to the sea. For the more recent period from 1993 to 2010, this component rises to $0.71\,W/m^2$ (shown as $\sim 1\,W/m^2$ in Fig. 9.4). Not only has this accumulation of energy increased with time, but its rate of storage has accelerated: the flux of energy that warms the ocean has increased from $0.55\,W/m^2$, when averaged over the past four decades, to $0.71\,W/m^2$ over the past 17 years. These estimations highlight the importance of the role played by the oceans in climate change.

25.2 Changes in precipitation, water vapour and extreme events

25.2.1 CHANGES IN PRECIPITATION

Precipitation (both snow and rainfall) falls in extremely variable amounts over the Earth's surface. It ranges from almost zero in certain deserts to more than 15 m/yr in some tropical areas. Moreover, precipitation can fluctuate irregularly. Estimating changes in global precipitation on the basis of local observations is therefore an even more challenging and complex task than changes in temperature.

IPCC 2013 shows the change in annual precipitation from 1900 to 2012, averaged over the globe (Fig. 25.5), as well as over land areas in four latitudinal bands. The conclusion of this report is that 'available globally incomplete records show mixed and non significant long term trends in reported global mean changes'. The report also includes figures for changes in annual precipitation over 19 large continental areas, from north to south. Here again, no single general trend emerges, and contrasting changes appear in various areas over the 20th century. Any noticeable trends depend on the latitude: 'Global satellite observation for the 1979–2008 period and land-based gauge measurements for the 1950–1999 period indicate that precipitation has increased over wet regions of the tropics and NH mid-latitudes, and decreased over dry regions of the sub-tropics. Such patterns of precipitation change are consistent with that expected in response to the observed increase in the tropospheric specific humidity' (IPCC 2013).

Fig. 25.5 Variation in annual precipitation averaged over the globe from five global precipitation data sets with respect to the 1981–2000 climatology. **Source:** IPCC 2013. *Climate Change 2013: The Physical Science Basis.* Working Group I Contribution to the Fifth Assessment Report of the Intergovernmental Panel on Climate Change, Figure 2.28. Cambridge University Press.

25.2.2 CHANGES IN ATMOSPHERIC WATER VAPOUR

Satellite observations of the water vapour content of the lower atmosphere over the oceans indicate considerable variability, which is linked to the sea surface temperature. Thus, strong El Niño currents have a

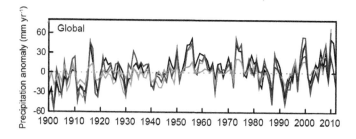

particularly high impact when the surface of the tropical Pacific Ocean is at its warmest. Temporary warming injects more water vapour into the atmosphere, as was observed during the strongest El Niño of the 20th century in 1997–8, when the water vapour content increased by nearly 4%. Despite the great variability of this signal, there is a slight long-term trend towards higher water vapour content (as indicated by radiosonde, GPS and satellite observations), since warmer air can contain more water vapour before reaching saturation. An increase in temperature therefore makes the water cycle more active (more water vapour in the atmosphere, more rain). We return to this point in Chapter 27, with respect to changes in salinity of ocean surface waters.

25.2.3 CHANGES IN 'EXTREME' EVENTS

Climate variables such as temperature, rain or wind can sometimes reach unusual values and produce natural catastrophes. These are dubbed 'extreme' events. Their definition in climatology is precise: they are the 10% least intense and 10% most intense events observed. They are relatively unlikely, yet more frequent than the popular view would suggest.

When such events occur in populated areas, they can cause human catastrophes. It should be said here that the cause of these catastrophes must be carefully analysed and a distinction made between what is due to the extreme climate event and what is due to our own management of the environment. *All too often, poor management amplifies the effects of an event that is not in itself extreme.* Flooding, for instance, is often due not so much to extreme rainfall as to a change in the use of the land and embankments, building in high-risk areas, soil waterproofing or other practices that lead to increased soil erosion. There is sometimes a bias in cause assessment that consists in blaming destruction and accidents on climate change rather than on recent environmental management, which can amplify the damage caused by climate events.

That said, let us return to changes in extreme events. These include cyclones, torrential rainfall, storms, heat waves, abnormally cold (or warm) winters, prolonged droughts and so on. We only touch on some of these.

More frequent tropical cyclones?

Tropical cyclones develop in the tropical areas of the Earth's three oceans (Pacific, Indian and Atlantic). They originate in the eastern parts of these oceans when the top 50 m of the ocean warms to at least 27°C. On moving west (driven by atmospheric circulation), their energy increases, fed by the ocean waters. When they touch land, this energy source vanishes and they lose power. The temperature of the water is thus one of the key factors in generating cyclones, which is illustrated by the fact that they are more frequent in the tropics of the NH than in the south (Fig. 5.6). Cyclone intensity is measured on a scale from 1 to 5, where 4 and 5 are the most intense cyclones. Since surface-water temperatures have increased over the past few decades in each of the three tropical oceans, the question arises as to whether this has had an impact on cyclone frequency or intensity. A study by Webster et al. (2005) analyses the number and intensity of tropical cyclones between 1970 and 2004. While the number of cyclones in the first three categories has barely changed, the number of force 4 and 5 cyclones has almost doubled. These observations appear to confirm the idea that recent warming has reinforced tropical cyclone activity. Finally, Katrina, the notorious cyclone that devastated the New Orleans (USA) area in 2005, which is often cited as an extreme climate event, helped to increase awareness in the United States that profoundly destructive events could become more frequent there due to global warming. The more recent typhoon Haiyan, which struck the Philippines in November 2013, is one of the most devastating tropical storms of the 20th and 21st centuries to date.

Record heat waves

An important change in the most recent decades is the emergence of a new category of so-called 'extremely hot' summers and *heat waves*, such as those of 2003 in Western Europe, 2009 in Australia, 2010 in Western Russia, 2011 in Texas and Oklahoma and 2013 in China. *How does a heat wave happen?* Take, for example, the heat wave in the summer of 2003, which affected much of Western Europe (Fig. 25.6). In France, the average summer temperature was approximately 19°C over the period from 1983 to 2002, and increased to 22–23°C in 2003 during this dramatic event. Its highly unusual character is illustrated in Fig. 25.7, which shows summer temperatures in Switzerland from 1863 to 2003. The average temperature in the summer of 2003 was approximately 22°C, much higher than the average temperature for the period from 1863 to 2003, which was about 17°C. The distribution of summer temperatures over this whole period is approximated by a Gaussian curve (the green line). In such a distribution, 66.6% of the events occur between $-\sigma$, and $+\sigma$, where σ is the standard deviation. In a warmer climate, the curve shifts towards a higher mean temperature and the probability of hot events increases. Since the beginning of the 20th century, the climate has warmed by about 1.5–2.0°C (in France, the summer temperature has increased from about 17°C to 19°C), which accounts for the higher frequency of hot summers and heat waves.

How did the heat wave affect the normal course of the summer? First, by its impact on the population. It was the cause of about 15,000 extra deaths in France (Rousseau 2006; Fig. 25.8) and some 70,000 extra deaths throughout Europe. In addition to its dramatic human consequences, it affected the soil and vegetation. The unusual warming, which went hand-in-hand with a drought, reduced agricultural output in France by approximately 20% for wheat, 30% for maize and 60% for animal feed. Vegetable production fell by 30%, an unprecedented drop since the beginning of the 20th century. It also released vast quantities of CO_2 from the soil in Western Europe (estimates vary from 0.2 to 0.5 *gigatonnes of carbon* (GtC) (Part V, Note 1), whereas at that time of year, owing to photosynthesis, soils are usually a sink for atmospheric carbon. These values should be compared with Europe's emissions from fossil fuels, about +1 GtC/yr, and with the amount normally stored each year in the soil (which acts as a sink), about −0.2 GtC/yr. All this suggests that, in the long run, plant productivity in temperate Europe may fall if such climate extremes (heat waves and droughts) become more frequent. Such a scenario is expected if greenhouse gas emissions from human activities continue their present rate of increase.

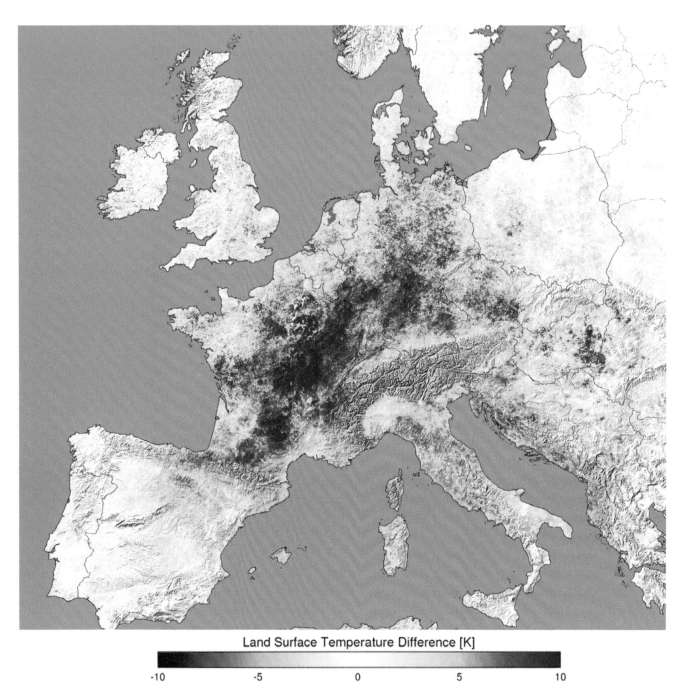

Fig. 25.6 The surface temperature anomaly over Western Europe in summer 2003, during the period 20 July to 20 August. The anomaly is calculated by subtracting from the 2003 measurements the average of observations made during cloudless days in 2000, 2001, 2002 and 2004.

Source: Reproduced with permission of Reto Stöckli.

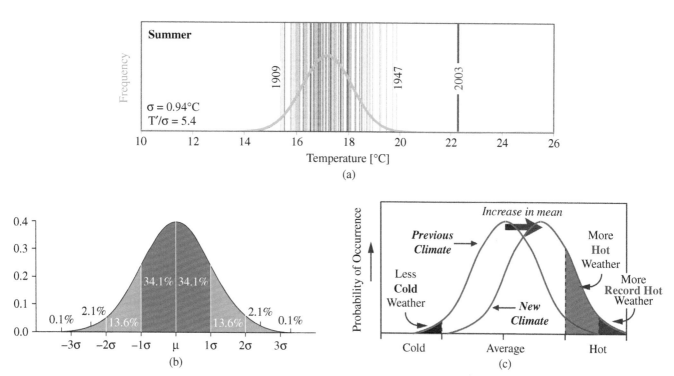

Fig. 25.7 (a) Summer temperatures in Switzerland between 1864 and 2003, on average 17°C, with fit to a Gaussian distribution (green curve). The extreme years in the record were 1909, 1947 and 2003. The standard deviation σ for 1864 to 2000, and the 2003 anomaly T'/σ, are indicated on the left (Schär et al. 2004). (b) The distribution (or probability of occurrence) of the temperature is approximated by a normalized Gaussian distribution (area equal to 1); μ, mean value. (c) A schematic showing the effect on the extreme temperatures when the mean temperature increases, for a normal temperature distribution. A shift towards a warmer climate brings less frequent extreme cold weather and more frequent extreme hot weather.

Sources: (a) and (c) from IPCC 2007. *Climate Change 2007: The Physical Science Basis*. Working Group I Contribution to the Fourth Assessment Report of the Intergovernmental Panel on Climate Change, FAQ 9.1 Figure 1 and Box TS.5 Figure 1. Cambridge University Press.

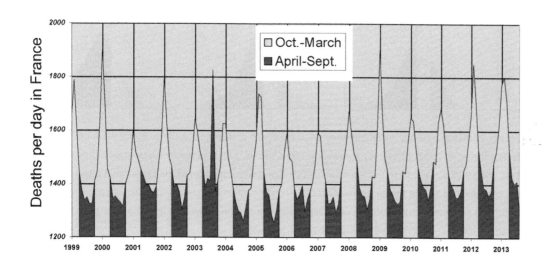

Fig. 25.8 Seasonal mortality (deaths/day) in France. The European heat wave of summer 2003 caused about 15,000 additional deaths in France, and more than 70,000 across Europe as a whole.
Source: Rousseau (2006). Reproduced with permission of D. Rousseau.

Changes in heat waves and extreme temperatures?

A quantitative study by J. Hansen (Hansen et al. 2012, 2013a) examines extreme climate events and their evolution since 1950. The authors investigated the distribution of average temperatures for each decade between 1951 and 2010 in specific regions of the world (250 km squares), for the three summer months and the three winter months. The study thus reviews the changes in local temperatures. For any given decade, the distribution forms a bell-shaped curve around an average value that can be approximated by a normal (Gaussian) distribution. The reference period is 1951–80, for which the local mean and the standard deviation σ are calculated. For each decade, the authors trace the difference between local temperatures and the reference average, divided by the standard deviation σ, thus obtaining a characteristic temperature distribution for that decade. They then compare the temperature histograms for the different decades by normalizing them with respect to the reference period parameters. In the absence of climate variations, the histograms for each decade should be constant and should superimpose on each other.

The study shows that the distributions for the first two decades do superimpose (Fig. 25.9), but a shift to the right occurs from the third decade onwards, and detaches itself quite clearly in the fourth, fifth and sixth decades. The local average temperature over the most recent decade (2000-10) has shifted by 1σ compared to the initial decade (1951–60). This observation is in line with the increase in average temperatures observed during the past few decades. Furthermore, the temperature distribution is gradually becoming broader, a trend that increases the probability of extreme events, because of both the average displacement and the greater width of the distribution. The most important change probably lies in the emergence of a new category of 'extremely hot' summers, where temperatures are more than 3σ higher than the average summer temperature of the period from 1951 to 1980. In those first three decades, during which the temperature distribution nearly coincides with the reference curve (shown in black in Fig. 25.9), such events are almost undetectable, their probability being close to 0.15%. In the most recent decade, however, because the curve has shifted by about 1σ, these same events only lie beyond 2σ, rather than beyond 3σ. The probability of their occurrence

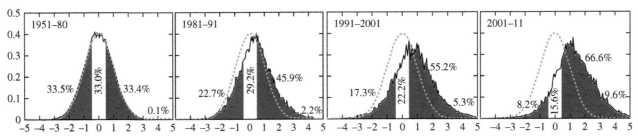

Fig. 25.9 The change in the frequency of local land June-August temperature anomalies (with respect to the 1951–80 mean) in the Northern Hemisphere in the past six decades (Hansen et al. 2012; Hansen et al. 2013a). The horizontal axis is in units of local standard deviation, σ. Temperature anomalies between 1951 and 1980 closely match the Gaussian distribution (green curve), which describes cold (blue), typical (white) and warm (red) seasons, each of probability 1/3. In the past three decades, the distribution has shifted towards higher temperatures, mirroring the regular increase in the global mean temperature. Extreme warm anomalies (more than 3σ warmer than the base-period mean) have increased from less than 1% to about 10%.

Source: Hansen et al. (2013a). Reproduced by permission of Makiko Sato.

has therefore risen to 2.5%. The authors thus conclude that over the past 10 years, these events have become more than ten times more likely than in the period from 1950 to 1980.

The changes in temperature over the past six decades clearly show how 'extremely hot' summers have become increasingly frequent, as a logical consequence of the upward shift in the distribution curve of the average temperature.

Storms and floods

Over the past few decades, various storms, extreme precipitation events and floods have struck the popular imagination on the different continents. Although media coverage of these extreme events may have led people to believe that they have become more frequent, the statistics show no significant increase in storm activity to date. The question is still topical. For floods and extreme precipitation, different trends emerge in different parts of the globe. With regard to flooding, however, it is important to re-emphasize that land development very often remains the principal cause of its increase.

25.3 An overview of the past few decades

Of the three parameters chosen to describe climate (temperature, rainfall and winds), over the past four decades only temperature exhibits a clear increase almost everywhere on the planet. This change has three features:

- a regularly increasing average;
- a consistent pattern of warming over the whole of the surface of the planet, as in all major climate changes – stronger on land than in the oceans, and greater in the higher NH latitudes than in the tropics; and
- changes in the extremes – more frequent hot events and fewer cold events.

The water cycle appears to have intensified slightly owing to increasing water vapour in the atmosphere. This is perfectly consistent with the increase in average temperature. It is confirmed by the change in salinity of ocean water masses, described in Chapter 27.

The agreement among all these observations justifies the notion of global warming over recent decades.

25.4 The impact of global warming: the key issue

The recent warming of our planet has brought about many changes to the environment. We concentrate on these in the next chapters, for two reasons. First, we identify the numerous environmental repercussions from changes in the temperature. Second, analysis of the observed facts gives a better understanding of the consequences of any future warming.

Climate modelling tells us that unless societies change their behaviour drastically, our climate in the 21st century will change far more than over the past few decades. It is by the yardstick of its impact on life in general (or, more exactly, on all the physical conditions in which life is sustained) that the magnitude of this change will be judged. To answer these questions, it is helpful to survey to what extent the recent warming, albeit still relatively modest, has already disrupted the world in which we live. Climate models are instructive. Observations of the recent past, however, not only complement them, but also provide irrefutable evidence. As we shall see in Part VI, they simply announce what the simulations predict. In the following chapters, we confine ourselves to a few observations relating to ice and oceans; we also note a few of the ways in which the biosphere has changed.

Chapter 26
The impact of global warming on the cryosphere

C hanges in sea ice, permafrost, the duration of seasonal snow and ice cover, glacier extension and ice sheets all illustrate the ongoing warming of the planet, with some nuances.

26.1 Sea ice, the 'canary' of our planet

26.1.1 SEA ICE

Sea ice forms in the winter when the seawater cools and freezes to an average depth of 3 m. On freezing, seawater releases salt, increasing the salinity of the underlying water column, which becomes denser and therefore sinks. We have seen that in certain regions, mostly in the North Atlantic, this mechanism contributes to the deep ocean circulation (*thermohaline circulation*, Chapter 5). In the Arctic Ocean, sea ice persists throughout the summer along the coastline of North America and Greenland, where the ice accumulates through the effect of currents. In this region, the lifetime of the sea ice can be as much as 10 years, and its thickness can be greater than 5 m. In summer, two mechanisms cause its partial disappearance: local warming of the ocean and atmosphere, and the influence of wind and ocean currents. The latter break up the thinning sea ice, which drifts to lower latitudes, where it melts faster. Any decrease of sea ice in the summer is due either to melting or to ice floes being carried away by the East Greenland current.

Climate Change: Past, Present and Future, First Edition. Marie-Antoinette Mélières and Chloé Maréchal.
© 2015 John Wiley & Sons, Ltd. Published 2015 by John Wiley & Sons, Ltd.
Companion website: www.wiley.com\go\melieres\climatechange

In the first half of the 20th century, the sea ice in winter covered the entire Arctic Ocean, some 16 million km^2. Over the three summer months, it would shrink to 11 million km^2. In the Southern Ocean, by contrast, while sea ice reached latitudes comparable to those in the Northern Hemisphere (NH) and covered up to 19 million km^2, in summer it would dwindle to 3 million km^2 along the coastline of Antarctica. Obviously, this dissymmetry is directly related to the geographical differences between the two hemispheres: in the north, the polar region is occupied by an ocean that freezes in the winter, whereas in the south, it is almost wholly occupied by a continent.

26.1.2 THE RETREAT OF ARCTIC SEA ICE: WHY IS IT SO IMPORTANT?

The state of the Arctic sea ice is a reliable indicator of global climate change. Its role is sometimes compared to that of a canary in a coal mine. If the canary stops singing and dies, toxic gases have seeped into the tunnel and the miners must immediately evacuate the mine! Our present danger, however, is not of an explosion, but instead that we are en route to global warming, and is illustrated by the large-scale and alarmingly fast retreat of the summer sea ice. The shrinking ice pack is the signal of the global climate change, but amplified, just as all global warming is amplified in the higher latitudes of the NH.

26.1.3 SEA ICE IN THE ARCTIC OCEAN OVER THE 20TH CENTURY

The reconstruction of the extent of the Arctic sea ice since the beginning of the 20th century shows that in the first half of the century the ice surface area did not change. The change in area averaged over each season (a 3-month average) shows that the shrinkage of the ice floe in autumn and winter started in the 1980s. In spring and summer, it had begun to recede 20 years earlier, in the 1960s. Figure 26.1 illustrates its decrease in area between 1979 and 2011, for September, when it is least extended.

Figure 26.2 shows the change in surface area since the early 1980s. Over that period, the maximum has decreased by about 1.5 million km^2, while its minimum has recently fallen from 5.5 to 3 million km^2. The summer of 2007, with its loss of nearly 1 million km^2 of sea ice with respect to the previous summer, appears outstanding. This melting was the result of exceptional weather conditions, with very high temperatures, especially in the Beaufort Sea. But this example was followed a few years later, with the further loss of almost a million square kilometres in the summer of 2012. In the most recent decades, the change has had a radical impact on the age (Maslanik et al. 2011) and thickness of the ice floe (Figs. 26.3 and 26.4): it is now younger and slimmer (which could make some envious!). Figure 26.3 illustrates how the age of the ice floe at the end of winter changed between 1985–2000 and 2011. Ice less than a year old – that is, formed over the most recent winter – occupies more and more of the area at the expense of older ice, which gradually disappears. At present, only a very small portion of the sea ice is older than 6 years, and first-year ice has become the dominant type in the Arctic. Similarly, the thickness of the sea ice has reduced, with a sharp decline in the area occupied by ice at least 5 m thick.

Fig. 26.1 The cryosphere in the Northern and Southern Hemispheres. The map of the Northern Hemisphere shows the sea ice cover during its minimum summer extent (13 September 2012). Yellow line: average location of the ice edge for the yearly minima from 1979 to 2012. Dark pink: areas of continuous permafrost; light pink: discontinuous permafrost. Green line: maximum snow extent; black line: contour for the 50% snow extent. The map of the Southern Hemisphere shows approximately the maximum sea ice cover during an austral winter (13 September 2012). Yellow line: average location of the maximum sea ice cover from 1979 to 2012. **Source:** IPCC 2013. *Climate Change 2013: The Physical Science Basis.* Working Group I Contribution to the Fifth Assessment Report of the Intergovernmental Panel on Climate Change, Figure 4.1. Cambridge University Press.

The reduction of sea ice in the NH over the past few decades may be due to various processes, such as warming of the Arctic Ocean, the influx of warm waters from the North Atlantic current into the Arctic and changes in the wind pattern. Ongoing melting creates positive feedback through the decrease in albedo: in summer, when the solar flux is strongest, the decreasing area of white sea ice reduces the reflection of the solar radiation and thus increases the warming of ocean water.

26.1.4 A COMPARABLE CHANGE IN THE SOUTHERN HEMISPHERE?

What information can be found in the Southern Hemisphere? There is hardly any change. Year-to-year variations are due to the strong influence of El Niño events, but the area of sea ice appears, on average, to be stable: no clear trend is discernible. Once again, this difference between north and south reflects the different configurations of the hemispheres at higher latitudes. The NH sea ice is surrounded by land and is influenced by their warming, which is always greater than that of the oceans. In the SH, the situation is completely different. The sea ice is surrounded on one side by a vast ocean, whose temperature at these latitudes

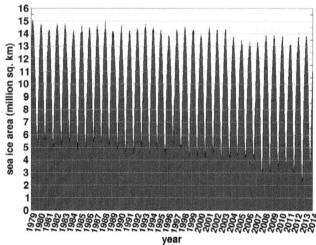

Fig. 26.2 The change in the area of sea ice in the Northern Hemisphere since 1979.
Source: Data provided by NSIDC: NASA SMMR and SSMI.

Fig. 26.3 The age of ice floes in the first week of March from 1988 to 2010. The multiyear sea-ice extent is continually decreasing.
Sources: Courtesy of National Snow and Ice Data Center, J. Maslanik and C. Fowler.

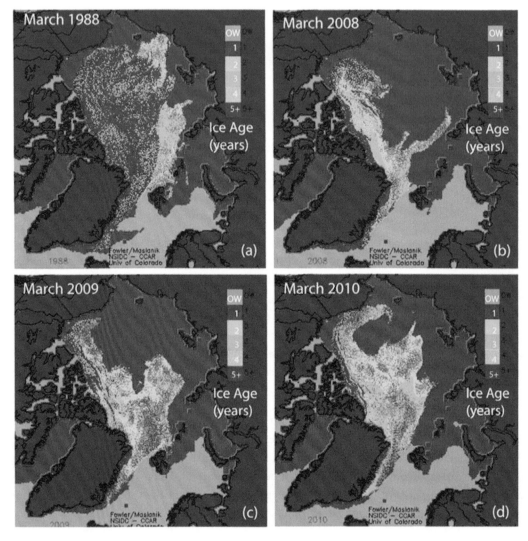

Fig. 26.4 Distribution of sea ice thickness in winter in the Arctic and the trends in average, first-year (FY) and multiyear (MY) ice thickness. Between 2004 and 2008 the average ice thickness decreased from around 3 to 2.5 m.
Source: IPCC 2013. *Climate Change 2013: The Physical Science Basis.* Working Group I Contribution to the Fifth Assessment Report of the Intergovernmental Panel on Climate Change, Figure 4.5. Cambridge University Press.

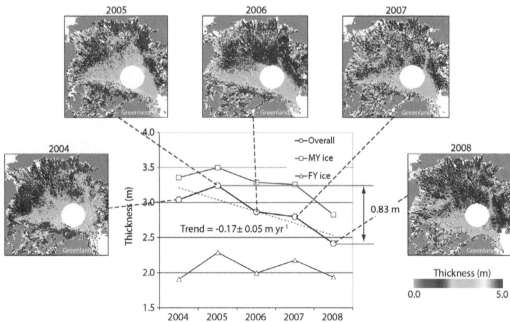

remains practically stable, and on its southern side by a continent covered by a permanent ice sheet.

26.2 Changes in glaciers

It is easy to see by eye whether a glacier is advancing or retreating over the years. *But what are the parameters that define its motion? Is it the position of its front, as recorded by the moraines it leaves behind, or by old maps and photos, historical archives or other more sophisticated techniques used in palaeoclimatology? Or its surface area?* Such evidence is valuable, but those estimates are not related directly to climate change; they depend on the particular shape of the glacier, its slope, its exposure and so on, which vary from one glacier to another. Above all, they hold the memory of past climate changes. Far more meaningful is the mass balance; that is, the annual variation of the mass of the glacier. This is the difference between the mass of ice accumulated and that lost by the glacier during the year. Nevertheless, because it takes far longer to measure the mass balance than the position of a glacier front, mass balance measurements are available for only a few glaciers, and only over recent times.

26.2.1 THE RELATIONSHIP BETWEEN GLACIERS AND CLIMATE

Unlike freeze-up (the annual freezing of water at the surface of lakes or rivers) and snow cover (annual snowfall), the dynamics of a glacier are not a direct response to the year's climate. While freeze-up and snow cover start each year from an unchanging initial situation (the absence of ice or snow) and therefore depend directly on the climate conditions of the given year, changes in the length of the glacier are the sum of past climate conditions. It follows that interpretation of the length of a glacier in terms of climate change is more complex. By contrast, the mass balance of a glacier, which is the resultant of snow accumulation and ice loss during the year, is directly related to the climate conditions of that year.

The mass balance of a glacier combines two independent factors: snowfall, which feeds the glacier during the winter, and energy income (governed by temperature and solar flux), which melts the ice during the summer. A glacier is defined by its *equilibrium-line altitude* (ELA), a line that divides the glacier into two zones. In the upper, or accumulation, zone, ice accumulates on average over the year. In the lower, or ablation, zone, ice disappears each year. In the middle latitudes, this disappearance is due to melting. In the tropics, melting is reinforced by sublimation (direct transformation of water from solid to vapour). At high latitudes, the ice flows directly into the open water of the ocean. A glacier front does not move when the mass balance of the glacier is zero; that is, when the mass of accumulated ice is equal to that lost. Despite the continuous and slow sliding motion of the whole glacier, the total amount of ice remains constant: the front is stationary and the glacier remains in equilibrium. But when the climate changes, the glacier evolves towards a new equilibrium.

To simplify, let us imagine an abrupt climate warming: the new climate is stable, with no transition period. The ELA now shifts to a higher altitude, and remains there; that is, the altitude at which the ice melts is now greater. The area of the accumulation zone is thus reduced, and that of the ablation zone increases. The glacier mass balance is now negative and the glacier loses ice – its front recedes and increases in altitude. With time, as the front rises, the new ablation zone shrinks, until balance is re-established with the mass of ice that has accumulated above the new ELA position. The annual balance then reverts to zero, and the position of the glacier front becomes stable once again. In the new climate it takes many years, decades or even centuries for the glacier to reach its new equilibrium.

Where is the ELA now? This obviously depends on the region: it is higher in warmer regions (at the Equator and in the tropics) and gradually becoming lower towards the poles. Within the same mountain range, it may vary by several hundred metres, depending on the orientation of the glacier and on the year. It is nevertheless possible to estimate its average value. At high latitudes it is close to sea level, while at the Equator, where very few glaciers subsist, it is much higher. In Irian Jaya (Papua New Guinea), the ELA is situated above 5000 m, while the island's tallest mountain culminates at 4884 m. All the glaciers in that region, which at the beginning of the century covered popular climbing areas, are thus located in the ablation zone and have receded at an extraordinary rate over the past few decades. All are doomed to disappear in the very near future, along with their ice records. In the Tropical Andes, the ELA is situated at 5100–5300 m: since the highest peaks rise to more than 6000 m, an accumulation zone survives. In Europe, in the Alps and Pyrenees, the ELA lies on average at around 3000 m. In the Alps (with summits above 4000 m), an accumulation zone remains, but not in the Pyrenees, where the highest summits reach only 3000 m. There, the glaciers are receding inexorably, from 40 km² at the beginning of the 20th century to 5 km² at the end of the century. In the Himalayas, the ELA varies between 5500 m in the north-west and 4800 m in the south-east.

26.2.2 OBSERVATIONS OF CHANGES IN GLACIERS

Over the 20th century, from the tropics to high latitudes, almost all glaciers have receded. We shall concentrate first on the middle latitudes, taking as an example the European Alps.

In the Alps, the mass balance has been estimated for six glaciers since 1950 (Vincent et al. 2005; Fig. 26.5). The data are normalized to allow comparison among the glaciers. The synchronization between the signals is remarkable. The six glaciers behave similarly, despite being located in areas 400 km apart, which implies that the Alps share a common climate, with only minor variations. *How has this common climate changed?* The retreat of Alpine glaciers in the 20th century was part of a worldwide process. In the Alps, it is estimated that this process started between 1820 and 1850 (Fig. 22.13). *What happened in the 20th century?* The changes can be expressed most succinctly and vividly by simply considering the thickness of these glaciers. On average, the thickness of Alpine glaciers shrank by ~30 cm/yr between 1850 and 1970–80 (this is the average thickness of the layer of ice lost, in units of liquid water). The average shrinkage rate accelerated to ~40 cm/yr in the 1980s, to ~80 cm/yr in the 1990s, and to ~1 m/yr or more since the beginning of the 21st century (Haeberli et al. 2007). In 2003, during the heat wave in Western Europe, the average loss amounted to 3 m. This shrinkage can be entirely attributed to the particularly warm summer of 2003, as the snow accumulation in the previous winter was comparable to that of a normal winter.

On the global scale, the ice loss is just as unmistakeable. Figure 26.6 recapitulates the mass balance since 1953 of 75 glaciers, located in 15 different regions. The general trend observed

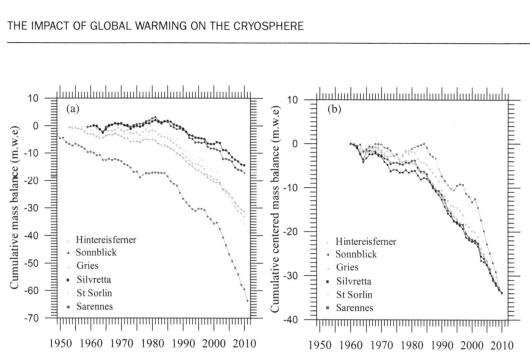

Fig. 26.5 (a): the variation in cumulative mass balance of six glaciers in the French, Austrian and Swiss Alps from 1950 to 2011. Over this period, the decrease in the St Sorlin glacier, for example, is equivalent to a layer of water 39 m deep, covering the whole of the glacier. (b): the balance difference of each glacier is normalized by the average loss of the six glaciers (34 m). The time variation is then comparable among the six glaciers, even though the most distant are separated by more than 400 km. This reflects a widespread change in climate on the scale of the Alpine mountain range.

Source: Vincent et al. 2004. Reproduced with permission of C. Vincent.

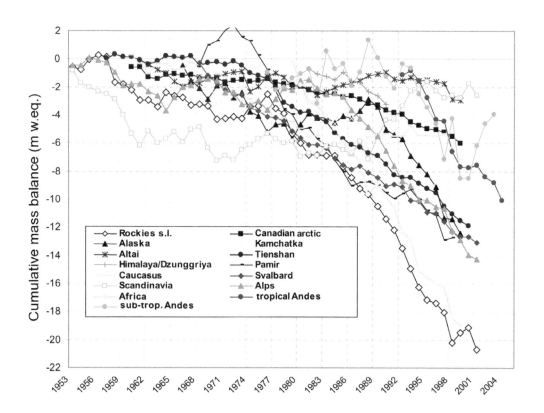

Fig. 26.6 The variation since 1953 of the average cumulative mass balance in 75 mountain glaciers of 15 large regions in the world for which information is reported regularly. The balance is almost always negative, indicating the general retreat of the world's glaciers. This retreat has accelerated since the 1980s.

Sources: Compilation by Ohmura (2004); completed and updated by Francou and Vincent (2007). Reproduced with permission of B. Francou.

in the European Alps appears quite clearly: almost all mass balances are negative over that period, signifying a generalized retreat of glaciers. As in the Alps, glacial retreat accelerated sharply in the 1980s. It is estimated that between 1993 and 2011, global glacier melting (excluding the ice sheets of the Antarctic and Greenland) has caused sea levels to rise on average by approximately 1 mm/yr.

The causes of the change

The general retreat of the glaciers, which affected the entire planet in the 20th century, is part of a global process that started before the beginning of that century. In the Alps, the retreat started at some time during the 19th century; in the tropical Andes, it started in the 18th century. This retreat marked the end of the Little Ice Age (LIA), during which the entire planet experienced a series of 'cold spells' at the end of the Middle Ages. The retreat during the 20th century is part of the natural climate change that concluded the LIA. Glacier retreat during the 18th and 19th centuries can hardly be attributed to human activity, since the greenhouse gases generated by humans at the beginning of the industrial era in the middle of the 18th century were still too limited. Glacier dynamics, which need time to reach equilibrium after a climate change, fully explain the retreat that began at the end of the LIA.

By contrast, the rapid acceleration of the phenomenon around the 1980s does indeed appear to bear the signature of the global warming that marked the second half of the 20th century. All indicators and analyses attribute this warming to the impact of human activity (Chapter 29).

26.3 Ice-sheet changes

The ice sheets are the largest reservoir of fresh water on Earth. If the Greenland Ice Sheet were to melt, the sea level would rise by 7 m; if the Antarctic Ice Sheet melted, it would rise by 58 m. Their partial melting, even by a few metres, would profoundly affect human life on Earth. This happened in a limited way during the last interglacial, some 125,000 years ago, when the sea level rose to about 7 m higher than now (Section 20.5). *What repercussions will the current global warming have on ice-sheet dynamics? Have they started to reduce in size?* Before describing the observations, we briefly recall the factors that govern the equilibrium of ice sheets.

Dynamics

As with glaciers, annual variations in the volume of the ice sheets stem from the balance between snow accumulation and ice loss. Loss is due to (1) melting of the ice-sheet surface in the warm season, and (2) discharge of ice into the ocean, forming icebergs. The ice sheets flow continuously into the ocean via outlet glaciers, which calve into icebergs – or, as in the Antarctic, large tabular icebergs that break off from the ice sheet. When snow accumulation compensates these mechanisms, the ice-sheet volume remains constant.

How can warming change the dynamics? We start by recalling that a warmer atmosphere contains more water vapour, thus allowing more precipitation. The distinction is particularly striking between cold ice ages and warm interglacial stages, in which ice accumulation doubles. When the climate warms, several mechanisms intervene, some increasing the volume of the ice sheets (greater ice accumulation due to more precipitation), while others decrease the volume (greater ice ablation, due to more melting and/or accelerated discharge from the outlet glaciers). When an ice age gives way to an interglacial, although more ice accumulates on the ice sheets, the ablation mechanism prevails. This is why the ice sheets in the NH almost disappear and the Antarctic Ice Sheet shrinks during glacial–interglacial transitions.

It is difficult to estimate the change in volume of the ice sheets, since they cover such vast areas. Nonetheless, much progress has been made over the past 10 years, thanks mainly to satellite data.

The Greenland Ice Sheet

Between 1979 and 2009, the area of Greenland affected at least one day per year by melting (about 550,000 km² in 2005) increased by 42%, and the average summer temperature increased by 2.4°C. This phenomenon occurs most dramatically in northern Greenland: since 2000, major ice-melt events have been observed at more than 1500 m in altitude, a situation that satellites had never detected previously. From year to year, the surface area affected by melting has increased. The map of Fig. 26.7 shows the increase in number of days of thaw per year, averaged over five consecutive periods between 1980 and 2012. Current estimates indicate that the shrinkage of the Greenland Ice Sheet is due in equal parts to surface melting and to discharge into the ocean.

Additional information is provided by a study of the outlet glacier flow velocity at Jakobshavn, a major outlet glacier on Greenland's west coast. Its velocity has increased significantly in the past two decades. This outlet alone drains close to 5% of the ice sheet (Fig. 26.8). Although its front is in constant retreat, the flow velocity almost doubled between 1992 and 2000 (from 5 km/yr to approximately 10 km/yr) and reached 15 km/yr in 2007. Similar acceleration is also observed on most glaciers of the west coast. These changes were not predicted by models for lack of spatial resolution and in the absence of the necessary physical mechanisms. The finding raises fresh questions. The acceleration is due partly to better lubrication between the glacier and the bedrock, as the increased meltwater infiltrates and flows along the glacier bed. Less friction means greater speed! However, this is not enough to explain the dramatic acceleration of the glacier. Another mechanism is at work. The ice platforms that cling to the

Fig. 26.7 The number of melt days per year of the Greenland Ice Sheet, averaged over five consecutive periods between 1980 and 2012. Surface thawing has increased in recent years.
Source: Fettweis et al. (2011). Reproduced with permission of Xavier Fettweis, Université de Liège.

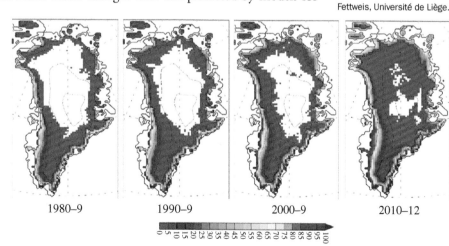

1980–9   1990–9   2000–9   2010–12

Number of melt days per year derived from satellites

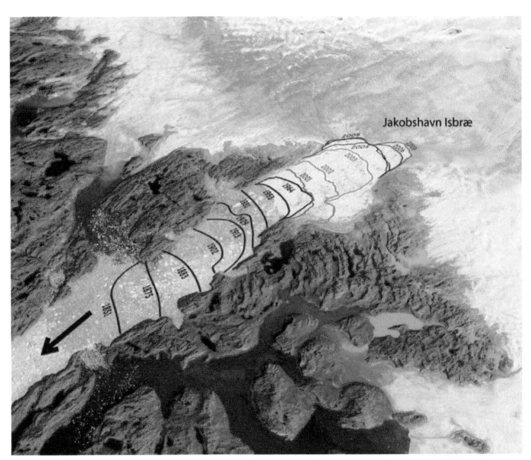

Fig. 26.8 The retreat of the Jakobshavn glacier between 1851 and 2009 is marked by the position of the successive fronts. The flow velocity (direction: black arrow) almost doubled between 1992 and 2000 (from 5 km/yr to approximately 10 km/yr) and attained 15 km/yr in 2007. **Sources:** NASA/Goddard Space Flight Center Scientific Visualization Studio. Historic calving front locations courtesy of Anker Weidick and Ole Bennike, Geological Survey of Denmark and Greenland.

coast at the front of the glacier are formed not from ice floes (ice floes are only a few metres thick and form when seawater freezes), but from the discharge of the glacier into the ocean. They can be several hundred metres thick and they create a barrier at the outlet of the glacier that slows its flow. Changing ocean currents have brought warmer water close to the glaciers, melting their frontal zones and destabilizing them, thus accelerating the outflow of the glacier.

The Antarctic Ice Sheet

The above mechanisms are not all equivalent: melting requires time, while discharge can be dependent on amplification mechanisms that release ice suddenly. The West Antarctic Ice Sheet, which, unlike the East Antarctic and Greenland Ice Sheets, rests on the seabed, has been the object of particular concern, since it may be liable to major destabilization.

For the Antarctic Ice Sheet, satellite measurements show that mass has been lost every year since the beginning of the 21st century. At the edges of the ice sheet, ocean warming and rising surface temperatures deepen the crevasses through melting and freezeback, and have caused spectacular disintegrations of certain platforms that had blocked the flow of ice. In the following years, the glacier flow accelerated in the region. The most spectacular example was that of the Larsen B ice shelf in the Antarctic Peninsula, more than 3200 km^2 of which collapsed in 2002. The first ice shelf, Larsen A, had already disintegrated in 1995. The third, Larsen C, seems to have resisted until now, but it cannot be ruled out that it may suffer a similar fate in the years to come. The warming of the ocean waters surrounding the ice sheets, whether in Greenland or in the Antarctic, appears to be caused by a change in the way warmer waters now circulate in the coastal areas.

In conclusion

Estimates of ice-sheet mass balance have improved significantly over recent years, by virtue of new techniques that provide a consistent record of the Greenland and Antarctic Ice Sheets over the past two decades. The ice loss in the ice sheets of Greenland and the Antarctic is shown in Fig. 26.9, expressed in terms of their contribution to the rise in sea level. Since 2000,

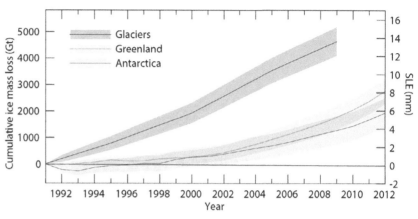

Fig. 26.9 IPCC 2013 assessment of the contribution to the rise in global sea level (Sea Level Equivalent, SLE, in mm) from 1991 to 2012 due to ice loss in terms of mass (Gt) from the ice sheets of Greenland and Antarctica, together with that from all glaciers (excluding those on the periphery of the ice sheets).

Source: IPCC 2013.*Climate Change 2013: The Physical Science Basis*. Working Group I Contribution to the Fifth Assessment Report of the Intergovernmental Panel on Climate Change, Fig TS.3. Cambridge University Press.

the mass balance of each of these two ice sheets has been negative: in the period 2002–11, the ice loss in Greenland was 215 ± 60 Gt/yr, and it was 147 ± 75 Gt/yr in the Antarctic. By 2012, the melting had raised the sea level by about 1 cm since 2000.

26.4 Changes in frozen soils

26.4.1 PERMAFROST AND FROZEN SOILS

In the high-latitude regions of the NH, the soil can be permanently frozen: this is called *permafrost*. Depending on the region, the soil may be frozen to a depth of several hundred metres, or merely a few metres. Permafrost covers 23 million km^2; that is, one quarter of all the land area in the NH (Fig. 26.3). The surface layer lying above the permafrost that freezes and thaws each year is called the 'active' layer. *What is the effect of warming in these regions?* It modifies both the thickness of the active layer, which increases (each year, a thicker layer of soil thaws), and the thickness of the permanently frozen soil, which decreases. The thickening of the seasonally thawed zone reduces the volume of permafrost and destabilizes the soil. In more southerly regions of the NH where there is no permafrost, the surface soil freezes to a certain depth every year. Warming in those regions decreases the thickness of the layer that freezes.

26.4.2 RECENT CHANGES

The retreat of permafrost

In the Arctic, where temperatures have increased by several degrees since the 1980s, the retreat of permafrost and warming of the soil are well documented. In the Russian Arctic, changes in permafrost and in the active layer have been monitored at 242 measurement stations (dis-

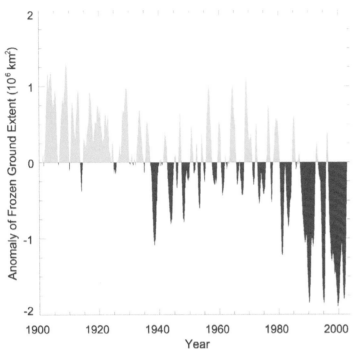

Fig. 26.10 Historical variations in the area of seasonally frozen ground in the Northern Hemisphere (including the active layer over permafrost). The figure shows the difference with respect to the average over the period 1901–2002. Blue: above-average monthly extent; red: below-average monthly extent.

Source: IPCC 2007. *Climate Change 2007: The Physical Science Basis*. Working Group I Contribution to the Fourth Assessment Report of the Intergovernmental Panel on Climate Change, Fig. 4.22. Cambridge University Press.

tributed from Western Russia to the Kuril Islands) during the second half of the 20th century. Data collected since 1955 show that the depth of the active layer has increased by approximately 20 cm, reflecting a retreat of the permafrost. In regions without permafrost, the bottom of the layer that freezes during winter has risen by almost 30 cm (averaged over 211 observation stations), proof that frost is not penetrating so deeply. In the Arctic, the area of permafrost shrank by about 2 million km^2 in the 20th century (Fig. 26.10). The impact of global warming has been especially acute in the most recent decades.

Another region of the NH that is particularly vulnerable to the retreat of permafrost is Tibet. This 'Roof of the World' is the most southerly region of the hemisphere possessing pockets of permafrost. Here again, observations reveal rapid change: the area of the 'islands' of permafrost has decreased by 36% over the past three decades.

Rockfalls

Above a given altitude, mountain faces remain permanently frozen from season to season. At high altitudes, distinct changes in mountain permafrost have been observed in recent decades: with the rising temperature, its lower boundary has climbed to higher altitudes and its internal temperature has risen. Warming can weaken the foundations of infrastructures at high altitude, such as cable cars and refuges, or generate far more sudden events: rockfalls. This process is well known to mountaineers and has always existed, but current conditions have exacerbated it. Under favourable geological conditions, three main factors trigger rockfalls: earthquakes, decompression of the mountain wall as glaciers retreat and deterioration of the permafrost. Permafrost warming is held responsible for the recrudescence of rockfalls in recent decades at high altitudes. Studies carried out in the Mont Blanc range (Chamonix, France), where mountaineering reigns unchallenged, have verified this hypothesis (Ravanel & Deline 2010). Recent rockfalls in the region have destroyed many legendary climbing routes (Fig. 26.11). A systematic inventory of rockfalls since 1855 (the end of the LIA in the Alps, and also when the first photographs of rock faces were taken) has been undertaken on the west face of the Dru and on the north face of the Aiguilles de Chamonix. Between 1850 and the middle of the 20th century, rockfalls were rare. No mention appeared in this category until the decade 1900–10, when only one event was recorded. This was followed by a second in the decade 1930–40. Since that date, the volume of collapsed rock has increased sharply, especially in the two most recent decades. A comparison of rock-

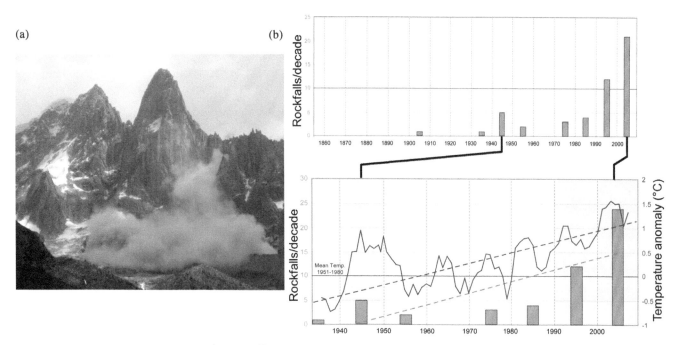

Fig. 26.11 (a) In June 2005, a rockfall of 265,000 m³ occurred on the west face of the Drus (summit at altitude 3754 m) in the Mont Blanc mountain massif (Chamonix, Western Alps), leaving a huge pale scar on the rock face. The fall destroyed the legendary Bonatti pillar climbing route. (b) The frequency of rockfalls on the west face of the Drus and the north face of the Aiguilles de Chamonix (orange), and the air temperature anomaly with respect to the period 1951–80 (red curve). The number of rockfalls, like the annual air temperature at Chamonix, has increased strongly in recent decades.

Sources: (a) Photograph by J. Malbert. Reproduced with permission. (b) Reproduced with permission of L. Ravanel.

fall frequency and changes in temperature at Chamonix exhibits a clear correlation between warming and rockfalls (Fig. 26.11): the greater the temperature anomaly, the greater is the number of rockfalls. They have been particularly noticeable in the past 10 years. It is notable that the end of the LIA was free of large rockfalls. This suggests that the recent warming cannot be compared with the warming that caused the retreat of the glaciers at the end of the LIA (Fig. 22.13). The average altitude at which rockfalls occur (3200–3600 m) reflects permafrost warming. It follows that as the temperature rises, rockfalls will occur at increasingly high altitudes.

Global warming has made rockfalls more frequent. Mountaineers are all too aware of it! Certain climbing routes have become more dangerous, but other routes can now open up on the newly virgin rock walls!

26.4.3 CONSEQUENCES

Degradation of permafrost affects the mechanical properties of the soil. Not only does thawing cause ground subsidence, weakening building and pipeline foundations and shortening the seasonal use of winter roads, but it also affects soil drainage systems and modifies Arctic lake landscapes. In the Arctic, thawing of the permafrost creates depres-

Fig. 26.12 Changes from tundra (left, 1978) to wet zones (right, 1998) subsequent to recent warming. The two photographs were taken at the same place – Tanana, central Alaska.

Source: Jorgenson et al. (2001). Photographs by Torre Jorgenson. Reproduced with permission.

Fig. 26.13 Sites of Siberian lakes that have vanished after three decades of rising soil and air temperatures (changes recorded by satellite imagery between the early 1970s and 1997–2004). The spatial pattern of lake disappearance suggests that permafrost thawing is the cause of the observed losses.

Source: IPCC 2007. *Climate Change 2007: Impacts, Adaptation and Vulnerability. Working Group II Contribution to the Fourth Assessment Report of the Intergovernmental Panel on Climate Change, Figure 15.4, Cambridge University Press.*

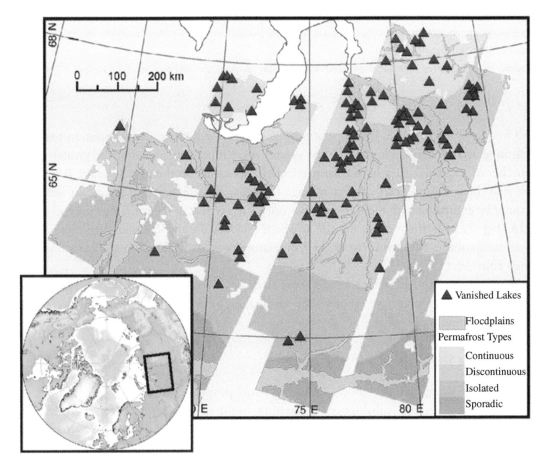

sions in which new wet zones form. Figure 26.12 shows the transformation of a central Alaskan landscape between 1978 and 1998, as tundra gives way to wetlands. As more land thaws, surface water drains into the subsoil, reducing the area of wetland and causing lakes to disappear (Jorgenson et al. 2001). Figure 26.13 illustrates satellite observations of Siberia over the past three decades (from the early 1970s to 1997–2004). Many of the lakes dotting the vast landscape have disappeared, affecting the entire area occupied by permafrost. More generally, the loss of wetland in the Arctic has disrupted the existing ecosystems.

26.5 Freeze-up and snow cover

Unlike glaciers, annual snow cover and the freeze-up of lakes and rivers constitute a direct response to the climate in a given year. Since there is no snow or ice during the summer season, the system is reset each year. Snow cover and freeze-up are therefore directly related to the annual snowfall and temperature changes. In January, snow covers 49% of the land in the NH. At present, the snow cover during the year varies from 45.2 million km² in winter to 1.9 million km² in summer.

In the NH, the reduction of snow cover is undisputed. Satellite data from the period 1966–2004 indicate a decrease of about 5%, with a particularly sharp decline around 1986–8. Between 1972 and 2000, for example, the maximum of the snow cover advanced from February to January, and the spring thaw shifted forward by 2 weeks. Figure 26.14 maps this retreat in the NH in April, showing that it affects both middle and high latitudes. Similarly, as the lakes generally freeze later and thaw earlier, the freeze-up season in the NH has shortened. In the SH, there is insufficient data available for a proper study of snow cover.

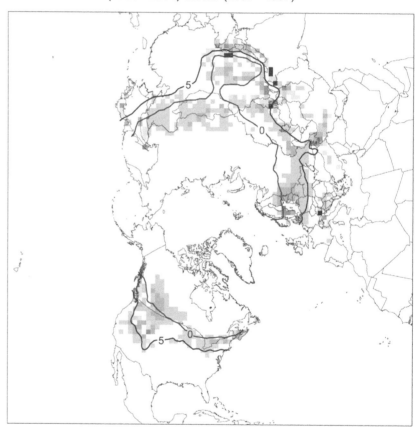

March — April Snow Departure
(1988 - 2004) minus (1967 - 1987)

■ -36 to -26 ■ -25 to -16 ▦ -15 to -6 ☐ -5 to 5 ▨ 6 to 15 ▨ 16 to 25 ■ 26 to 38

Fig. 26.14 Decrease in Northern Hemisphere snow cover at the end of the 20th century. The map shows the difference in March–April average snow cover (expressed in %) between the periods (1967–1987) and (1988–2004). Negative values indicate greater extent in the earlier portion of the record (data from NOAA/NESDIS maps). Red curves show the average 0°C and 5°C isotherms for March and April between 1967 and 2004. **Source:** IPCC 2007. *Climate Change 2007: The Physical Science Basis.* Working Group I Contribution to the Fourth Assessment Report of the Intergovernmental Panel on Climate Change, Figure. 4.3. Cambridge University Press.

The source of this reduction: less snow or more heat?

The worldwide observation of reduced snow cover is the result of warming rather than a decrease in winter precipitation. In France, observations by the Centre d'Etudes de la Neige de Grenoble (Météo-France–CNRS) illustrate this mechanism at the Col de Porte ski station (altitude 1325 m), at the gateway to the Alps in the Chartreuse mountain range. This station is representative of the changes occurring in the mid-altitude range of the Western Alps, which are affected by weather systems coming from the Atlantic. Over the past five decades, the average snow cover has decreased by half. The decrease cannot be attributed to lower precipitation (which has on average remained stable) but, rather, to a local rise of about 2°C in winter temperatures. The higher temperatures mean that the ratio of snowfall to total precipitation has diminished, and that the winter snow cover is thinner.

The changes in snow cover in the Western Alps since 1958 are reconstructed on the basis of weather data, observations and meteorological re-analyses. The reconstruction uses both the SAFRAN model for improved spatial resolution (25 km rather than 120 km) and the CRO-CUS model, which yields the snow cover at various altitudes. These results show that in recent decades, snow cover has decreased over the entire range of altitude up to 2000 m.

Chapter 27
The impact of warming on the ocean

The understanding of how and why the oceans have changed in recent decades is extremely important, but also, owing to their coupling with the atmosphere, particularly complex. It is important since both the oceans and the atmosphere transport heat from the Equator to the polar regions. Any modification of ocean circulation patterns can therefore be a source of climate change, especially at high latitudes. It is also important because any increase in sea level directly affects highly populated coastal areas. Moreover, the ocean is home to rich ecosystems where the lowest links in the food chain, which are the foundation of all marine life, can be affected by changes in ocean acidity. Finally, the ocean plays an important role as a sink for anthropogenic carbon, by absorbing the carbon dioxide that accumulates in the atmosphere. *How will this system evolve?* It is particularly challenging to answer these questions, since response times in the ocean are much longer than in the atmosphere. They are longer because water is far denser than air (~830 times denser), its specific heat capacity is four times higher and the mass of the oceans is far greater (~250 times greater). The question is also challenging because the density of water, a key parameter that plays a role in the formation of deep ocean currents, depends on two variables: temperature (cold water is denser, down to 4°C) and salinity (the higher the salinity, the denser the water). Either of these factors may fluctuate for a variety of reasons, further complicating the dynamics of the system.

Consequently, it can take several decades for natural regional trends to run their course, making it all the more difficult to distinguish natural variations from those caused by recent global warming. To discriminate between natural fluctuations from long-term trends, we must first identify the global patterns.

Climate Change: Past, Present and Future, First Edition. Marie-Antoinette Mélières and Chloé Maréchal.
© 2015 John Wiley & Sons, Ltd. Published 2015 by John Wiley & Sons, Ltd.
Companion website: www.wiley.com\go\melieres\climatechange

Various mutually consistent trends, perceptible at a large scale, reveal the impact of recent warming. One issue that strikes public opinion and has raised grave concerns about our future is the rise in sea level: a few dozen centimetres would profoundly disrupt highly populated regions located close to sea level (through flooding, coastal erosion, cyclones and storms, and so on). But other global changes have also made their appearance, including changes in salinity (which can modify ocean circulation), and in dissolved carbon dioxide (which can interfere with the carbon cycle and ocean acidity and therefore affect marine life).

27.1 Change in sea level

27.1.1 WHAT CHANGES THE AVERAGE SEA LEVEL?

Over several decades or over a century, the global average sea level may change for two reasons: the amount of water in the oceans may increase (or decrease) and its volume may expand (or contract). At a regional or local level, sea levels may also be affected by other factors that affect the relative position of the sea and land surfaces. These movements can have various causes, including:

- Plate tectonics: movements of the Earth's crust (uplift or subsidence) and deformation of the lithospheric plates.
- Glacio-isostatics: viscoelastic relaxation of the Earth's crust and mantle, induced by melting of the ice sheets that covered the NH 18,000 years ago. Currently, this process is causing the seaboards of Western Europe and the eastern United States to subside, and those of Scandinavia and Canada (e.g. Hudson Bay) to rise.
- Hydrology: stream sediments in large river deltas can make the ground subside.
- Human activity: groundwater pumping can cause the ground to subside by 1–3 m (e.g. Shanghai). Exploitation of oil and gas fields (e.g. in the Gulf of Mexico) can have similar consequences.
- Atmospheric pressure at the sea surface can produce a local or regional change in sea level: a rise (or drop) of 1 hectopascal (1 hPa = 1 millibar or 0.76 mmHg) causes the sea level to fall (or rise) by one centimetre.
- Changes in ocean circulation.

27.1.2 OBSERVED CHANGES

Various measurements carried out during the 20th century show that the mean sea level has risen on average by approximately 1.7 mm/yr. More recently, since 1993, altimetry satellites (TOPEX/Poseidon and Jason) have been providing high-resolution coverage of all the oceans. Data collected over the past 15 years indicate that in that period the mean sea level rose faster, at a rate of about 3.3 mm/yr. These very precise figures illustrate the regularity of the rise from year to year. *Is this rise not just another fluctuation among the variations of the recent millennia?* Reconstructions of the global sea level since 1700 show that from that date, the level remained stable until the middle of the 19th century, at about 10–20 cm below its present level. In the past two millennia, the level appears to have hardly moved. Figure 27.1 illustrates these various measurements.

What is the change due to?

Studies based on the Argo float network data (contributions of temperature and ocean salinity) and spatial gravimetry measurements (ocean and polar ice-sheet mass variations, the GRACE mission) indicate that the rise is the result of:

- Accelerated ice melting. Melting glaciers contribute about 1 mm/yr, which is comparable to that of the ice sheets (~1 mm/yr, including ~0.6 mm/yr for Greenland and 0.4 mm/yr for the Antarctic Ice Sheet). Altogether, ice melting contributes about 2 mm/yr.
- Ocean warming. Thermal expansion contributes about 1 mm/yr.

27.1.3 IS SEA-LEVEL RISE UNIFORM OVER THE PLANET?

The mean sea level has gradually risen by an amount that appears to be consistent with the steady global warming over the past decades. *But is the change uniform over the planet? And what will happen to coastal populations over the years?*

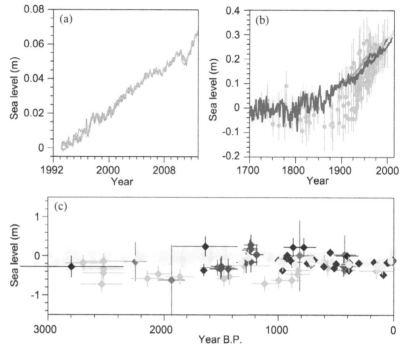

Fig. 27.1 The change in global mean sea level. (a) Since 1993: a reconstruction from satellite altimetry data sets from five groups, with the mean shown as a bright blue line. (b) A comparison of three types of data sets: palaeodata from salt marshes from Northern and Southern Hemisphere sites (purple); tide gauge data sets (dark blue, green and orange) and altimetry data (light blue). Shifts have been applied to adjust the mean values of these data sets. (c) Palaeo sea-level data for the past 3000 years from Northern and Southern Hemisphere sites. The effects of glacial isostatic adjustment have been subtracted from these records.

Source: IPCC 2013. *Climate Change 2013: The Physical Science Basis.* Working Group I Contribution to the Fifth Assessment Report of the Intergovernmental Panel on Climate Change, Figure 13.3a, d, e. Cambridge University Press.

The TOPEX/Poseidon satellite has been mapping changes in the sea level since 1993. Figure 27.2(a) shows the rate of change of sea level between 1993 and 2003 in different regions. The change is anything but uniform. In some regions, the sea level has risen by more than 10 cm during the decade (more than three times the global average), while in others it has dropped sharply. *Why?* The phenomenon has to do with the thermal structure of the first 1000 m in the ocean. Figure 27.2(b) is a simulation of the rate of change of the sea level calculated only from variations in the two variables that affect water density, namely the heat content and salinity of the first 1000 m of the ocean. The similarity between these two maps justifies the assumption that the regional differences in sea level are, on the timescale of these observations, due to density differences. *Does this also apply in the longer term?*

Fig. 27.2 (a) Geographical distribution of short-term linear trends in mean sea level (mm/year) from 1993 to 2003, based on TOPEX/Poseidon satellite altimetry. (b) The geographical distribution of mean thermal expansion (mm/year) from 1993 to 2003, based on temperature data from the ocean surface to a depth of 700 m.

Source: IPCC 2007. *Climate Change 2007: The Physical Science Basis*. Working Group I Contribution to the Fourth Assessment Report of the Intergovernmental Panel on Climate Change, Figure 5.15. Cambridge University Press.

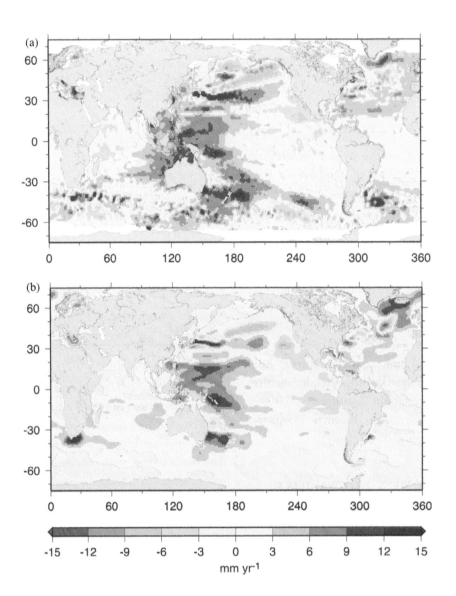

A similar comparison was carried out for the period from 1950 to 2003. Again, the two maps show clear similarities, confirming the interpretation. However, the trends observed in the two periods (1950–2003 and 1993–2003) are not the same. This indicates that the regional thermal content of the ocean varies, as is illustrated below by the example of atolls in the tropical western Pacific. In addition to these fluctuations, however, the global sea level is steadily changing: under the influence of global warming, the mean sea level has risen regularly.

To illustrate the concept, consider a swimming pool that is gradually filling with water, with its average level steadily rising. As it fills, waves on the water's surface make a floating cork bob up and down. The amplitude of this movement is far greater than the slow rise in the water level, and can at first mask it. However, if we measure the average position of the water surface in the pool, the oscillations cancel out and the slow steady rise due to the incoming water becomes apparent. In our case, the rise-and-fall processes of the sea level in various regions of the planet are not water waves (as in the pool), but are due to changes in deep thermal structures, which change with different timescales. The oscillations are superimposed on the slow rise of the average sea level caused by global warming.

Atolls and rising sea level

The gradual rise in sea level due to global warming poses a particular threat to low-lying atolls. *What does this imply for coastal populations of tropical Pacific islands?* First, on the timescale of a couple of years, they may suffer the consequences of El Niño/La Niña events, which affect sea levels by modifying atmospheric circulation patterns. During an El Niño event, the movement of warm waters from west to east can make sea levels fall in the west and rise in the east by as much as 20–30 cm (see Figs. 17.1 and 17.2). Figure 27.3 illustrates this variation at Kwajalein, located in the tropical western Pacific region (8°N, 167°E, in the Marshall archipelago), where the two strongest El Niño events of the past few decades (1982 and 1997–8) caused the sea level to drop by about 25 cm.

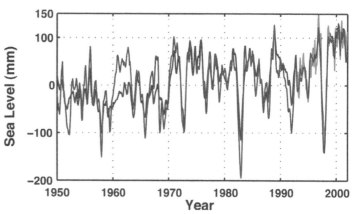

Fig. 27.3 Local variations in sea level. Monthly mean sea level curve for 1950 to 2000 at Kwajalein (8°44'N, 167°44'E, Marshall Archipelago). Sea level observed from tide gauge measurements is in blue, reconstructed sea level in red and satellite altimetry record in green. Annual and semi-annual signals have been subtracted. The impact of El Ninō events is clearly illustrated by the two large falls in sea level in 1982 and 1998 (about 25 cm), years of strong El Niño.

Source: IPCC 2013. *Climate Change 2007: The Physical Science Basis.* Working Group I Contribution to the Fourth Assessment Report of the Intergovernmental Panel on Climate Change, Figure 5.18. Cambridge University Press.

In the longer term, three factors may cause sea levels to rise or fall:

- Ocean warming and loss of continental ice. This increase is attributed to global warming.
- Thermal subsidence (subsidence of the Earth's crust under volcanic islands), causing an apparent rise in sea level.
- Low-frequency variability (fluctuations over several decades). This variability stems from regional changes in the thermal structure of the ocean. It is superimposed on the sea-level rise due to global warming, either amplifying or masking it.

What do we learn from recent decades? We compare the changes in sea level for two Pacific atolls, Funafuti (8°S, 179°E) in the Tuvalu archipelago, and Yap (9°N, 138°E) in the Caroline Islands. These atolls lie several thousand kilometres apart. Detailed estimates exist of sea-level changes since 1950 in the tropical western Pacific (Becker et al. 2012). Figure 27.4 illustrates the sea-level changes between 1950 and 2009, and between 1993 and 2009. The mean rise between 1950 and 2009 in the Funafuti region is distinctly faster than in the region of Yap. By contrast, between 1993 and 2009, it is slower. This illustrates the variability of sea-level change: the chronology of sea-level rise varies from place to place, and in one place it varies in time. The study also shows that on some islands the 'total' sea level (as perceived by the inhabitants) has risen significantly over the past 60 years. This is particularly true in Tuvalu, where the 'total' sea level has risen on average by 5.1 (±0.7) mm per year, nearly three times faster than the global average rate over the same period (1950–2009). Subsidence accounts for 10% of the rise, and global warming over the period for only about one third.

Fig. 27.4 Sea-level
changes since 1950
in the tropical western
Pacific: maps of
reconstructed sea-level
trends between (a) 1950
and 2009 and (b) 1993
and 2009. The two
maps illustrate how the
rise in sea level varies
in space and time. The
average rise in sea
level at Funafuti (8°S,
179°E, in the Tuvalu
archipelago) between
1950 and 2009, for
example, appears to be
much faster than in the
Yap region (9°N, 138°E,
in the Caroline Islands).
By contrast, between
1993 and 2009 it is
slower.
Source: Becker et al.
(2012). Reproduced with
permission of Elsevier.

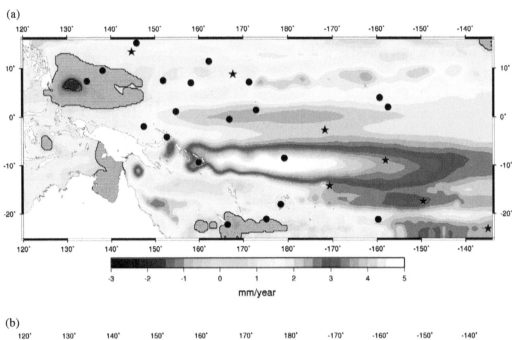

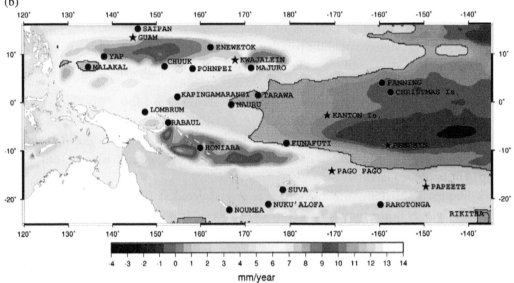

27.2 Regional changes in ocean salinity

The differences in ocean salinity have already been discussed in Chapter 5. Regional ocean salinity can vary for different reasons:

- an increased influx of fresh water from rain, rivers or continental ice melt, which reduce the salinity;
- evaporation of water from the marine reservoir, which increases the salinity;
- an influx of fresh water from melting sea ice, which consists mainly of fresh water; and
- movement of water of differing salinity from one region to another, its density depending on the salinity.

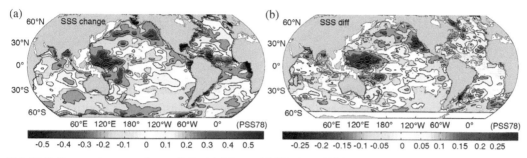

Fig. 27.5 Changes in sea surface salinity. (a) 58-year (2008 minus 1950), with seasonal and ENSO signals removed. (b) 30-year (2003–2007 average centred at 2005, minus 1960–1989 average, centred at 1975). Salinities measured using PSS-78 (Practical Salinity Scale 1978) do not have units. White areas are marginal seas where calculations are not carried out.

Source: IPCC 2013. *Climate Change 2013: The Physical Science Basis.* Working Group I Contribution to the Fifth Assessment Report of the Intergovernmental Panel on Climate Change, Figure 3.4. Cambridge University Press.

Sufficient data has become available since 1955 to provide a relatively detailed picture of changes in surface salinity in the three major basins of the Atlantic, Pacific and Indian oceans (Fig. 27.5). *Can the global scale of these changes be explained by the effects of climate warming?* One possible explanation is that intensification of the water cycle due to global warming has increased evaporation, and hence salinity in the tropical Atlantic, by exporting fresh water into the tropical Pacific. The inflow of excess fresh water into the tropical Pacific decreases its salinity. The reduced salinity at middle and high latitudes of the NH, both in the Pacific and the Atlantic, may also be a consequence of the intensification of the water cycle. The meridional atmospheric circulation transfers warm wet air masses from the tropics to the middle and high latitudes, where the water vapour condenses and precipitates. Greater evaporation in the tropical belt implies more precipitation in the middle and high latitudes and, accordingly, lower ocean salinity there. *The consistency of these changes suggests that the global water cycle is strengthening, with greater evaporation in the tropics.*

27.3 Is deep ocean circulation slowing?

The Meridian Overturning Circulation (MOC) is the meridian ocean circulation cell that transfers tropical waters to the high latitudes. In the North Atlantic, this cell also transfers the surface waters downwards, initiating deep-water circulation. Moreover, in transporting heat from the Equator to the poles, it moderates the climate of Northern Europe. The variability of the MOC in the North Atlantic is an important issue, because any change affects both deep-water formation and the climate in Europe. Intensive campaigns have been carried out over the past few years to measure changes in this cell between Greenland and Portugal. At present, it appears that the surface waters are sinking more slowly, which implies slowing down of the deep circulation.

Is this phenomenon the consequence of natural fluctuations over several decades, or of the recent global warming? As the North Atlantic dynamics are particularly complex, it is not yet possible to establish whether the change has been happening steadily over the past few decades or if it is the effect of decade-to-decade fluctuations. Opinions still diverge.

27.4 Changes in dissolved carbon dioxide and ocean acidification

Atmospheric gases dissolve in the ocean in such a way that their partial pressures in the air and water are at equilibrium. When the partial pressure of a gas increases in either reservoir, molecules are transferred from one to the other to rebalance the partial pressures.

27.4.1 CHANGES IN DISSOLVED CARBON DIOXIDE

Between the beginning of the 19th century and the 21st century, the concentration of CO_2 in the atmosphere has risen from about 280 ppm to about 400 ppm due to human activity (Chapter 29). Re-equilibration of the surface water with the atmosphere has steadily increased the amount of CO_2 (measured in ppm) dissolved in the surface layer. Measurements carried out over recent decades indicate that dissolved CO_2 has increased in the surface layer by 1.6–1.9 ppm/yr, in line with the increase in the atmosphere (1.5–1.9 ppm/yr).

By absorbing part of the CO_2 discharged into the atmosphere, the ocean acts as a CO_2 sink. The gas enters the water at the surface and propagates downwards as the water layers mix, thereby gradually penetrating into the ocean mass via ocean circulation. Chlorofluorocarbons (CFCs) enter into the ocean mass in the same way. As these gases do not exist naturally, their concentration in the various ocean layers provides useful information regarding the path followed by the ocean circulation. The same applies to the radioactive isotope of carbon, ^{14}C, released during the nuclear tests of the early 1960s. Its penetration into the ocean directly reveals that of CO_2, since it appears in the form $^{14}CO_2$.

IPCC 2013 provides estimates of the amount of anthropogenic CO_2 that has entered the oceans. Between the middle of the 18th century (the onset of the industrial era) and 2011 (the date of the inventory), an estimated 155 GtC of anthropogenic CO_2 was injected into the oceans, as shown in Fig. 27.6. This penetration is essentially confined to the surface layers,

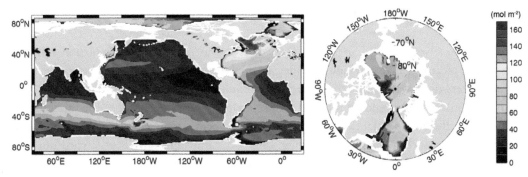

Fig. 27.6 Inventory of anthropogenic CO_2 in the ocean water column (mol/m²) in 2010. For the global ocean (excluding marginal seas) the estimated amount is 150±26 GtC (marginal seas: 5.7 to 8.0 GtC). By far the most anthropogenic CO_2 is injected into the North Atlantic Ocean by deep water formation.

Source: IPCC 2013. *Climate Change 2013: The Physical Science Basis.* Working Group I Contribution to the Fifth Assessment Report of the Intergovernmental Panel on Climate Change, Figure 3.16. Cambridge University Press.

since mixing with the lower layers is slow. Nevertheless, in places where deep waters form (mainly in the North Atlantic), it has reached the ocean floor and is gradually propagating from the north to the south of the Atlantic. That is the region of the ocean where the water column contains the greatest amount of dissolved anthropogenic CO_2.

The Global Carbon Project (GCP n.d.) recently published a review for the period 2003–13 (see Fig. 29.7). On average, anthropogenic emissions amount to 9.4 GtC/yr, while ocean uptake is 2.6 GtC/yr; that is, 27% of all annual anthropogenic emission. *Will this percentage of injection increase or decrease in the future?* The question is crucial, because it affects the amount of anthropogenic CO_2 that accumulates in the atmosphere.

27.4.2 OCEAN ACIDIFICATION

Increased amounts of dissolved CO_2 raise the acidity of seawater. The average pH of the ocean surface water has already decreased by 0.1 since the beginning of the industrial era, which corresponds to a 30% increase in H+ ions in the water (assuming constant ocean alkalinity and temperature). Despite this acidification, the ocean remains alkaline (pH > 7), varying between 7.9 and 8.3, with an average surface pH about 8.1. Because CO_2, like all gases, is more soluble in cold water, the decrease in pH is greater in the colder water at high latitudes (−0.12) than in the tropics (−0.06). When the water becomes undersaturated in carbonates, this increased acidity can reduce the concentration of carbonate ions. Carbonate ions are essential to many organisms for skeleton formation and for calcium carbonate shells (corals, shellfish, algae etc.). Such organisms grow in regions where water is saturated in carbonates. The boundary of these regions is defined by the *carbonate compensation depth* (Box 27.1).

BOX 27.1 THE CARBONATE COMPENSATION DEPTH IN THE OCEAN

Changes in the physical and chemical characteristics of water at different depths, along with deep ocean circulation, divide the water column into two regions, separated by the lysocline, *saturation horizon* or *Carbonate Compensation Depth* (CCD). Above the lysocline, the water is saturated with carbonates; below this level, pressure and temperature conditions make the carbonates unstable, and the water becomes undersaturated.

Marine organisms and corals that form their shells or skeletons from calcium carbonate (calcite or aragonite) can do so only if they develop in an environment saturated in carbonate ions; that is, above the saturation horizon. In deeper waters, it is more difficult for them to form strong shells, as the calcium carbonate gradually dissolves. The saturation horizon varies in depth between a few hundred and a few thousand metres. For calcite, it lies at a depth of about 3500 m in the Atlantic, but it can rise to about 200 m in certain high-latitude regions. Figure B27.1 shows the saturation of aragonite in the North Pacific, together with the partial pressure pCO_2.

Over geological time, the CCD rises and falls. It rises slightly from a glacial to an interglacial period, as atmospheric CO_2 increases and the pH of the surface water decreases. A rise in the lysocline reduces the thickness of the saturated layer between the ocean surface and the lysocline. The present warming has increased the CO_2 partial pressure in the ocean and raises the saturation horizon.

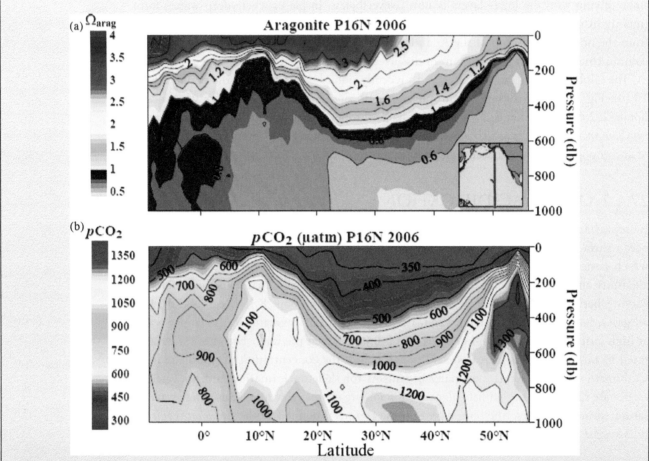

Fig. B27.1 The dependence on depth of: (a) aragonite saturation Ω_{arag}; and (b) the partial pressure of CO_2 in seawater (pCO_2) in March 2006 (P16N transect along 152°W in the North Pacific). In this transect, the saturation horizon (where $\Omega_{arag}=1$) descends from 100 m depth at 55°N and 20°N to about 500 m depth at 30°N. The solvation capacity of water increases with increasing dissolved CO_2. A pressure increase of 1000 decibar (db) indicates a depth of 1000 m.

Source: Fabry et al. (2008). Reproduced by permission of Oxford University Press.

What are the consequences of ocean acidification? Some authors claim that the increased CO_2 content has caused the aragonite saturation horizon to rise by 30–200 m in the east Atlantic (between 55°S and 15°N), in the north Pacific and in the northern Indian Ocean. According to the same authors, the calcite saturation horizon has risen by 40–100 m in the Pacific, north of 20°N. The oceans of the SH and those of the high northern latitudes are particularly sensitive to this acidification because the saturation horizon is naturally closer to the surface there than in other oceans. We illustrate the effect of acidification on living organisms in Chapter 28.

How does this recent decrease in ocean pH compare with past records? Studies show that over the past million years, atmospheric CO_2 has oscillated between the pre-industrial level of 280 ppm, typical for interglacial periods, and 200 ppm, typical for glacial periods. Its pH is therefore higher by 0.1 during a glacial period (lower atmospheric CO_2, less acidic oceans) than during an interglacial. The recent 30% increase in atmospheric CO_2 has lowered the pH of the ocean surface by a further 0.1 compared to typical interglacial levels. In the past million years, the average surface-water pH has never been as low as at present.

27.5 In summary: consistency over the globe

Changes in the ocean over recent decades have become visible in various ways, but the patterns are similar across the globe:

- Sea-surface temperatures have increased almost everywhere, except in the Southern Ocean, where they appear still to be stable.
- The average sea level has risen steadily: two thirds of the rise is due to melt water and one third to thermal expansion of the ocean surface layer.
- CO_2 from the increased atmospheric content has dissolved and penetrated into the ocean. Its transport is governed by ocean circulation.
- The increase in dissolved CO_2 has gradually acidified the ocean surface waters,
- The changes in salinity are consistent with an intensification of the water cycle (increased evaporation in the tropics and increased rain at middle and high latitudes).

All these changes are consistent with the observed increase in atmospheric CO_2 and global warming over the recent decades. They are illustrated in Fig. 27.7.

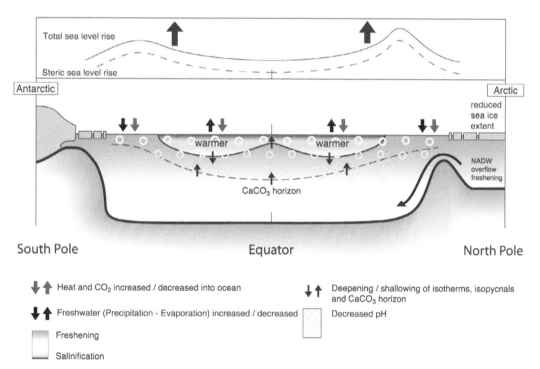

Fig. 27.7 A schematic diagram of observed changes in the state of the ocean. These include ocean temperature, salinity, sea ice, sea level, and biogeochemical cycles. The legend identifies the direction of the change in these variables. **Source:** IPCC 2007. *Climate Change 2007: The Physical Science Basis.* Working Group I Contribution to the Fourth Assessment Report of the Intergovernmental Panel on Climate Change, Figure 5.23. Cambridge University Press.

Chapter 28
The impact of warming on the biosphere

We emphasize again that the ultimate concern of climate change studies is the effect of global warming on the biosphere and on ecosystems. The sheer magnitude of the topic is far beyond the scope of this book! We shall limit ourselves to a few examples showing how numerous links in the chain of life have been disturbed in recent decades. Life in all its variety may be understood as part of a sequence: genes, individuals, species, populations, ecosystems, biomes and so on. Each of these entities can be affected or modified by the present global warming, and in turn affect the structure to which they belong.

28.1 Ongoing changes

To respond to the global warming of the past few decades, living organisms can develop three strategies:

- migration (tracking the climate change);
- adapting to the earlier arrival of warm seasons (changes in the life cycle both of plants, by advanced germination, leaf emergence, flowering and fruiting, and of animals, by earlier nesting, changes in hibernation dates, increase in annual insect reproduction cycles etc.); and/or
- adapting to a warmer climate by changes in physiology (size of individuals, of their eggs etc.).

Climate Change: Past, Present and Future, First Edition. Marie-Antoinette Mélières and Chloé Maréchal.
© 2015 John Wiley & Sons, Ltd. Published 2015 by John Wiley & Sons, Ltd.
Companion website: www.wiley.com\go\melieres\climatechange

If the cost of the required changes (migration–adaptation) is too high for the individual, its survival is no longer guaranteed and an entire species may become extinct. This can weaken the ecosystem to which they belong. Such changes may occur gradually, or, in the event of large-scale extreme events, precipitously.

Territories may expand or contract, or relocate. There is, however, a fundamental difference between land and sea. In the oceans, flora and fauna generally move freely, without spatial constraint. On land, by contrast, plants and animals are restricted by natural barriers, such as mountain ranges and oceans. Moreover, a significant portion of land is already used by agriculture and human settlement, limiting the space that plants and animals can occupy. Living organisms can thus respond and adapt to climate changes more easily in the oceans than on land. Here, we discuss a few examples.

28.2 Oceans

28.2.1 MIGRATION IN THE NORTH ATLANTIC

Changes in plankton habitat in the North Atlantic in recent decades strongly suggest that they adapt to the observed surface warming (between 1°C and 1.5°C). Since 1958 the Planktonic Recorder has monitored copepods, small crustaceans at the bottom of the food chain, over a large region of the North Atlantic. Copepods can be subdivided into four categories according to the temperature of the water: warm waters (warm to temperate waters and temperate waters) or cold waters (temperate to cold waters and subarctic waters). Figure 28.1 shows the observed changes (Beaugrand et al. 2002, 2009). Over recent decades, the number of copepod species living in temperate environments has increased and these species have moved to higher latitudes. Conversely, the number of copepod species living in cold environments has fallen, and their habitat has contracted northwards. Similarly, global warming has changed the distribution of various fish species in the North Sea. Figure 28.2 illustrates the decrease in abundance of cod (a boreal species that is also affected by fishing) in the southern part of its distribution area and the increase of mullet (a Portuguese species) and anchovy (a subtropical species) in the northern part of their distribution area. *Several decades of observation of a very large portion of the North Atlantic thus reveal the large impact of climate warming on fauna.*

28.2.2 CORAL BLEACHING

Corals form complex marine ecosystems and, although they occupy less than 0.1% of the ocean surface, coral reefs are home to more than 25% of the world's marine biodiversity. Most coral reefs are located in shallow tropical oceans, the three largest being in Australia (the Great Barrier Reef), New Caledonia and Belize. Coral is a product of symbiosis between an animal, the polyp, and an alga, zooxanthella, which lives in the polyp and gives it its colour. In response to stress (pollution, change in temperature etc.), the polyp expels the alga and its cells lose their colour, leaving only the white limestone structure of the coral: this phenomenon is known as coral bleaching. If the condition persists, the animal may die. Under moderate stress, the alga later re-enters the coral cells, and the initial colour returns.

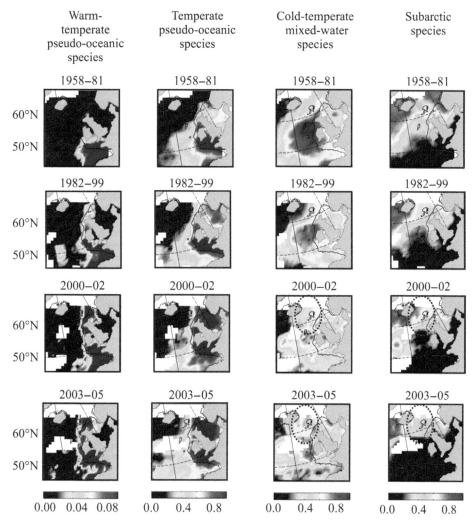

Mean number of species per CPR sample per assemblage

Fig. 28.1 Changes in distribution of copepods (marine zooplankton) in the north-east Atlantic from 1958 to 2005. The different species are grouped into four assemblages, characteristic of habitat. The scale, from 0 to 1, indicates the mean number of species per assemblage. The period from 1958 to 1981 was relatively stable, while between 1982 and 1999, rapid northward shifts took place (Beaugrand et al. 2002). Black dotted ovals denote areas of pronounced change. For some species assemblages, a shift by as much as 10° in latitude over 30 years was observed (260 km per decade). Warm-temperate and temperate pseudo-oceanic species increased, whereas cold-temperate mixed-water and subarctic species decreased. The irregular black line denotes the edge of the continental shelf (depth 200 m).

Source: Modified from Beaugrand et al. (2009). Reproduced with permission of Gregory Beaugrand.

Prior to the 1970s, bleaching occurred in restricted areas only, but since the beginning of the 1980s, major bleaching episodes have been observed over regions of several thousand square kilometres. This mass phenomenon occurs when the temperature becomes unusually warm, notably during El Niño events, since they cause pronounced warming over large areas of the Pacific Ocean. Such heat-related episodes have increased in frequency in response to

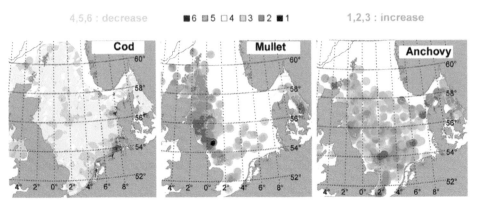

Fig. 28.2 Changes in the distribution of Atlantic cod, red mullet and anchovy in the North Sea between 1977–89 and 2000–5. A decrease in abundance (yellow to red) of Atlantic cod (left) occurred in the southern part of the distribution area, and an increase (green to dark blue) of red mullet (middle) and of anchovy (right) in the northern part. Atlantic cod (*Gadus morhua*) is a boreal species, the abundance of which has decreased by about 90% in the North Sea. Red mullet (*Mulus surmuletus*) is a Lusitanian species (southern limit, Senegal, Canaries; northern limit, the coasts of Scotland and Norway) that was not recorded in the North Sea prior to 1989. Anchovy (*Engraulis encrasicolus*) is a species with subtropical affinity, today found in 80% of the North Sea, but only occasionally reported in 1977–89.

Source: Reproduced by permission of the International Council for the Exploration of the Sea, Denmark.

the ongoing global warming. As surface water temperatures rise in the tropical oceans, these episodes are liable to become more frequent and widespread.

28.2.3 OCEAN ACIDIFICATION

As mentioned in Section 27.4, the average pH of surface water has decreased by 0.1, from 8.2 to 8.1. The negative impact of surface-water acidification on marine organisms has been widely documented by laboratory studies of fish and calcifying organisms. The latter produce calcium carbonate to form their skeletons (corals) or shells (oysters, mussels and other shellfish, as well as certain algae at the bottom of the food chain, e.g. cocco-lithophores).

Experiments have been carried out on cod (*Gadus morhua*) at levels of atmospheric CO_2 four to ten times higher than current values. Certain organs showed signs of damage, with anomalies to the pancreas, liver and kidneys when the larvae develop in waters that are more acidic than present levels. Other studies were carried out at high CO_2 levels (600 ppm and 1000 ppm) on silversides (*Menidia beryllina*), a species of fish that is very common off the north-east coast of the USA. The studies find that the survival rate of larvae exposed to such levels of acidity decreases by 50% at 600 ppm (and by 75% at 1000 ppm). It should, however, be recalled that these experiments involve future scenarios in which atmospheric CO_2 levels are higher than those of today.

How does increased acidity affect the development of calcium carbonate shells and skeletons (calcite or aragonite)? In the same way as a drop of vinegar damages a marble table. Calcium carbonate

(a)   (b)   (c)

Fig. 28.3 In the reefs of the D'Entrecasteaux Islands (Papua New Guinea), the pH falls from 8.1 (the pH of normal seawater) to about 7.4 within a distance of a 100 m, due to underwater volcanic vents that eject CO_2 (Fabricius et al. 2011). Bubble tracks mark CO_2 emission. This natural situation illustrates how increasing acidity of the seawater (due to the increase in pCO_2) interferes with coral growth. (a) The control site, pH 8.1. (b) Between pH 8.1 and 7.8 only large *porites* remain, since complex corals with branches and tabular forms cannot develop. (c) Below pH 7.7 (> 1,000 ppm of CO_2), the reef ceases to grow.

Sources: Photographs by Katharina Fabricius/Australian Institute of Marine Science.

that is synthesized in sea water can be affected when the acidity increases. If, due to the increased acidity, the water becomes undersaturated in carbonate ions, shells start to erode and dissolve. When a living membrane protects the skeleton, as is the case in corals, the damage is minimized, but the skeleton may still be weakened. Experiments have shown the negative effect of increased acidity on calcifying organisms, but until recently no observation has been reported of how organisms adapt over time to changes in pH. Recent *in situ* observations of the coral reefs of Papua New Guinea help fill this gap, and confirm the loss of biodiversity caused by acidification (Fabricius et al. 2011; Fig. 28.3). Although this investigation involves a natural situation in the absence of human intervention, it illustrates the large-scale effect of increasing the content of dissolved CO_2. Coral reefs offer a home to tens of thousands of species; in view of the biodiversity that they support, the impact of acidification on coral skeletons is a particularly critical issue. The reefs of the D'Entrecasteaux Islands (Milne Bay Province, Papua New Guinea) are located in clear shallow waters at temperatures of 28–29°C. Observations were made in a small area within which, owing to underwater volcanic vents that emit CO_2, the pH falls from 8.1 to almost 7.4 over a distance of 100 m. *These experimental observations demonstrate that coral diversity diminishes as acidity increases.* Between pH 8.1 and 7.8, the reef community gradually changes. The coral cover remains constant, but only large porites remain, since complex corals with branches and tabular forms cannot develop: the taxonomic richness of hard corals is reduced by 39% (Fig. 28.3b). Below pH 7.7 (> 1000 ppm of CO_2), the reef ceases to develop – no reef has been observed below this pH (Fig. 28.3c).

28.3 Land

In the terrestrial biosphere, changes in the range of species and fauna density have been observed in most regions of the world and in numerous taxonomic groups (insects, amphibians, birds, mammals, and so on). We offer a few illustrations of these changes.

28.3.1 WARMING AND MIGRATION

With higher temperatures, several species of insects and butterflies, including predatory butterflies such as the gypsy moth, became far more abundant in Central Europe at the beginning of the 1990s, and the range of several species of damselflies, dragonflies, cockroaches, grasshoppers and locusts (crickets) shifted northwards.

The case of the *pine processionary moth* is a good example of the relation between climate change and expansion of habitat. Its urticating caterpillar attacks pines and cedars. Its range adapts rapidly to climate change, as the caterpillars develop over the winter and are sensitive to increases in temperature during that season. The progression of its habitat has been widely studied in France (Battisti et al. 2005). Before the 1990s, winter temperatures acted as a barrier south of the Paris basin. When this barrier failed in the 1990s, northward expansion started and the range of the species shifted north by an average of 87 km between 1972 and 2004, while winter temperatures rose by an average 1.1°C in that area. Since the beginning of the 1990s, the rate of northward advance has been 5.6 km/yr. Climate warming could be tracked even more readily by this species, through accidental transport when planting infested trees (Robinet et al. 2012). In the Italian Alps, the upward expansion of the pine procession moth, also related to warming, varies between 7 and 3 m/yr, depending of the exposure of the slope (south or north) (Battisti et al. 2005).

Our last example is the European corn borer, which 20 years ago had been observed only as far north as Orléans, is now found in Belgium, some 300 km further north. We recall that the temperature gradients are approximately 1°C per 250 km in latitude and 1°C per 150 m in altitude.

Disease-bearing parasites and insects are no exception to this trend, and the ranges of many of them have also shifted to higher latitudes and altitudes. The patterns of distribution of such vector-borne diseases as malaria, dengue and so on, and food- or water-borne infectious diseases such as diarrhoea, have also been affected by climate change. Finally, with developments in transport, certain diseases and pests can settle in regions where today's warmer climate conditions allow them to thrive and develop. Various animal diseases have thus emerged in new regions: West Nile virus in horses in the Camargue (the South of France); sheep bluetongue first observed in Corsica (an island in the northern Mediterranean Sea) in 2000; leishmaniasis in dogs and humans in the Mediterranean (temperature is one of the main barriers to its northward progression in Europe); and dengue-carrying mosquitoes first observed in France in 1999. The plant world is also affected by these emerging diseases: the subtropical whitefly *Bemisia tabaci*, first identified in Europe in the last decade, now poses a serious threat to greenhouse crops in Southern Europe. Ticks can also carry many pathogens, such as Lyme disease (borreliosis) and viral encephalitis: borreliosis cases are reported wherever ticks are to be found, whereas encephalitis occurs only in certain areas. Recent warming seems to have caused a northward shift of encephalitis outbreaks in Europe to Scandinavia, prompted by milder winters (cases have multiplied in Sweden). In France, Lyme disease is strongly on the rise in many regions, but climate change may not be the only factor.

Finally, plants have also colonized new habitats. Northward shifts of boreal forests, however, are increasingly disturbed by infestation and wildfires that obscure the longer term trends.

Birds and butterflies

Can some species act as an indicator of climate change over a wide distance scale? What 'climate debt' (delay in the response of a species to changes in temperature), if any, is accumulating? Birds and butterflies, which disperse easily, should be capable of tracking climate change more easily than other taxonomic groups. Recently, the climate debt accumulated by bird and butterfly communities in Europe was measured at 9490 and 2130 sample sites, respectively, and compared over two decades (1990–2008) (Devictor et al. 2012). Observational data were used to assess whether certain groups respond more or less quickly than others over large areas. Researchers calculated the *Community Temperature Index* (CTI) for each bird and butterfly community from 1990 to 2008. The map in Fig. 28.4 shows the changes with time in bird and butterfly CTI for each country.

Given the temperature increase between 1990 and 2008, species should have shifted northwards by 249 ± 27 km. The study shows that changes in the composition of the communities are rapid, but different between birds and butterflies: bird communities shifted northward by only 37 ± 3 km and butterflies by 114 ± 9 km, thus accumulating climate debts corre-

Fig. 28.4 European trends (1990–2008) in the Community Temperature Index (CTI) for birds (brown arrows, 9490 sampled sites) and butterflies (green arrows, 2130 sampled sites). The size of the arrows is proportional to the change in CTI. The increase in CTI is greater for butterflies (114 ± 9 km) than for birds (37 ± 3 km). This figure indicates that neither birds nor butterflies keep up with temperature increase.

Source: Modified from Devictor et al. (2012). Reproduced with permission from Nature Publishing Group and Vincent Devictor, ISEM Université de Montpellier.

sponding to a lag of 212 km (for birds) and 135 km (for butterflies) over two decades. It seems that not only are birds and butterflies not tracking climate change fast enough on a large distance scale, but that the gap between the two groups is widening. This may disrupt interactions between the species.

The finding that butterflies in Europe respond on average more rapidly may be due to their relatively short life cycle and to the fact that, as cold-blooded animals, they can track changes in temperature very closely. This finding is echoed in palaeoclimate studies: the palaeo-archives reveal that during large-scale climate changes of the past, such as glacial–interglacial transitions, beetles were the first migrators to colonize a warmer environment.

28.3.2 WARMER, EARLIER

In middle-to-high latitudes, as global warming causes spring to arrive earlier, the growth cycles of various crops, grasslands and forests change and accelerate. In Western Europe, in the agricultural sector in France, observations show that dates of flowering and fruiting have advanced throughout the country. Since 1970, maize sowing dates have advanced by almost one month at the four sites monitored by the Institut National de la Recherche Agronomique (France).

The impact of climate change on cereal growth varies according to the crop. Under similar farming techniques and for similar water availability, maize yields have increased while wheat yields have stagnated, due to the sensitivity of wheat to heat stress (shrivelled wheat grains yield less flour). Over the past 20 years, maize crops have shifted northward within Europe: the surface area of cultivated land has exploded in Denmark, while cultivation areas in France have slowly shifted to the north-west.

The effects of warming on prairies vary from one region to another. In Northern Europe, the main effect is to increase biomass. In the south, while rain patterns have not changed significantly, the dry period with no plant growth has lengthened in water-scarce regions (south-eastern Europe) by 8–10 days per decade owing to increased evapotranspiration (plant transpiration, which increases with warming). A dividing line appears around 45°N in Western Europe, separating the 'Winning North' from the 'Losing South', with areas above an altitude of 1000 m being included in the winning regions.

Vine and wine

The impact of warming on grapevine provides a good illustration of its impact on vegetation in general. Although it is only a modest facet of the world's economy, we dwell on this example because good wine is a universally recognized nectar. The history of vines and wine is closely linked to that of humankind. The beginnings of wine-growing and winemaking date back 8000 years in the Caucasus and Mesopotamia (6000 BC), and continued through Egypt and Phoenicia (3000 BC), to Greece (2000 BC), then to Italy, Sicily and North Africa (1000 BC) before reaching Spain, Portugal and southern France (1000 to 500 BC). Vineyards were planted in Northern Europe and the British Isles between 500 BC and the Middle Ages (AD 1000). Finally, with colonization, vines were imported into South Africa, Mexico, Chile and Argentina in the 16th and 17th centuries, and into California and Canada in the 18th century. At present, with favourable climate and soil conditions, they are grown on five continents, in a band of latitude centred around 40°N (between 30°N and 50°N) and 30°S (between 20°S and 40°S). In these regions, the average yearly temperature varies between 10°C and 20°C.

Wine quality is influenced mainly by climate, grape variety and soil type. About 10,000 grape varieties exist, of which 3500 are conserved in the Domaine de Vassal (Institut National de Recherche Agronomique) in the South of France (Part V, Note 2). Varieties suited to a given region are largely determined by the local climate. Four climate types, with growing season temperatures between 13 and 20°C, are well suited to the 20 most widespread varieties: cool (13–15°C), intermediate (15–17°C), warm (17–19°C) and very warm climates (19–24°C), illustrated by the Jones diagram in Fig. 28.5. On average, each variety acclimatizes to a temperature range of 2–3°C, in some rare cases up to 4°C. Certain varieties are particularly sensitive to the climate: Pinot noir (2°C range) is restricted to 'cool-intermediate' growing areas centred around 15°C, while Chardonnay can adapt between 14°C and 17°C, spanning from cool to warm areas. When warming exceeds this range, the varieties struggle to adapt!

How does warming affect wine-growing? Earlier flowering brings earlier harvests. Also, under warmer conditions, grapes become sweeter and less acidic. Sugars are transformed into alcohol, producing a less acidic wine with a higher alcohol content. Acidity gives bite and an edge to wine and can often make it feel lighter in the mouth, while alcohol gives body and richness. As temperatures rise, wines will feel fuller and heavier in the mouth. Higher temperatures

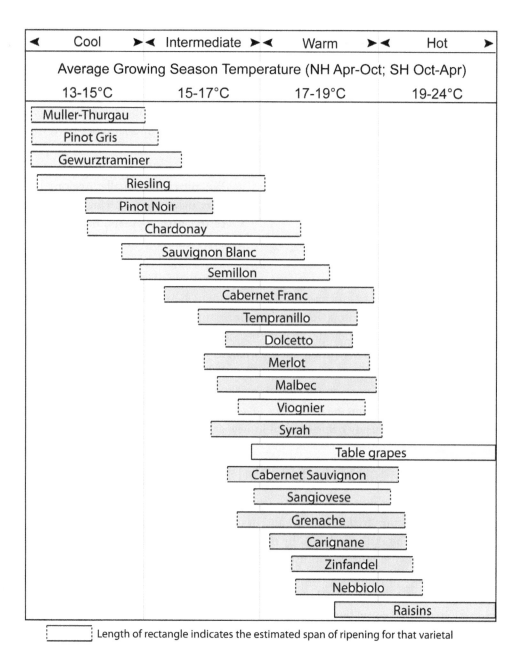

Fig. 28.5 The relation between climate and grape maturity for the major grape varieties (Jones 2006). The temperature range of the growing season (tolerance about 2°C) is indicated for various grape varieties. The climate maturity groupings are based on the relationship between phenological requirements and climate for high to premium quality wine production in the world's benchmark regions for each variety. The temperature limits (dashed lines at the end of the bars) are only indicative, but changes of more than ±(0.2–0.6)°C are highly unlikely. Green, white variety of grapevine; pink, red variety grapevine; blue, dessert grapes. Source: Jones (2006). Reproduced with permission of Gregory V. Jones.

during the growing season can have consequences that are positive when the grape variety reaches optimum conditions, or negative when the optimum temperature is exceeded. The cultivated variety is then less suited to the warmer climate.

In 2005, a study was carried out by G.V. Jones on 27 high-quality wine-growing regions around the globe, documenting changes in climate between 1950 and 1999 and their consequences on the industry (Jones 2006; see also Jones et al. 2012). In 17 of the 27 regions, mostly in the USA and Europe, growing-season temperatures increased by an average of 1.2°C. (In the Bordeaux area (France), temperatures warmed by about 2°C from 1967–89 to 1990–2005). In 2 of the 27 regions, growing-season temperatures had exceeded the optimum conditions

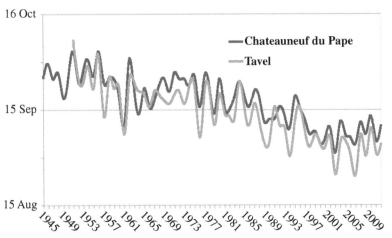

Fig. 28.6 Variation in the starting date of grape harvesting in southern Côtes du Rhône, France, for Tavel *Appellation d'Origine Contrôlée* (AOC) over the period from 1951 to 2012, and for Châteauneuf du Pape AOC over the period from 1945 to 2012. For these two vineyards, grape harvesting is now 3 weeks earlier than in the 1950s/1960s.
Source: Inter-Rhône data. Reproduced with permission of the Observatoire National sur les Effets du Réchauffement Climatique (ONERC).

above which vintage ratings tend to decline. In the wine-growing regions of the Southern Hemisphere (SH), by contrast, no significant changes were detected. This is consistent with the global temperature changes: in the SH, the mid-latitudes have warmed less than in the NH owing to the presence of oceans (Fig. 25.3).

We turn to a smaller region, France, with its rich variety of wine-growing areas. Over the past 50 years, the warming has brought earlier bud burst and flowering and ripening, harvest dates have advanced by 3–4 weeks (Fig. 28.6) and the temperature of the growing season has increased. Alcohol contents have risen by 1–2° depending on the region, while acidity has dropped, yielding consistently high-quality wines over the past 20 years. The Jones diagram indicates that with such warming, some varieties are becoming less and less well-adapted to the current local climate.

In conclusion, global warming has already affected vineyards in many places in the world, often beneficially. But, as temperatures continue to rise, in the words of Jones et al., *'some traditional wine making regions are scrambling to adapt, while others see themselves as new wine frontiers'*. Vineyards are already being planted in new areas in the British Isles.

28.3.3 CHANGES IN DEVELOPMENT AND LIFE CYCLES

In recent decades, many changes have been recorded in the animal world (changes in size of individuals, reproduction frequency etc.) and in the plant world. In the latter, however, certain changes cannot be easily explained by the recent climate change alone. Thus, forests are unquestionably growing more vigorously at middle and high latitudes, such as in Canada. Several factors can contribute to the increase of photosynthesis that has caused this higher productivity; for example, increased atmospheric CO_2, increased nitrogen compounds from human activity (which contribute a further fertilizing effect) and warming. It is still difficult to assess the impact of each of these factors. At the same time, however, boreal forest fires at high latitudes have increased, which is also a consequence of the observed warming: over the past 20 years, the area burned annually in western North America by forest fires has doubled.

As regards insects and diseases, certain signs, such as changes in the reproduction cycle of various insects, may be linked directly to climate change. In France, for instance, more frequent reports have been recorded of three generations per year of the *codling moth*, a pest of apples and stone fruits that causes significant damage to orchards during hot, humid summers. There has also been an increase in the diversity of *aphids*, the small sap-sucking insects involved in various phenomena studied in *phenology* (Part V, Note 3). Conversely, the *phomopsis fungal pest* of the sunflower died out in south-west France after the 2003 heat wave.

28.4 Portents of dysfunction

To what extent will the seasonal life cycles of living organisms respond fast enough to adapt to the new calendar? Some collateral damage has already occurred, and dysfunction is bound to take place when the supply and demand for food fall out of step during the reproduction period.

The following classic example, drawn from observations in the Netherlands, illustrates the underlying mechanisms (Fig. 28.7). The great tit, the caterpillar and the oak tree coexist in a close relationship. Great tits generally lay their eggs at a time of year when food is plentiful, in the form of newly hatched caterpillars feeding off tender oak leaves that have just opened. In the past two decades, early warming has disrupted these patterns: oak buds are developing a few days earlier, and caterpillars are hatching 2 weeks ahead of schedule – too soon to access the new leaves – whereas the great tits have not yet changed their date of laying. Warming has depleted the resources for caterpillars and birds alike, weakening their reproduction cycles.

Many other instances of loss of synchronization have been observed in recent decades, whether in freshwater systems or in pelagic (surface-water) ocean communities, between primary (phytoplankton), secondary (copepods and zooplankton) and tertiary producers (meroplankton, or plankton consisting of larval-stage animals). Lastly, disruptions at the top of the Arctic food chain must be mentioned, where polar bears and seals now struggle to survive against early melting of the ice floes. Polar bears mostly hunt ringed seals on the ice floe. Their demand for food is particularly high at the end of their hibernation period, which runs from October to February, when the female must feed her newly born cubs. The earlier onset of ice floe melt (Hudson Bay, Canada now has three more ice-free weeks than 30 years ago) and ice fragmentation disrupt and reduce the bears' hunting territory, making it more difficult for them to find food. Recently, more cases of bear cannibalism, drowning and weight loss have been reported; these are clear indications of their struggle to survive. This last example illustrates the vulnerability of life cycles when an abrupt climate change modifies the timing between living resources and the demand for food in the crucial reproduction period. Such disruptions in the ecosystem have been detected at many levels.

In summary, warming acts on life cycles through two types of change:

- Increased average temperatures cause animal and plant migration (terrestrial plants, birds, insects, mammals, algae, fish, germs, viruses, bacteria etc.) and adaptation (changes in size, reproduction frequency etc.).
- Earlier warm seasons shift animal and plant life cycles, possibly generating a mismatch between reproduction dates and food availability dates.

When a habitat is no longer viable and/or reproduction is no longer possible, failure to adjust or adapt ends in demographic collapse and extinction. This, in turn, can weaken an entire ecosystem.

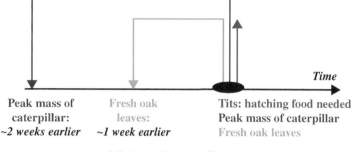

Peak mass of caterpillar: ~2 weeks earlier

Fresh oak leaves: ~1 week earlier

Tits: hatching food needed
Peak mass of caterpillar
Fresh oak leaves

Mistiming due to earlier spring

Fig. 28.7 An illustration of ecosystem disruption caused by recent warming. Oak budding, caterpillars and egg laying by tits were previously synchronized, providing food resources for both caterpillars and chicks. The recent warming has advanced the date of caterpillar hatching and budding, but not that of egg laying. Since hatching of caterpillars precedes opening of buds and hatching of chicks, the food resources of both caterpillars and chicks are reduced.

Chapter 29

Warming in the 20th century: natural or human-induced?

What is the origin of recent climate changes: nature or man? Natural causes may be divided into two categories: external causes, which arise outside the climate system and affect its functioning, and internal causes that are inherent in the dynamics of the system. On the timescale of 10 years, the main external factors are variations in solar activity (more or less the energy delivered to the Earth's surface) and atmospheric disruptions due to volcanic activity (cooling the Earth's surface). We examine these external factors in Section 29.4.

Internal causes arise from interactions among the components of the climate system. On our timescale, it is mainly the interaction between the atmosphere and the ocean that affects the climate. Here, the present warming trend could be the result of multi-decade fluctuations in this interaction, involving the natural variability of the climate system. Extensive research is currently under way to estimate multi-decade variability in the regional climate systems. It is unlikely, however, that such effects could cause the world's climate to change as uniformly and to such an extent as is observed.

Of course, there are also effects due to human activity. Of all the human-related changes that could influence the climate, the major factor affecting the entire planet is the drastic

Climate Change: Past, Present and Future, First Edition. Marie-Antoinette Mélières and Chloé Maréchal.
© 2015 John Wiley & Sons, Ltd. Published 2015 by John Wiley & Sons, Ltd.
Companion website: www.wiley.com\go\melieres\climatechange

increase in atmospheric greenhouse gases (GHGs), primarily carbon dioxide (CO_2). As the atmospheric CO_2 concentration is defined by the global carbon cycle, we start with a brief introduction to this cycle.

29.1 The carbon cycle prior to the industrial era

29.1.1 THE KEY ROLE OF CARBON IN THE UNIVERSE

The carbon atom occupies an exceptional position in the universe. Its synthesis is essential to all processes that have occurred since the first stars formed. It is the fourth most abundant element after hydrogen (H, 90.8%), helium (He, 9.0%) and oxygen (O, 0.08%). The elements oxygen (O), carbon (C), neon (Ne) and nitrogen (N) make up 0.13% of all the atoms in the universe, while all the others, with the exception of hydrogen and helium, represent only 0.012%. Although estimates vary, the order of magnitude remains the same. Since helium and neon are chemically inert, the overwhelming majority of all chemical reactions in space produce molecules made up of H, O, C and N atoms. These reactions generally produce small molecules, mainly water (H_2O), methane (CH_4) and carbon dioxide (CO_2). The stability of carbon bonds and carbon's ability to bond with up to four atoms at a time allow it to form an extremely rich variety of complex multi-atom molecules, such as amino acids, or long molecular chains as found in biopolymers. Thus, for reasons linked both to nuclear physics and to its electronic properties, carbon and its compounds are the backbone of life.

Carbon chains are both very strong and very flexible; they can be recombined with other chains, changing the order of their sequences. What other atoms have the same features? The natural abundance in the universe of silicon, another atom that can form four bonds, is an order of magnitude smaller than that of carbon. Its bonds require less energy to break, and silicon does not offer the same variety of chemical possibilities as carbon. Out of 110 molecules recorded in the interstellar medium, 83 include carbon atoms, while only seven include silicon atoms. Throughout the universe, carbon appears as the most favourable element to form the basis of life.

Photosynthesis is the process by which organisms stay alive: it allows the energy contained in the photons to be stored. In photosynthesis, the carbon from CO_2 stores this energy in the form of sugars. Through respiration, the reverse reaction, this energy is released and used by the organism itself, or by another organism (Section 29.1.3). The carbon in the sugars is then transferred and recombined with oxygen to regenerate CO_2. In the carbon cycle, biomass is a reservoir through which the carbon atom actively transits.

29.1.2 CARBON RESERVOIRS

The carbon cycle (Fig. 29.1) describes the exchange of carbon between the Earth's various reservoirs, in which the atom may be stored in various forms. The various reservoirs on land exchange CO_2 mainly with the atmosphere, over a range of different timescales. Box 29.1 explores their characteristics.

BOX 29.1 THE IMPORTANCE OF CO$_2$ ON EARTH

CO$_2$ and organic matter

CO$_2$ is one of the key components in the life cycle on Earth. It is one of the three building blocks, along with water vapour and sunlight, that constitute the starting point of all living matter. As well as initiating life, it is also the final degradation product once the organic matter is completely oxidized. Methane (CH$_4$) is also part of this cycle; in the absence of oxygen, organic matter degrades into methane when carbon cannot completely oxidize. This means that methane may still oxidize and release heat (i.e. supply energy), while CO$_2$, which contains a fully oxidized carbon atom, cannot burn. CO$_2$ is therefore the end product of the decomposition of all organic matter, whether this matter has burned (forest, savannah or peatland fires, or combustion of organic matter such as oil, coal or gas) or is decomposed in the soil through oxidation by microbial organisms. When organic matter is buried rapidly in an anaerobic environment (wetlands, the seabed), carbon chains synthesized by living organisms transform into compounds in which carbon atoms remain bound to hydrogen atoms, such as fossil fuels. The present reserves of fossil fuels were formed between millions and hundreds of millions of years ago, during the Carboniferous era (360–300 Myr ago).

CO$_2$ and carbonate rocks

Just like fossil organic matter, most limestone mountain ranges are also the result of the cycle of life. As well as organic matter, life produces carbonates in the form of shells, coral and so on that are deposited and buried in the seabed, slowly transforming into limestone. Tectonic plate movement later exhumes the limestone, as can be seen almost everywhere on Earth. *How were these carbonates synthesized by life?*

Over hundreds or thousands of millions of years, the water cycle, with its rain slightly acidified by CO$_2$ dissolved from the secondary atmosphere (mainly water vapour, CO$_2$ and N$_2$), leached the soil and supplied the ocean with carbonate and calcium ions. From these ions, calcium carbonate was synthesized by organisms and, on their death, deposited in the form of carbonate rock sediment. This created a spectacular transformation in the composition of our atmosphere by slowly stripping it of its main component, CO$_2$. In comparison, on our similarly sized neighbouring planet, Venus, which lacks both liquid water and water cycle, the atmosphere still contains nearly 200,000 times more CO$_2$ gas than ours. This is the same order of magnitude as would be found here had our CO$_2$ not been transformed into carbonates. The greenhouse effect caused by the atmospheric CO$_2$ on Venus is the main reason for its scorching surface temperature (~460°C).

CO$_2$ and volcanic activity

Another source of CO$_2$ is limestone calcination. This process requires higher temperatures than are naturally found at the surface of the Earth, except in volcanoes, which are therefore a long-term source of atmospheric CO$_2$. In the manufacture of cement from limestone, the same mechanism emits CO$_2$ into the atmosphere. In a given year, volcanoes sometimes contribute only very small amounts of CO$_2$ to the atmosphere, but this process is crucial, as it compensates for the slow decrease in atmospheric CO$_2$ from precipitation and storage in ocean and sedimentary reservoirs. Over very long times, volcanic emissions thus help to maintain the equilibrium level of CO$_2$ in the Earth's atmosphere.

Carbon stocks are expressed in gigatonnes (1 GtC = 10^{15} grams of carbon). Transfers between reservoirs are expressed in GtC/yr. Carbon may be stored in gas, liquid or solid form in the atmosphere, in the sea and on land. On land and in the oceans, further distinctions may be drawn between inorganic and organic carbon stocks. Estimates of carbon stocks in these three main reservoirs, together with the exchange of flux are shown (black and blue) prior to

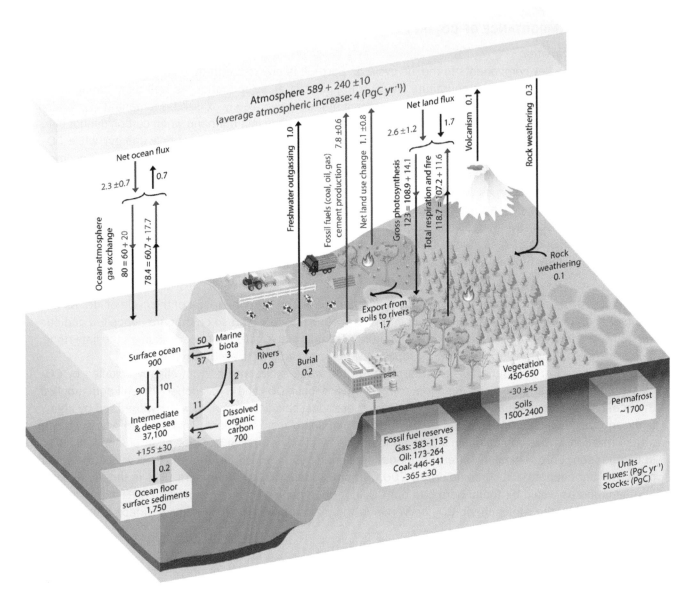

Fig. 29.1 A simplified schematic diagram of the global carbon cycle connecting the three reservoirs (atmosphere, ocean and continent). Numbers represent reservoir size (in PgC) and carbon exchange flux (in PgC/year). Black: reservoir size and natural exchange estimated for times prior to the Industrial Revolution (about 1750). Red: anthropogenic fluxes averaged over 2000–9, and cumulative changes in the reservoir over the period 1750–2011 caused by human activity (1 PgC = 1 GtC).

Source: IPCC 2013. *Climate Change 2013: The Physical Science Basis.* Working Group I Contribution to the Fifth Assessment Report of the Intergovernmental Panel on Climate Change, Figure 6.1. Cambridge University Press.

the industrial era in Fig. 29.1, as published in the IPCC 2013 report. Only the orders of magnitude are significant, as estimates vary depending on definitions and sources.

Four main continental reservoirs can be distinguished:

- An inorganic reservoir of carbonate rock (not included in Fig. 29.1). Depending on the source, it is estimated at between 50 and 100 million GtC.

- A 'fossil' reservoir of organic matter, including all fossil fuels (coal, gas and oil). This organic matter was buried and, lacking oxygen, failed to degrade into inorganic matter. In the IPCC 2013 report, the *reserves* indicated (1000–2000 GtC) predate the Industrial Revolution. Human activity has now depleted these reserves by ~365 GtC.
- A reservoir of terrestrial organic matter, which enters the cycle of living matter (~2500 GtC) either as living matter (mainly plants; the animal kingdom constitutes less than 1 GtC) or as decaying dead matter. This reservoir can be subdivided into the 'above ground part' of plant growth (including living roots) (450–650 GtC), and the underground part (excluding living roots), which includes decaying organic matter and the chain of life that its decomposition sustains (1500–2400 GtC).
- A reservoir in the permafrost (~1700 GtC)

In the oceans, a similar distinction may be drawn between the following:

- Living matter, found mainly in the ocean surface layers (~3 GtC).
- Carbon dissolved in the ocean mass (~38,000 GtC), by far the majority of all carbon present in the oceans. This reservoir includes dissolved inorganic carbon (CO_2 dissolved in the water and the corresponding ions, HCO_3^- and CO_3^{2-}), as well as dissolved organic carbon. This can be further subdivided into two reservoirs, one in the ocean surface layer, which extends to a depth of about 100 m, and in which most of the marine photosynthesis occurs (~900 GtC), and the other in the lower layer (~37,100 GtC).
- Sediments, mainly containing carbonates.

Prior to the Industrial Revolution, the atmospheric reservoir contained approximately 590 GtC. At present, it contains about 830 GtC. Note that the atmosphere contains almost 50 times less carbon in the form of CO_2 than the oceans. Any change in oceanic carbon content can therefore have a significant impact on atmospheric CO_2.

29.1.3 EXCHANGE BETWEEN RESERVOIRS

The rates of exchange between these main reservoirs possess very different timescales. The highest rates are between the atmosphere and the biosphere on land, on the one hand, and between the atmosphere and the ocean, on the other.

Atmosphere–biosphere exchange on land

The plant world generates living matter by synthesizing organic matter, which stores energy through photosynthesis. *Photosynthesis* is the reaction that converts H_2O, CO_2 and solar energy into sugars or carbohydrates of generic formula (CH_2O), releasing an oxygen molecule. It is a means of producing fuel (carbohydrates) and storing energy:

$$6H_2O + 6CO_2 + \text{solar energy} \Rightarrow C_6H_{12}O_6 \text{(glucose)} + 6O_2 \quad \text{(photosynthesis)}$$

The organism then uses the stored energy in complex processes, notably respiration. In respiration (where oxygen is consumed), the fuel is oxidized in a succession of biochemical reactions and decomposed into $H_2O + CO_2$, thus releasing the stored energy as and where needed in the organism:

$$C_6H_{12}O_6 \text{(glucose)} + 6O_2 \Rightarrow 6H_2O + 6CO_2 + \text{energy} \quad \text{(respiration)}$$

Photosynthesis and *respiration* are the two motors that drive the exchange of CO_2 between the atmosphere and the land-based biosphere.

Respiration as described above is said to be *autotrophic*; that is, carried out by the living organism itself (the same process exists in the animal world, the only difference being that the individual does not synthesize the organic matter directly, but takes it from the plant world). Another form of respiration is set in motion on the death of the individual or the plant: the compounds that make up the dead organic matter are still a source of energy. They are slowly decomposed by bacteria that use oxygen from the air to burn the organic matter and retrieve the energy stored within. This process is called *heterotrophic respiration*, because it is performed by organisms other than that to which the organic matter originally belonged. In the carbon-cycle balance, the flow resulting from these two types of respiration (both of which emit CO_2 into the atmosphere) are of the same order of magnitude.

Currently, photosynthesis by land plants absorbs approximately 120 GtC/yr of atmospheric CO_2. The living organism uses part of this stored energy through respiration (~60 GtC/yr of CO_2 emitted in *autotrophic respiration*). The rest goes to increase the store of organic matter on the Earth. This store is slowly decomposed by micro-organisms into inorganic molecules (*heterotrophic respiration*), thereby releasing CO_2 (~60 GtC/yr). In all, about 120 GtC of CO_2 returns to the atmosphere each year.

Photosynthetic activity and respiration vary according to the seasons of the planet, which are in opposite phase in the two hemispheres. Worldwide, about 120 GtC are exchanged in this way each year between atmosphere and land masses.

Atmosphere–ocean exchange

This exchange is driven by the difference in CO_2 partial pressure between the atmosphere and the ocean. The partial pressure of CO_2 dissolved in the ocean varies both with temperature and with marine photosynthesis, which removes dissolved CO_2. It decreases when the temperature increases (which is why a bottle of carbonated water loses gas more quickly when it is warm). Various regions of the ocean can thus become either carbon sinks or carbon sources, depending on the season and local photosynthesis activity. Averaged over a year, input and output balance out: some 80 GtC are exchanged each year with the atmosphere.

All in all, some 200 GtC of CO_2 are exchanged each year between the atmosphere and the Earth's surface.

29.1.4 THE CARBON CYCLE IN GLACIAL PERIODS

During glacial–interglacial transitions, large variations take place in the atmospheric concentration of CO_2. It oscillates between a high (280–300 ppm) in interglacials and a low (180–200 ppm) during glacial periods (Fig. 19.2). The fastest change occurs during deglaciation:

the 100 ppm increase takes only about 10,000 years. These variations are due mainly to the reorganization of ocean circulation during the transitions. The ocean contains far more carbon than does the atmosphere. Any change in ocean circulation that affects the carbon cycle in the ocean reservoir therefore has an amplified effect in the atmosphere. For example, consider the onset of a glacial period, in which the deep ocean circulation becomes slower. Exchange between surface and deep water is reduced, causing greater stratification in the ocean. Nevertheless, the deep water continues to be enriched by organic residue falling from the surface layer (where biomass is synthesized), and thus contains a greater amount of organic matter and carbon compounds. As deep-water formation decreases and the deep-water current slows, its carbon stock increases at the expense of that of the atmosphere. During deglaciation, the opposite phenomenon occurs: CO_2-rich water resurfaces more often, returning CO_2 to the atmosphere: the CO_2 reservoir of the ocean diminishes, and the atmospheric reservoir again increases.

Two other mechanisms also decrease atmospheric CO_2 at the onset of a glacial period, but more moderately. First, since gases are more soluble in cold seawater, the ocean extracts CO_2 from the atmosphere as the partial pressures in the atmosphere and the ocean come to equilibrium. This mechanism increases the amount of CO_2 dissolved in the ocean, enhancing the ocean carbon stock. This occurred at the onset of the last glaciation, which culminated towards 20,000 years ago, when ocean temperatures were lower on average by 2°C. Second, biomarine activity is assumed to be higher in glacial periods, activating the *biological carbon pump* of the ocean (Chapter 5). By contrast, changes in the land-based biosphere, which is less active during glacial periods, do not increase carbon storage in the soil and thus do not decrease the atmospheric CO_2.

29.1.5 YEAR-TO-YEAR VARIATIONS IN CARBON SINKS AND SOURCES

It is of interest to identify the years in which atmospheric CO_2 departs substantially from its trend, and to determine the causes, bearing in mind that fluctuations in the average global temperature are due mainly to El Niño/La Niña events and volcanic eruptions.

Since 1980, various groups have estimated CO_2 exchanges by dividing the Earth's surface into three regions: the tropics (25°S–25°N), the Southern Hemisphere (>25°S) and the Northern Hemisphere (>25°N) (IPCC 2013). CO_2 exchanges were further subdivided into ocean–atmosphere and land–atmosphere fluxes (Fig. 29.2). These assessments show, first, the scale of the disruption caused by an El Niño event. Such events generate emission of CO_2 from land in the tropics, the band where the El Niño events occur. The emissions were very pronounced in 1998, the strongest El Niño of the century, but were also detected during the events of 1982 and 1987–8. The year 1998 is memorable for the severe drought, accompanied by violent, devastating wildfires in tropical regions such as New Guinea, the Philippines and so on that are normally wet. A strong El Niño generally brings drought to parts of the tropics (Fig. 17.4), thereby contributing to wildfires and increased CO_2 emission. By contrast, El Niño events have no detectable impact on the ocean flux, indicating that they do not affect the CO_2 sinks and sources in the ocean. *What is the effect of volcanic eruptions?* The eruption of Mount Pinatubo (1991) caused average

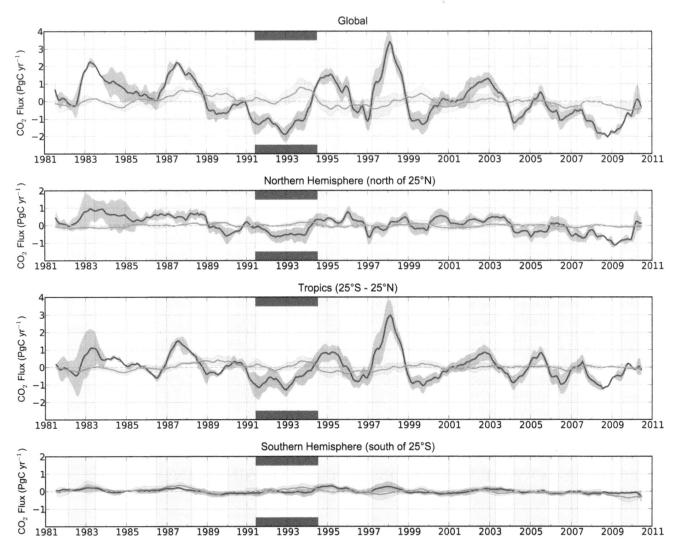

Fig. 29.2 Variability of the surface CO_2 flux over the globe, in the Northern Hemisphere, in the Tropics and in the Southern Hemisphere, from 17 atmospheric inversion models. Long-term mean CO_2 flux was subtracted and seasonal signal removed. Green, land–atmosphere fluxes; blue, ocean–atmosphere fluxes. Grey shaded regions indicate El Niño episodes, black bars indicate the cooling period following the Mount Pinatubo eruption. Positive flux means a larger than normal source of CO_2 feeding the atmosphere (1 PgC = 1 GtC).

Source: IPCC 2013. *Climate Change 2013: The Physical Science Basis.* Working Group I Contribution to the Fifth Assessment Report of the Intergovernmental Panel on Climate Change, Figure 6.9. Cambridge University Press.

global temperatures to drop by a few tenths of a degree. Following the eruption, the flux of CO_2 between land and the atmosphere became slightly negative, while that with the ocean was unaffected.

Obviously, to these natural variations we must add those related to human activity, which depend on the consumption of fossil fuels and the rate of deforestation, and hence vary from year to year.

29.2 The impact of human activity on the carbon cycle

29.2.1 THE INCREASE OF ATMOSPHERIC CO_2 AND CH_4 IN THE 'ANTHROPOCENE'

All existing climate records relating to the atmosphere show that, during the current inter-glacial period – that is, for about 10,000 years – atmospheric CO_2 remained fairly stable at around 280 ppm. From the beginning of the 19th century, atmospheric CO_2 has slowly increased, betraying an imbalance in the carbon cycle. The present situation shows that since then the atmosphere has already stored an additional 240 GtC, and now contains approximately 830 GtC (Fig. 29.1). For more than a century, the carbon cycle has been perturbed each year by the injection of additional CO_2 emanating from human activity. At present, human activity emits almost 10 GtC/yr, of which about half remains in the atmosphere, and the other half is absorbed more or less equally by the ocean and by the land biosphere. Let us examine more closely the history of this impressive change, and the situation at the beginning of the 21st century.

Atmospheric CO_2 has been measured directly and continuously at Mauna Loa (Hawaii) since 1958, due to the initiative of C.D. Keeling. Against all odds, Keeling found the resources to carry out these measurements. Since then, an observation and measurement network has been set up that now covers a large part of the planet. Figure 29.3 traces these changes since 1970, along with measurements of O_2, $\delta^{13}C$ and CH_4. For earlier periods, the composition of the atmosphere can be reconstructed by analysing air bubbles trapped in ice. These unique records enable us to track atmospheric changes over almost 800,000 years (Fig. 19.2). Figure 29.4 shows changes in CO_2 and CH_4 over the past 10,000 years, corresponding to the relatively stable and warm climate of the Holocene. Until the 19th century, the atmospheric concentrations of both gases remained approximately constant.

The atmospheric CO_2 and CH_4 contents in the present interglacial are comparable to those of each interglacial period of the past hundreds of thousands of years. In the warm inter-glacial climate, a characteristic equilibrium is reached in the atmospheric carbon cycle. During a glaciation, atmospheric CO_2 falls by about 30% (from 280 to 180 ppm), and atmospheric methane by about 50% (from 720 to 350 ppb) (Fig. 29.4). Until the 19th century, atmospheric CO_2 oscillated between 180 ppm and a maximum value of 300 ppm, in phase with the glacial and interglacial oscillations.

Disruption, the Anthropocene

Since the mid-19th century, however, this balance, established over the past 10,000 years, has been abruptly disturbed by a massive increase of all GHGs: by 2013, CO_2 had increased from 280 to 395 ppm (+115 ppm), CH_4 from 0.72 to 1.82 ppm (about +1 ppm) and N_2O from 0.27 to 0.32 ppm (+0.05 ppm).

The amplitude and rate of increase in GHGs over the past 150 years has profoundly dis-rupted the equilibrium conditions on Earth. The disruption is unequalled over the past several

Fig. 29.3 Atmospheric concentration of CO_2 and O_2, the stable isotope ratio $^{13}C/^{12}C$ in CO_2 ($\delta^{13}C$) and CH_4, recorded over the recent decades at representative stations in the Northern and Southern Hemispheres. Oscillations are seasonal variations. The anti-correlation between the average change of CO_2 and that of O_2 demonstrates that the increase in CO_2 stems mainly from combustion of organic matter. (a) CO_2 from Mauna Loa (MLO) and South Pole (SPO) atmospheric stations; (b) O_2 from Alert (ALT) and Cape Grim (CGO) stations; (c) $\delta^{13}C$ from Mauna Loa and South Pole stations; (d) CH_4 from Mauna Loa and South Pole stations. **Source:** IPCC 2013. *Climate Change 2013: The Physical Science Basis.* Working Group I Contribution to the Fifth Assessment Report of the Intergovernmental Panel on Climate Change, Figure 6.3. Cambridge University Press.

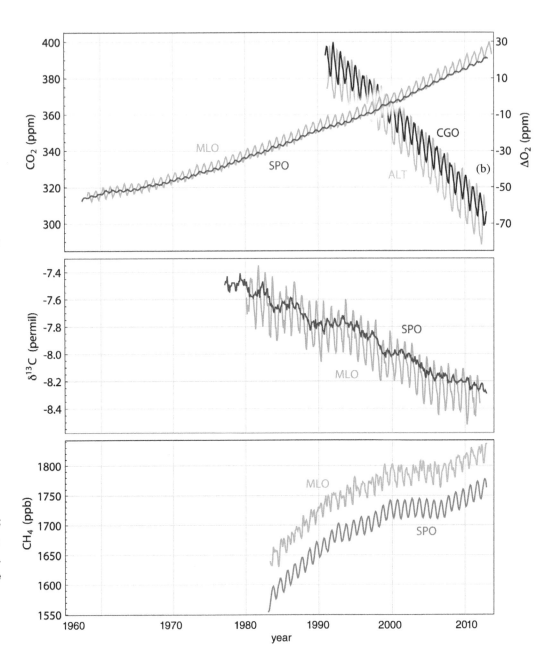

hundred thousand years or even, in all likelihood, over the past few million years. This change of direction, starting in the 18th century, sends the planet on a new route. Increased CO_2 and CH_4 (as well as other minor GHGs, such as N_2O) are not its only symptoms; all aspects of life on Earth are deeply affected. With the Industrial Revolution, the demographic explosion and large-scale exploitation of planetary resources, we have entered what P. Crutzen has dubbed the *Anthropocene*, a brand-new geological era.

29.2.2   CHANGES IN THE CARBON CYCLE

Figure 29.1 shows the carbon cycle at the beginning of the 21st century, where the red numbers in the reservoirs denote cumulative changes over the Industrial Revolution. Red

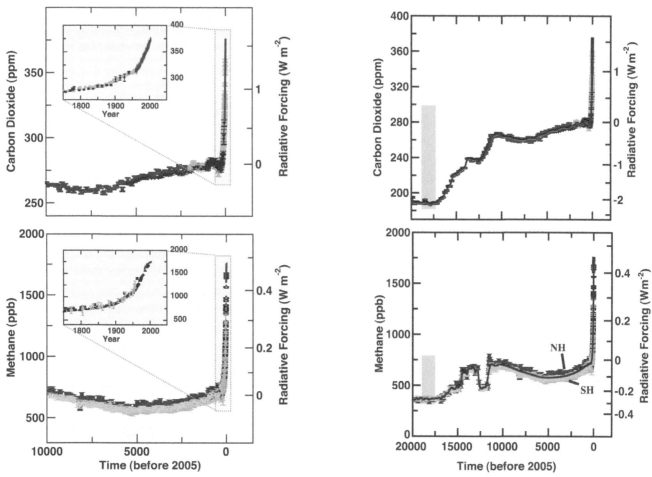

Fig. 29.4 Left panels: Atmospheric concentrations of carbon dioxide (CO_2) and methane (CH_4) over the past 10,000 years (large panels) and since 1750 (inset panels). Measurements shown are from ice-core (symbols with different colours for different studies) and atmospheric samples (red lines). The corresponding radiative forcings are shown on the right-hand axes of the large panels. The last 10,000-year time period corresponds to the present warm interglacial. Right panels: Over the past 20,000 years; that is, including the last glacial maximum and the present interglacial. Grey bars show the reconstructed range of natural variability for the past 650,000 years. Present concentrations far exceed the highest naturally occurring concentrations within this timescale in the past.

Source: IPCC 2007. *Climate Change 2007: The Physical Science Basis.* Working Group I Contribution to the Fourth Assessment Report of the Intergovernmental Panel on Climate Change, Fig SPM1 and Figure 6.4. Cambridge University Press.

arrows and numbers indicate fluxes arising from human activity, averaged over the period 2000–9. Total fluxes exchanged between the different reservoirs are shown according to their origin, natural or human. *Do not be misled by the indicated precision of the values! The uncertainty in the estimated exchange flux is about 20%: the values in this figure are adjusted to balance the flux.*

Over the past 10 years, anthropogenic emissions have increased atmospheric CO_2 by about 2 ppm/yr (4.25 GtC/yr), and atmospheric CH_4 more unevenly by up to 0.015 ppm/yr (0.03 GtC/yr). These two rates differ by a factor of 100. The same factor applies to the increase in atmospheric content of these two gases between the pre-industrial era and the 21st century (+100 ppm versus +1 ppm). It is therefore almost exclusively the increase in CO_2 that

Fig. 29.5 Fraction of fossil fuel emissions remaining in the atmosphere (airborne fraction) each year (bars), and five-year averages (solid black line). This fraction has been fluctuating around 50% for several decades.
Source: IPCC 2007. *Climate Change 2007: The Physical Science Basis.* Working Group I Contribution to the Fourth Assessment Report of the Intergovernmental Panel on Climate Change, Figure 7.4. Cambridge University Press.

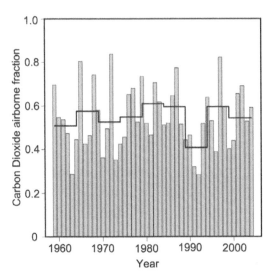

governs the changes in carbon content of the different carbon reservoirs. The changes discussed below are therefore restricted to the flux of CO_2.

Accumulation in the atmosphere of anthropogenic CO_2

The present human activity emits about 10 GtC into the atmosphere each year. *How does this figure compare with the natural flux between the Earth's surface and the atmosphere?* The latter is about 200 GtC/yr. It might be hoped that the transfer mechanisms between the atmosphere and carbon sources and sinks (biosphere and oceans) could easily absorb the 5% excess percentage of CO_2, without noticeably changing the atmosphere's contents. However, the mechanisms that developed over many thousands of years have in fact little leeway, and the carbon sinks can absorb, at best, only half of the anthropogenic CO_2 emissions. This is revealed by estimates of year-by-year CO_2 emissions and their accumulation in the atmosphere since 1960. Although the emissions are increasing, roughly 50% still accumulate in the atmosphere (Fig. 29.5).

Absorption of anthropogenic CO_2 by the terrestrial biosphere and the ocean

The oceans and land together absorb more than half the CO_2 generated by human activity. On land, it is directly absorbed by the biosphere. In the oceans, the excess atmospheric CO_2 is gradually absorbed through equilibration of the CO_2 partial pressure between the air and the water (Chapter 27). This gradual penetration takes place differently in the different oceans. In most oceans, the process is limited to the surface mixing layer, whereas in the North Atlantic it slowly sinks into the depths due to deep-water formation that takes place in that region (Fig. 27.6). The excess CO_2 in the ocean does not necessarily increase marine photosynthesis, as the amount of dissolved CO_2 is not a limiting factor for life. On land, the opposite happens. The land biosphere benefits from the increased CO_2 level, as it stimulates photosynthesis (plants grow better) and acts as a sink, extracting part of the excess CO_2 from the atmosphere.

Changes in sinks and sources resulting from human activity

Figure 29.6 (bottom panel) shows diagrammatically how carbon sources resulting from human activity (upper half) and corresponding sinks (lower half) have evolved since 1850. In this figure, the atmosphere is considered as a sink, since it accumulates CO_2. Emissions from fossil fuels and cement factories are shown in grey, and those due to deforestation in yellow. The sum of these sources (the envelope curve in the upper half of the figure), which is the total annual anthropogenic emission, is compensated by accumulation in the different sinks, represented by the symmetric curve in the lower half of the figure. The accumulated atmospheric

CO_2, which is a measured quantity (light blue), fluctuates significantly. The contribution of the ocean sink (dark blue) varies regularly. That of the land sink (green) is the complement of these two quantities. The figure illustrates how year-to-year fluctuations in the increase in atmospheric CO_2 are due to the variability of the land sink, which is often related to El Niño/La Niña events. Figure 29.7 illustrates the different sources and sinks of anthropogenic emission, averaged between 2000 and 2008, as estimated by the Global Carbon Project. On average, 45% of the CO_2 emitted by human activity accumulates each year in the atmosphere, while 29% is absorbed by the biosphere on land and 26% by the oceans.

29.2.3 THE SIGNATURE OF ANTHROPOGENIC EMISSIONS

How can we be sure that the changes in atmospheric CO_2 are caused by human agency? After all, each year, nature emits nearly 20 times more CO_2 into the atmosphere than we do. This is where our latest scientific tools come into play, in particular carbon isotopes. The various approaches give similar estimates.

First, we investigate the radioactive isotope of carbon, ^{14}C. Its signal can be used for dating prior to the 1950s, when thermonuclear testing severely disrupted atmospheric levels of ^{14}C. This isotope (half-life 5730 years; see Part III, Note 2) is present naturally in atmospheric CO_2, but not in fossil fuels (coal, gas and oil), as these were formed such a long time ago (millions of years) that all ^{14}C atoms have disappeared through radioactive decay. Carbon emitted by these fuels, therefore, does not contain any ^{14}C. The ^{14}C

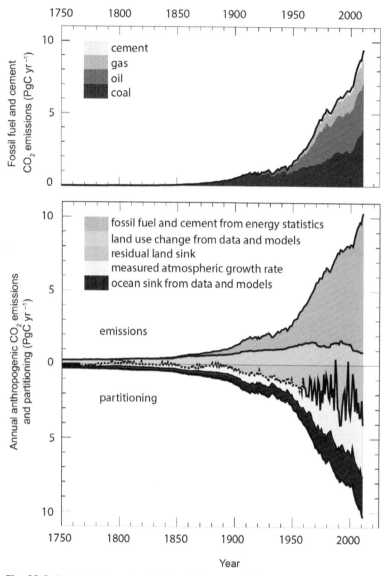

Fig. 29.6 Annual anthropogenic CO_2 emissions and their partitioning among the atmosphere, land and ocean (PgC/year), from 1750 to 2011. (Top) fossil fuel and cement emissions by category. (Bottom) upper part, Sources: emissions from fossil fuel and cement (grey) and from changes in net land use, mainly deforestation (beige); lower part, Sinks: land (green), atmosphere (light blue), ocean (dark blue). The land sink is estimated from the residual of the other terms (1 PgC = 1 GtC).

Source: IPCC 2013. *Climate Change 2013: The Physical Science Basis.* Working Group I Contribution to the Fifth Assessment Report of the Intergovernmental Panel on Climate Change, Figure 6.8. Cambridge University Press.

content of atmospheric CO_2 is recorded in plants that fix this gas, notably the trunks of century-old trees. Prior to 1950, the ^{14}C content in the atmosphere decreased with time while the amount of CO_2 increased. This observation indicates that the vast majority of excess CO_2 that

8.6±0.4 GtC/yr   92%

0.8±0.5 GtC/yr   8%

+

4.3±0.1 GtC/yr
45%

2.6±0.5 GtC/yr
27%

2.6±0.8 GtC/yr
27%

Calculated as the residual
of all other flux components

Fig. 29.7 The net destination of emissions generated by human activity (2003–12 average): 45% enters the atmospheric reservoir, 27% the continental reservoir and 27% the ocean. **Source:** Global Carbon Budget and trends 2013, based on Le Quéré et al. (2014) and CDIAC data. Reproduced by permission of the Global Carbon Project.

has been injected into the atmosphere does not contain ^{14}C, and can therefore be attributed to fossil fuels ... *or could it be volcanic emissions?*

Another observation relates to changes in atmospheric oxygen (O_2) and $\delta^{13}C$ (Fig. 29.3). If the increase in atmospheric CO_2 were the result of combustion of organic matter (fossil fuels and deforestation), atmospheric oxygen levels should decrease, because combustion consumes oxygen and emits CO_2. However, if the increase in atmospheric CO_2 results from volcanic emissions or from transformation of carbonate rocks, atmospheric oxygen levels should not be affected, as these processes do not consume oxygen. The measurements show that O_2 concentrations decrease continuously as CO_2 concentrations rise, signifying that the additional CO_2 emissions are the result of combustion and not of volcanism. A drop in O_2 and a simultaneous increase in CO_2 is thus the signature of combustion-related CO_2 emission. Fortunately, this decrease is very small compared to the size of the atmospheric reservoir, and we won't be running out of oxygen any time soon!

Differences in $\delta^{13}C$, the ^{13}C isotopic content of atmospheric CO_2, confirm that the changes in this gas stem from organic matter (which is enriched in ^{12}C), and not from volcanic emissions. The $\delta^{13}C$ of atmospheric CO_2 has decreased over time, indicating that atmospheric CO_2 has become enriched in ^{12}C.

Finally, we compare changes in the amount of CO_2 in the Northern and Southern Hemispheres (Fig. 29.3). The increase in atmospheric CO_2 in the Southern Hemisphere lags behind that in the north by about a year. This is consistent with the fact that most emissions are generated in the more industrialized Northern Hemisphere, and the transfer of excess CO_2 to the south is determined by the mixing time between the hemispheres; that is, about one year.

29.3 Changes related to human activity

29.3.1 GREENHOUSE GAS EMISSIONS FROM HUMAN ACTIVITY

Atmospheric GHG concentrations have increased steadily over the past two centuries. Some of these gases are present naturally in the atmosphere (water vapour, H_2O; carbon dioxide, CO_2; methane, CH_4; nitrous oxide, N_2O; and ozone, O_3). These contribute naturally to the warming of the planet via the greenhouse effect. But human activity has considerably increased the concentrations of several of these gases, thereby enhancing their effect. Other gases, such as chlorofluorocarbons (CFCs), are man-made. The increase of these gases introduces an additional greenhouse effect. It is not a simple matter to estimate their respective contributions to the greenhouse effect, since this calls the following into play:

- their individual properties (selective absorption in the emission spectrum of the Earth's surface: see Chapter 7);
- their combined effects when the absorption lines overlap;
- their lifetime in the atmosphere; and
- chemical reactions that transform them into other molecules (methane in the atmosphere, for example, gradually transforms into carbon dioxide with a lifetime of 12 years).

To account for these different factors, the emission of these gases is expressed in equivalent CO_2: the amount of CO_2 to be added to the atmosphere to generate the same warming as that amount of the gas. Anthropogenic GHG emissions are shown in Fig. 29.8 for the period between 1970 and 2004, with the different sectors of emission in 2004. Before the industrial era (the 19th century), atmospheric concentrations were stable at about 280 ppm for CO_2, about 720 ppb for CH_4 and about 270 ppb for N_2O. Since then, because of anthropogenic emissions, these concentrations have increased by ~40%, ~150%, and ~20%, respectively. Compared to the timescale of exchange between the hemispheres (one year), their lifetimes in the atmosphere are sufficiently long (100–300 years, ~12 years and ~120 years, respectively) for them to be homogenized. However, owing to local temperature variations and photosynthetic activity, their atmospheric concentrations vary regionally from one season to the next.

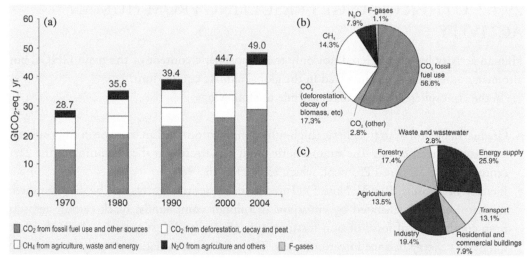

Fig. 29.8 (a) Annual greenhouse gas emissions of human origin in the world between 1970 and 2004. (b) Respective contributions of the different anthropogenic greenhouse gases in the 2004 emissions (expressed in CO_2 equivalent). (c) Contribution of the different sectors to the total anthropogenic greenhouse gas emission in 2004 (forestry includes deforestation).

Source: IPCC 2007. *Climate Change 2007: Synthesis Report*. Contribution of Working Groups I, II and III to the Fourth Assessment Report of the Intergovernmental Panel on Climate Change, Figure 2.1. IPCC, Geneva, Switzerland.

Natural versus anthropogenic emissions

The three principal human activities that generate CO_2 emissions are energy production from fossil fuels (coal, gas and oil), deforestation (Part V, Note 4) and cement production (Fig. 29.6). The total amount of CO_2 generated by these three activities is currently about 10 GtC/yr (2012), of which 8–9 GtC/yr come from fossil fuel combustion and cement production (the latter constituting about 3%); the remainder, approximately 1–2 GtC/yr, from deforestation. The main sources of natural emissions are the continental biosphere and the oceans. *Today, human emissions contribute about 5% of all CO_2 emissions on the surface of the Earth.*

The sources of methane emission are numerous and diffuse. The main anthropogenic sources are biomass fires and ruminants, followed by wetlands such as rice paddies, and by mining operations, landfills and gas transport. The main natural sources are wetlands (tropical wetlands and those at high northern latitudes), followed by termites (10% only). Estimations of these emissions have a large degree of uncertainty and only their order of magnitude is significant. The increase of atmospheric CH_4 slowed momentarily between 1995 and 2007, before resuming again (Fig. 29.3). The brief reduction could be related to a temporary decrease in natural emissions at high latitudes (fewer wetlands) or to a decrease in anthropogenic emissions from the fossil fuel industry (fewer gas losses). *Today, anthropogenic emissions constitute more than 60% of all CH_4 emissions from the Earth's surface.*

Anthropogenic emissions of N_2O are mainly the result of increased microbial production from the expansion of agricultural areas and the use of fertilizers. The main natural sources are vegetation-covered land surfaces and the oceans. *At present, anthropogenic emissions amount to 40% of all N_2O emissions from the surface of the Earth.*

29.3.2 OTHER CHANGES RESULTING FROM HUMAN ACTIVITY

Human activity has also affected not only the atmospheric content of the main GHGs, but also many other parameters involved in the global climate equilibrium.

In the atmosphere, these changes include the following:

- Gradual depletion of stratospheric (high-atmosphere) ozone and an increase in tropospheric (low-atmosphere) ozone. The stratospheric trend was arrested at the beginning of the 21st century by the Montreal Protocol, which took effect in 1989.
- Increase of aerosols (Part V, Note 5). Three main classes of aerosols can be distinguished: sulphate aerosols, generated by emissions of sulphur compounds; black carbon aerosols from combustion of fossil or non-fossil organic matter; and dust particles. Aerosols affect the Earth's energy balance in various ways. They alter solar radiation directly by reflecting it (mainly sulphate aerosols) or by absorbing it. They also alter the Earth's radiation by absorbing it (black carbon aerosols). Indirectly, they affect cloud formation (through water droplet nucleation), as well as the reflection and absorption properties of the clouds. It was aerosols, mostly sulphates from the combustion of coal, which contains sulphur, that caused the appalling acid rains of the 1970s. Their increase in the atmosphere at that time is well documented. Stringent regulation effectively contributed to a decrease in atmospheric aerosol concentrations from the 1980s (Fig. 29.9).

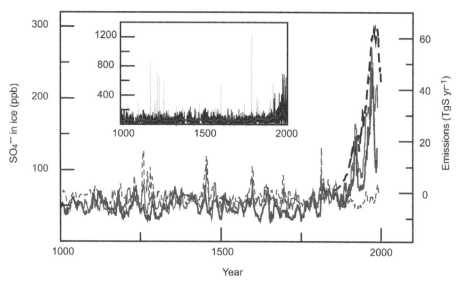

Fig. 29.9 Measurements of sulphate (SO_4^{2-}) concentrations in ice cores from Greenland (red and blue lines) and Antarctica (dashed purple) during the last millennium. Also shown are the estimated anthropogenic sulphur (S) emissions for the Northern Hemisphere (dashed black line). Unlike the ice of Greenland, deposits in Antarctica are unaffected by anthropogenic emissions. The ice core data have been smoothed by 10-year averaging thereby removing the peaks of major volcanic eruptions. The inset illustrates the influence of volcanic emissions over the last millennium. Green, measured monthly sulphate concentration; black, the same, but with identified volcanic spikes removed (most recent spikes were neither assigned nor removed); red: result with 10-year averaging. These records are illustrative examples and can be influenced by local deposition events.
Source: IPCC 2007. *Climate Change 2007: The Physical Science Basis*. Working Group I Contribution to the Fourth Assessment Report of the Intergovernmental Panel on Climate Change, Figure 6.15. Cambridge University Press.

- An increase in fires in vegetation-covered surfaces (forests etc.), which release not only CO_2 and aerosols, but also many other volatile compounds.
- Changes in surface coverage of the continents (desertification, reservoir construction, cultivation, deforestation, urbanization etc.), which modify the albedo and the water cycle.

All these environmental changes affect the energy balance of the planet.

Now that we have identified the main changes attributable to human activity that can disrupt the climate system, let us examine which natural variations could have contributed to the observed warming over the past century. Over this timescale, two stand out: increased solar activity and/or decreased volcanic activity.

29.4 Natural causes: solar and volcanic activity

Solar activity

Solar activity has been directly measured only since the 1980s. For earlier times, it must be reconstructed from various proxies (Chapter 13). Solar activity did increase to a certain degree in the first half of the 20th century, which may have contributed to the warming observed during the period from 1920 to 1940. However, no measurable increase of this activity has been detected over the past four decades, during which the global temperature has continued to rise (Figs. 13.1 and 22.14). At present, solar activity is not considered to be a major factor in recent climate change.

Volcanic activity

Major volcanic explosions cause overall cooling of the planet's surface by a few tenths of a degree, for a period of one or two years. Volcanic activity could only have contributed to the observed global warming if its impact had decreased over the past few decades. But history does not confirm this scenario (Fig. 23.3). Unlike previous decades, there have been three major volcanic eruptions in the past 50 years: Mount Agung (1963), El Chichón (1982) and Mount Pinatubo (1991).

29.5 An overview of all the causes: the major role of human activity

29.5.1 HUMAN-INDUCED RADIATIVE FORCING

The major disturbances that could have affected the climate balance since 1750 are displayed in Fig. 29.10, in terms of *radiative forcing* (see Chapter 10). This figure includes anthropogenic disturbances (mainly GHGs, aerosols and surface albedo) and natural causes (solar irradiance). The principal anthropogenic GHGs have generated a positive radiative forcing of more than $3\,\text{W/m}^2$. CO_2 is the main contributor ($1.7\,\text{W/m}^2$), followed by methane (about $0.7\,\text{W/m}^2$). These estimations are very reliable (low uncertainty). Such is not the case for estimates of ozone and aerosol forcing, the reliability of which ranges from poor to doubtful (high uncertainty). Taken together, net anthropogenic radiative forcing amounts to approximately $+2.3\,\text{W/m}^2$ (estimates vary from +1.1 to $3.3\,\text{W/m}^2$).

This radiative forcing generates surface warming, and feedbacks amplify it by a factor of about 2.5 (1.6–3.8). We saw in Chapter 10 that a radiative forcing of $1\,\text{W/m}^2$ leads, ultimately (i.e. including feedbacks), to an average warming of about 0.8°C (0.5–1.1°C), when the equilibrium temperature is reached. Human-induced radiative forcing of $2.3\,\text{W/m}^2$ will thus produce a final warming ΔT_{equil} of about 1.8°C (1.1–2.5°C).

Is this result consistent with the estimate of about +1 °C warming observed since the end of the 19th century? The answer is yes if we consider the range of uncertainty of the radiative forcing, and also the fact that we are currently in a transient state where the transient temperature is smaller than T_{equil}.

What about temperature changes in the 20th century? While GHG emissions increased steadily (Fig. 29.3), average temperatures stopped increasing from the 1950s until the 1970s (Fig. 25.1). Anthropogenic sulphate aerosol emissions in the troposphere may have played an important role in this trend. Through negative radiative forcing (by increasing reflection of solar radiation), their presence leads to cooling. Their increase in the second half of the 20th century, followed by a sharp decrease from the 1970s, when they became regulated, are clearly recorded in the Greenland ice cores (Fig. 29.9). This episodic increase helped temporarily to mitigate anthropogenic greenhouse warming until the 1980s. The effect was more pronounced in the Northern Hemisphere, where most aerosols are emitted (Part V, Note 5).

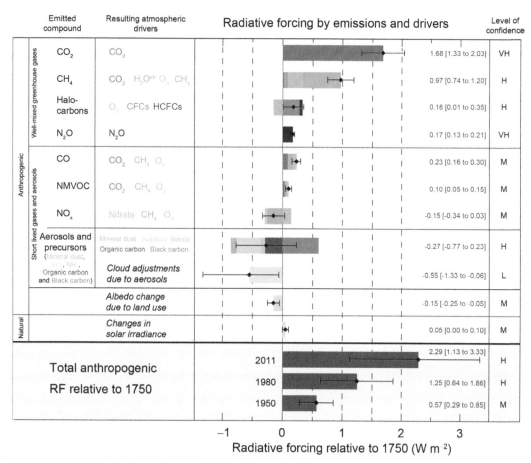

Fig. 29.10 Radiative forcing estimates in 2011 relative to 1750 and aggregated uncertainties for the main drivers of climate change. Values are global average radiative forcing (RF) partitioned according to the emitted compounds or processes that result in a combination of drivers. The best estimates of the net radiative forcing are shown as black diamonds with corresponding uncertainty intervals; the numerical values are provided on the right of the figure, together with the confidence level in the net forcing (VH – very high, H – high, M – medium, L – low, VL – very low). Albedo forcing due to black carbon on snow and ice is included in the black carbon aerosol bar. Small forcing due to contrails (0.05 W/m² including contrail induced cirrus), and HFCs, PFCs and SF$_6$ (total 0.03 W/m²) are not shown. Concentration-based RFs for gases can be obtained by summing the like-coloured bars. Volcanic forcing is not included as its episodic nature makes it difficult to compare to other forcing mechanisms. Total anthropogenic radiative forcing is provided for three different years relative to 1750.
Source: IPCC 2013. *Climate Change 2013: The Physical Science Basis.* Working Group I Contribution to the Fifth Assessment Report of the Intergovernmental Panel on Climate Change, Figure SPM 5. Cambridge University Press.

29.5.2 TWENTIETH-CENTURY GLOBAL WARMING: NATURAL VERSUS ANTHROPOGENIC ORIGIN

We discuss two types of study, covering, respectively, the 20th century and the recent decades. In these studies, different groups evaluated the relative contributions from natural and anthropogenic causes to the current changes in the global surface temperature.

The studies of the first type regard the period as a whole. Four of them are considered in IPCC 2013 and are shown in Fig. 29.11. Their conclusions are similar. To distinguish between

Fig. 29.11 (a) Variations of the observed global mean surface temperature anomaly (i.e. deviation from the 1961–90 mean value, black line), and (b–f) results of multivariate analyses performed by 4 groups (coloured bars). Contributions from (b) ENSO, (c) volcanoes, (d) solar forcing, (e) anthropogenic forcing and (f) other factors (Atlantic Multidecadal Oscillation, semi-annual Oscillation, Arctic Oscillation).

Source: IPCC 2013. *Climate Change 2013: The Physical Science Basis.* Working Group I Contribution to the Fifth Assessment Report of the Intergovernmental Panel on Climate Change, Figure 10.6. Cambridge University Press.

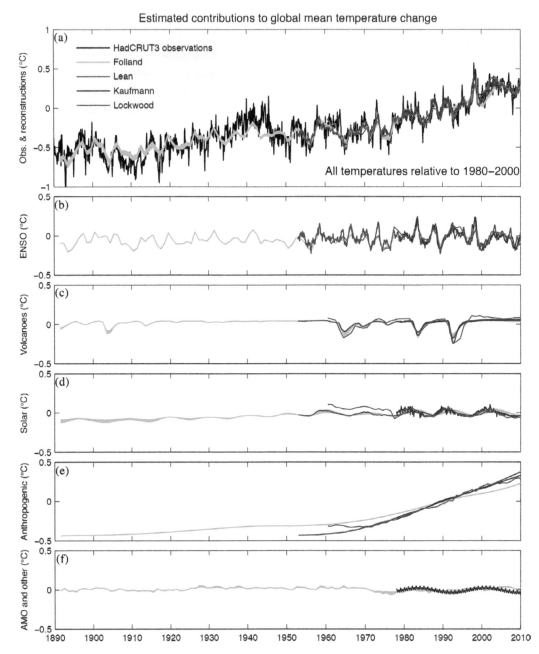

simultaneous natural and anthropogenic impacts on surface temperature, multivariate analyses were performed using the best available estimates, together with the observed surface temperature record. Figure 29.11 shows how the different natural (ENSO, volcanic and solar) and anthropogenic components contribute to the evolution of net global surface temperatures. Natural influences produce as much as 0.2–0.3°C warming during major ENSO events, and about 0.25°C cooling following large volcanic eruptions. The 11-year solar cycle gives rise to a cyclic variation (of no more than 0.1°C), and a slight warming (~0.1°C) took place in the first half of the 20th century due to the increase in background solar activity. Anthropogenic radiative forcing comes principally from increasing GHG concentrations and the effects of tropospheric aerosols. This forcing produces a net warming of about +1°C, the major part of which has occurred in the past four decades.

From 1890 to 2010, the temperature trends produced by all three natural influences are at least an order of magnitude smaller than the observed surface temperature trend. According to this analysis, solar forcing has contributed negligible long-term warming in the past 25 years and 10% of the warming in the past 100 years.

The second study, by Foster and Rahmstorf (2011), focuses on the period from 1979 to 2010, seeking to extract the human contribution in global warming. It casts light on the causes of the slowdown of warming during the first decade of the 21st century (Box 29.2).

BOX 29.2 SLOWING OF GLOBAL WARMING SINCE 2000?

In the first decade of the 21st century, the rate of increase of the global mean temperature became slower. This immediately raises the following question: *have future projections not overlooked some important parameter? Is the future warming not therefore overestimated?* The work of Foster and Rahmstorf (2011) dashes this hope, showing that the slowdown is merely the temporary result of natural fluctuations. This study is based on five reconstructions of the global temperature, three from the surface temperature and two from the lower troposphere temperature (Fig. B29.1a). These authors filter out the short-term variability attributable to the El Niño Southern Oscillation (ENSO), to the solar cycle and to volcanic activity. The largest overall impact comes from ENSO, with volcanic aerosols having the second-largest influence and solar variation exhibiting the smallest influence.

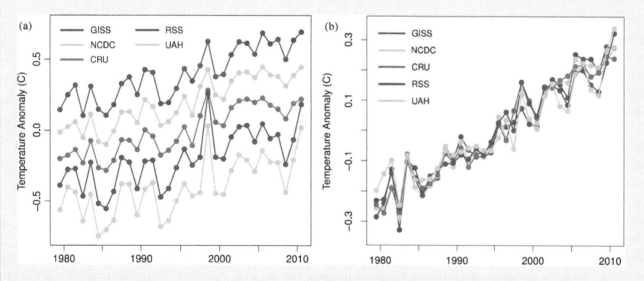

Fig. B29.1 (a) Global temperature records from five different sources. Successive records are shifted vertically by an arbitrary constant. (b) Adjusted data sets for each of the five sources after subtracting the influence of El Niño, volcanic eruptions and solar variations. The upward trend, due only to the anthropogenic contribution, is linear.

Source: Foster and Rahmstorf (2011). © IOP Publishing Ltd. CC BY-NC-SA. Reproduced with permission.

After subtracting these influences, Foster and Rahmstorf show that over the past three decades (1979–2010), the global surface temperature continues to rise at a remarkably steady rate of about 0.15°C per decade (Fig. B29.1b). In the words of Foster and Rahmstorf, 'because the effects of volcanic eruptions and of ENSO are very short-term and that of solar variability very small, none of these factors can be expected to exert a significant influence on the continuation of global warming over the coming decades'.

29.5.3 IS ANTHROPOGENICALLY INDUCED WARMING NOW DETECTABLE? IF SO, SINCE WHEN?

Modelling the 20th-century climate change

As part of the initiatives requested by IPCC 2013 to determine the extent of human activity in recent warming, some 15 Atmosphere–Ocean General Circulation models were launched to simulate average annual temperature changes over the 20th century under two different scenarios. The first excluded human activity: it took into account only the natural conditions observed during the 20th century (solar activity and volcanism). To these natural conditions, the second scenario added various disruptions related to human activity (GHG and aerosol emissions, changes due to land use and changes in the ozone layer). In order to identify regional warming trends, Fig. 29.12 displays the results of these simulations for different regions. The simulations included the following:

- The global average temperature (oceans and land), the average temperature on land and the average ocean temperature (oceans warm more slowly than land).
- The average temperature for each of the seven large continental regions located at various latitudes (North America, South America, Europe, Africa, Asia, Australia and Antarctica), and the six oceanic regions. Since climate change is more pronounced at high latitudes, these continents exhibit different temperature 'signatures' as warming progresses.
- The area covered by sea ice (Arctic and Antarctic) and the ocean heat content.

In each case, changes in the simulated average temperatures were compared to observed changes (black curve): human activity was included (pink shading), or not (blue shading), in the simulation. *The results demonstrate that the recent warming over at least the past two decades cannot be simulated unless human activity is taken into account.*

Detection of the anthropogenically induced warming

These numerical simulations confirm that the recent warming is caused mainly by the anthropogenic GHG emissions of the past few decades. Its effect is now detectable in the mean temperatures of the continents and of the oceans, and in the majority of regions. Its effect is strongly suspected in two other domains, the global ocean heat content and the arctic sea-ice area, where the uncertainty in the two curves (with or without human influence) still overlaps slightly. Other investigations, such as that shown in Fig. 29.11, lead to a similar conclusion: for the past 30 years or so, natural forcing has been clearly outstripped by anthropogenic forcing. It is therefore legitimate to conclude that since the 1980s, the mean climate of the planet has been warming as a result of human activity.

We recall that this increase in average temperature reflects an increase in available energy at the Earth's surface (as explained in Part I), the proxy for which is the mean temperature.

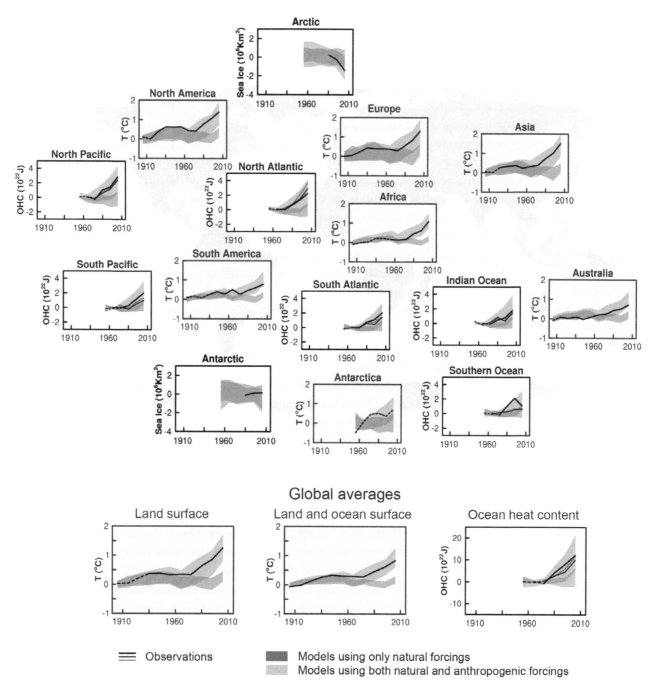

Fig. 29.12 A comparison of observed and simulated climate change based on three large-scale indicators in the atmosphere, the cryosphere and the ocean: change in continental land surface air temperatures (yellow panels), Arctic and Antarctic September sea-ice extent (white panels) and upper ocean heat content in the major ocean basins (blue panels). Global average changes are also given. Anomalies are given relative to 1880–1919 for surface temperatures, 1960–80 for ocean heat content and 1979–99 for sea ice. All time-series are decadal averages, plotted at the centre of the decade. For temperature panels, observations are dashed lines if the spatial coverage of areas being examined is below 50%. For the ocean heat content and sea-ice panels, the solid line is where the coverage of data is good and higher in quality, and the dashed line is where the data coverage is only adequate, and thus, uncertainty is larger. Model results shown are Coupled Model Intercomparison Project Phase 5 (CMIP5) multi-model ensemble ranges, with shaded bands indicating the 5–95% confidence intervals. **Source:** IPCC 2013. *Climate Change 2013: The Physical Science Basis.* Working Group I Contribution to the Fifth Assessment Report of the Intergovernmental Panel on Climate Change, Figure SPM 6. Cambridge University Press.

PART V SUMMARY

The variation in the climate over the past few decades is clearly visible all over the world. Of the three parameters that serve as broad measures of climate, namely temperature, precipitation and wind, only the temperature exhibits a pronounced change with time when averaged over the planet. This does not mean that the changes in the other two parameters are non-existent – their effects are felt on a more regional level. The warming shows up regularly and consistently at the surface of the continents, in the oceans to a depth of several hundred metres and in the atmosphere to an altitude of several kilometres. The warming exhibits features of global climate changes that have occurred in the past: (i) it is more pronounced on the continents than in the oceans; (ii) it is more pronounced at high latitudes in the Northern Hemisphere than in the tropics; and (iii) it is more pronounced in the Northern Hemisphere than in the Southern Hemisphere, which is mainly occupied by ocean. Since the beginning of the 20th century, the annual average temperature on the whole of the Earth's surface (the global mean temperature) has risen by about 1°C. In the Northern Hemisphere, warming has reached about 1°C on the continents at low latitudes, between 1 and 2°C at mid-latitudes and more than 2°C over the whole of the high latitudes.

What is the impact of this warming? Its consequences are numerous. Some of the most outstanding are to be found in the changes in the cryosphere (snow cover, permafrost, glaciers, ice sheets and sea ice), in the oceans, in the water cycle and in the biosphere. All the observations about the general regression of the cryosphere concur. They are clearly illustrated by the melting of the Arctic sea ice in summer, which has accelerated in the past few years, as well as by the advance of about 2 weeks in the date of thaw for all the mid-latitudes in the Northern Hemisphere. The gradual increase in sea level has been detected in the course of the 20th century by several techniques, the results of which are mutually consistent. In recent decades, the average rate of increase has risen to its present value of about 3 mm/yr. This is due in part to thermal expansion of the surface water and in part to melting of glaciers and ice sheets. The average rise is nevertheless difficult to detect on a local scale, because local or regional variations are frequently much greater, whether they are due to natural causes or to human activities. A change in the water cycle is also beginning to appear that is consistent with the observed warming: intensification of the cycle accompanied by stronger evaporation from the tropical oceans. This changes the salinity of the surface water of the oceans on a worldwide scale (higher salinity in the tropics, reduced salinity at mid-latitudes). All these repercussions disturb the equilibrium of the biosphere.

The impact of the warming on the biosphere (flora and fauna) is multiple and varied, with two main driving forces: (i) the increased average temperature and (ii) the shortened winter period. These perturbations modify the areas in which species are distributed as well as the equilibrium of the ecosystems. Migration towards higher latitudes is well documented in the Northern Hemisphere, and retreat to higher altitudes is observed, both in the animal and in the vegetable kingdom, in oceans and on continents. The advance in the annual growth cycle perturbs the established equilibrium of ecosystems: dysfunctions arise, for example, when the cycle of food resources is out of phase with the animal reproduction cycle, thereby jeopardizing reproduction. Lastly, heat waves are becoming more frequent and more intense. Their regular occurrence reduces agricultural productivity.

What is the cause of the warming? According to the latest information, the IPCC 2013 report concludes that it is *extremely likely* that human activities have caused most of the observed increase in global average temperatures since the 1950s. All the analyses at present converge in attributing, with very high probability, most of this rise to the increase in anthropological greenhouse gases. The most important contribution is from CO_2, due mainly to combustion of fossil fuels, half of which accumulates each year in the atmosphere. On the different continents and on the oceans, all the simulations show that natural causes (variability of solar radiation, volcanic eruptions) are incapable of replicating the order of magnitude of the observed warming. By contrast, when the effect of the increase in greenhouse gases produced by humankind is incorporated, the observed warming is accounted for.

The changes in global climate during the 20th century – in particular, in the second half and at the beginning of the 21st century – thus appear to be fairly well understood. We may therefore conclude that the warming of the mean climate at the Earth's surface since the 1980s is caused principally by human activity.

The results of the models are sufficiently in agreement with observations for their simulations of future changes to be credited as meaningful (Part VI).

PART V NOTES

1. A gigatonne of carbon (or GtC) is a measurement based on carbon atoms. To calculate the equivalent quantity of CO_2, it must be multiplied by the ratio of the mass of a CO_2 molecule (44 g/mole) to that of a carbon atom (12 g/mole); that is, 3.66.

2. The Vassal Conservatory contains the largest collection of grape varieties in the world. Started in 1876 in Montpellier (France) to combat the destructive effect of *phylloxera*, the collection was transferred in 1949 to Vassal, beside the sea, where the sandy soil was free of phylloxera.

3. Phenology is the study of the influence of climate conditions on phenomena that punctuate the life of plants and animals throughout the year: germination, flowering, fruiting, the arrival and departure of migrating animals, the breeding season, hibernation dates and so on.

4. Savannah fires, practised regularly for many centuries to encourage regrowth for pasture, do not constitute a source of emissions. The carbon emitted yearly by these fires in the form of CO_2 is fixed again through regrowth of the grasslands. The annual carbon balance is therefore zero.

5. Anthropogenic aerosols are emitted mainly at the middle latitudes of the Northern Hemisphere, where most industrialized countries are located. The aerosols are washed out by precipitation. Atmospheric circulation patterns confine them essentially to the latitude band where they were emitted, since the water cycle brings them back to the surface within about 10 days. This is why, unlike greenhouse gases, which are evenly distributed over the planet, such aerosols remain mostly at the latitudes where they were emitted; that is, the middle latitudes of the Northern Hemisphere.

PART V FURTHER READING

Archer, D. (2012) *Global Warming – Understanding the Forecast*. John Wiley & Sons, Hoboken, NJ; see Part 2.

Climate Change 2013: The Physical Science Basis. Working Group I Contribution to the Fifth Assessment Report of the Intergovernmental Panel on Climate Change (IPCC), Cambridge University Press; available at http://www.ipcc.ch/report/ar5/wg1/-; see Chapters 2, 3, 4, 6 and 10.

Fletcher, C. (2013) *Climate Change – What the Science Tells Us*. John Wiley & Sons, Hoboken, NJ; see Chapters 2 and 3.

Houghton, J. (2009) *Global Change – The Complete Briefing*, 4th edn. Cambridge University Press, Cambridge, UK; see Chapter 1.

Ruddiman, W.F. (2008) *Earth's Climate – Past and Future*, 2nd edn. W.H. Freeman, New York; see Part 5.

PART VI
CLIMATE IN THE 21ST CENTURY: DIFFERENT SCENARIOS

Chapter 30
Two key factors

Humankind is disrupting the planet's balance in a multitude of ways. The climate systems of the Earth are so complex that any number of human activities could affect the natural climate balance to a certain extent. The principal culprit, however, which drastically disrupts the composition of the atmosphere and the way in which the surface of the planet is heated at a global level, is emission of greenhouse gases (GHGs). We saw in Chapter 29 how observations and calculations identify with almost absolute certainty these anthropogenic emissions as being the cause of the recent climate change. The range of future climate change is set by the socio-economic scenario of future emissions.

If emissions go unchecked, what levels will be reached, and what will the conditions on Earth be within a century? How will these levels compare with those of several million years ago?

But first, let us look at what stabilizing the temperature means for GHG emissions. For this, we restrict ourselves to CO_2, and place ourselves in the economic and social context corresponding to these emissions.

30.1 Greenhouse gas emissions

30.1.1 THE LEADING ROLE OF CO_2

Figure 29.8 shows the increase in each major anthropogenic GHG since 1970 and the contribution of various human activities in these emissions. The greatest impact on global warming comes from changes in the concentration of CO_2 (Fig. 29.10), and the mechanisms causing its

Climate Change: Past, Present and Future, First Edition. Marie-Antoinette Mélières and Chloé Maréchal.
© 2015 John Wiley & Sons, Ltd. Published 2015 by John Wiley & Sons, Ltd.
Companion website: www.wiley.com\go\melieres\climatechange

steady increase appear to be well defined. To stabilize our climate, we must therefore stabilize atmospheric CO_2. The emissions are related directly to the world economy by its energy requirements, and to the composition of human society (demographics): stabilizing GHG levels will necessarily involve both factors.

30.1.2 CO_2 EMISSIONS AND GLOBAL ENERGY PRODUCTION

Anthropogenic CO_2 emissions vary from year to year (Fig. 29.6), but over several years they exhibit a slow but steady upward trend. In 2010, average emissions stood at about 10 giga-tonnes of carbon per year (GtC/yr), of which about 1 GtC/yr were generated by deforestation, and the remaining 9 GtC/yr by combustion of fossil fuels (oil, gas and coal); cement manu-facture also contributes, but only marginally. The main source of CO_2 emission in the world is clearly the production of energy.

The problem of reducing CO_2 emissions takes on an economic dimension when we con-sider the place of fossil fuels in global energy production. If we include biomass in these sources (i.e. wood, peat, plant products etc.), they amount to approximately 80% of the world's primary energy sources (Part VI, Note 1). Figure 30.1 plots energy production in recent decades. Between 1970 and 2004, energy production more than doubled (+110%), which corresponds to an average annual increase of about 2%. This was accompanied by a similar surge in CO_2 emissions: over the same period, these increased by 90% on average.

This simple observation shows how difficult it is to reduce GHG emissions, since the world economy depends almost entirely on energy production. Various economic scenarios have been developed to estimate future GHG emissions and their consequences, from the most

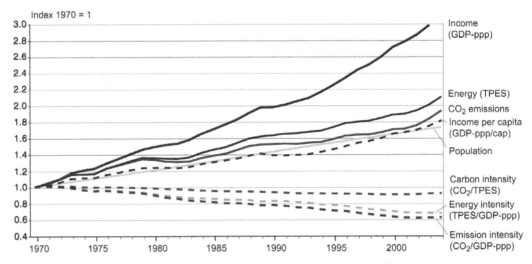

Fig. 30.1 Relative global changes in income, total primary energy supply (TPES), CO_2 emissions and population with respect to 1970. CO_2 emissions stem from fossil fuel combustion and gas flaring, with a much smaller contribution from cement production.

Source: IPCC 2007. Climate Change 2007: Mitigation of Climate Change. Working Group III Contribution to the Fourth Assessment Report of the Intergovernmental Panel on Climate Change, Figure 1.5. Cambridge University Press.

'innovative', involving early action to reduce emissions, to the most *laissez-faire*, in which current trends continue unabated.

30.1.3 EMISSIONS VARY FROM COUNTRY TO COUNTRY

The amounts of GHGs emitted worldwide can be expressed on a national and also on a per capita basis. This enables us to compare the impact of various lifestyles and to identify the main producers of atmospheric GHGs. Figures vary substantially from one country to another, depending on local economic conditions. Figure 30.2 provides an overview for 2010 of total carbon emissions and per capita

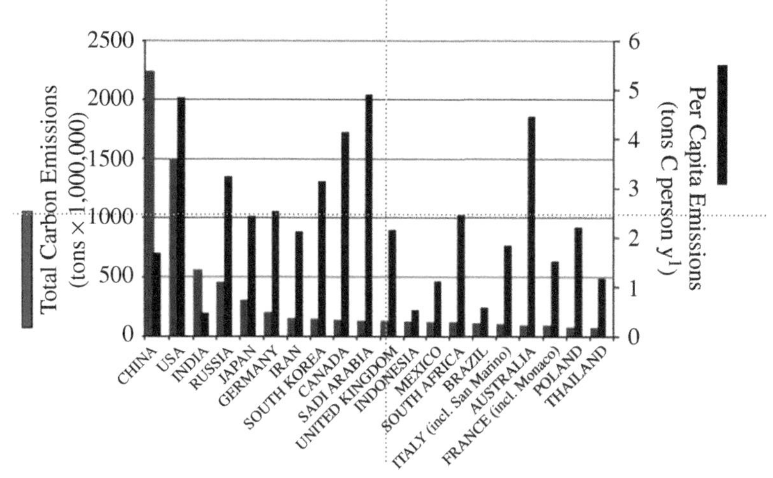

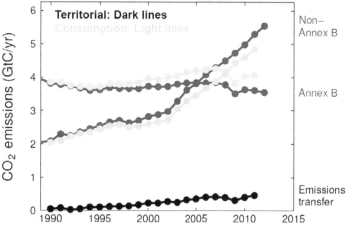

Fig. 30.2 The 20 countries that generated the most CO_2 fossil fuel emissions in 2010: total carbon emission (red bars) in megatonnes of carbon, and per capita emission (black bars) in tonnes of carbon per year.

Sources: Global Carbon Budget 2010; data from Boden et al. (2011) and CDIAC (2011).

emissions for 20 countries that generate the most CO_2 emissions. The Kyoto Protocol of 1992 distinguished different groups of countries. Among them are two groups, Annex B countries (Part VI, Note 2), which are mostly industrialized and are sometimes also called 'Developed Nations', and all the others, the non–Annex B group, sometimes called the 'Developing Nations'. Figure 30.3 shows the CO_2 emissions for the Annex B group and the non–Annex B group between 1990 and 2012. A question arises: *is it fair to portray a nation's contribution to global CO_2 emissions by those that it generates directly?* Surely not, because we must also take into account goods manufactured abroad and consumed by the inhabitants of that country. The direct cause of CO_2 emissions generated in manufacturing such goods is the consumer. We can calculate a nation's *true contribution* to global emissions by subtracting emissions associated with exported goods and adding emissions related to imported goods. The adjusted figures are displayed in Fig. 30.3. As Annex B nations import more consumer goods than they export (in terms of CO_2 emission), their CO_2 'consumption' is higher than their 'production', in contrast to other nations. Despite this adjustment, a general trend appears clearly: over the past 10 years, emissions have stagnated overall in the group of Annex B nations and doubled overall in the group of non–Annex B nations.

Fig. 30.3 CO_2 emissions from Annex B nations of the Kyoto protocol (so-called 'Developed Nations') (blue curve), and from non–Annex B nations ('Developing Nations') (red curve) from 1990 to 2011. Dark lines: territorial emissions. Light lines: consumption-based emissions (territorial emission + imported emission – exported emission). In Annex B, production-based emissions have had a slight decrease. Consumption-based emissions have grown at 0.5% per year, and emission transfers have grown at 12% per year (1 PgC = 1 GtC).

Sources: Global Carbon Budget and trends 2013, based on Le Quéré et al. (2013) and Peters et al. (2011). Reproduced by permission of the Global Carbon Project.

30.1.4 ACCUMULATION OF CO_2 AND CH_4 IN THE ATMOSPHERE

After their emission, how do GHGs accumulate in the atmosphere? In the case of CO_2, common sense, as well as the trends that have been observed for more than a century and a half, provide a clear indication of how atmospheric CO_2 is likely to change over the next few decades. On average, half of the CO_2 emitted each year as a result of human activity remains in the atmosphere, while the other half is absorbed in equal parts by the terrestrial biosphere and the oceans (Part V). It is therefore easy to estimate roughly how atmospheric CO_2 will evolve with human activity. This relationship seems to be changing slightly, however, as the terrestrial CO_2 sink of the planet has become somewhat less efficient in recent years. Current calculations clearly suggest that both CO_2 sinks will gradually become slower, in which case more CO_2 will accumulate in the atmosphere.

Atmospheric CH_4 levels are more complex. Unlike CO_2, the CH_4 molecule is chemically reactive and can be oxidized. Atmospheric CH_4 levels therefore depend on several parameters. We recall that after a temporary pause (from 2000 to 2007), CH_4 levels in the atmosphere have recently started to rise again.

30.2 Population growth

Figure 30.1 shows how between 1970 and 2004, the world's population grew by almost 70%. The economies of one particular group of countries, the *transition economy* countries led by China, India, Brazil and South Korea, are booming. The Kyoto Protocol classification ought therefore be adjusted to distinguish not two, but three, groups: *More Developed Countries* (MDCs), *Emerging Countries* (ECs) and *Developing Countries* or *Less-Developed Countries* (LDCs), within which GHG emissions per capita are more or less comparable, although certain extremes remain. Global GHG emissions can then be formulated in the following way:

$$\text{Emitted GHG} = \text{Em}_{\text{MDC}} \times N_{\text{MDC}} + \text{Em}_{\text{EC}} \times N_{\text{EC}} + \text{Em}_{\text{LDC}} \times N_{\text{LDC}}$$

where Em and N are the average emission per capita and the population of each group. Global emissions will thus depend on changes in per capita emission and in the population. In varying proportions and for different reasons, each of the three terms of this equation is increasing. The main parameters contributing to higher emissions are as follows:

- in More Developed Countries, increasing consumption per capita, since the population is approximately stable;
- in Developing Countries, population increase; and
- in Emerging Countries, increasing individual consumption and population increase.

Global population control is therefore as important a factor in controlling GHG emissions as consumption control.

Chapter 31
Projections: economic scenarios and climate models

31.1 Successive steps in a projection

To construct an economic scenario, estimates of the following constraints are needed for the coming decades of the 21st century:

- Per capita emission. In this step, we are concerned mainly with: (i) energy consumption and energy sources (fossil fuels, renewable energy etc.); (ii) deforestation (whether replanted or not); and (iii) agricultural management.
- Population growth.

Obviously, a multitude of scenarios is possible. The challenge is to circumscribe the possible by its two extremes, and to select a few conceivable intermediate situations.

To estimate the degree of climate change that human activity could generate, we must:

1. Estimate the year-by-year emission of the different gases and aerosols (CO_2, CH_4, N_2O, sulphate aerosols etc.).
2. Calculate the increase in concentration of these gases and aerosols.
3. Calculate the change in temperature, including feedback.
4. Calculate the impact of temperature changes on the water cycle (precipitation, drought, cyclones, sea level etc.).

Climate Change: Past, Present and Future, First Edition. Marie-Antoinette Mélières and Chloé Maréchal.
© 2015 John Wiley & Sons, Ltd. Published 2015 by John Wiley & Sons, Ltd.
Companion website: www.wiley.com\go\melieres\climatechange

5. Estimate the effect of the changes in temperature and water cycle on the biosphere (from individuals to species, both plant and animal).

6. Last but not least, estimate the impact of these changes on the structure of each ecosystem, and on the interactions between ecosystems. This latter is the basic principle that ensures the balance of life on Earth. Agriculture and human health depend on this balance.

Point [1] is addressed in various economic scenarios. It engages the scientific community of economists, sociologists, demographers, political scientists and so on, who project demographic and consumer trends. They are in a position to quantify the impact of human activity on the environment: atmospheric emissions (GHGs and others), modifications to the Earth's surface (e.g. deforestation and agriculture) and so on.

Points [2], [3] and [4] are addressed by climate modelling after human activities (point [1]) have been estimated. This is where climatologists intervene, by studying the changes in the atmosphere, oceans, continental surfaces, bio-physicochemical cycles and energy transfer. Figure 31.1 summarizes this process.

Since all these variables are interdependent and involve feedback, climate models must address a complete set of interacting processes simultaneously. This is a complex task. Each parameter that changes leads in turn to a new climate balance. The model must integrate all the changes brought about by the initial disturbance in order to simulate the resulting climate balance properly. However, although albedo and water vapour concentration are relatively easy to identify, many other types of feedback – changes in cloud cover, soil status, impact on vegetation and so on – are not. This is where interactive models become complex (Chapter 11).

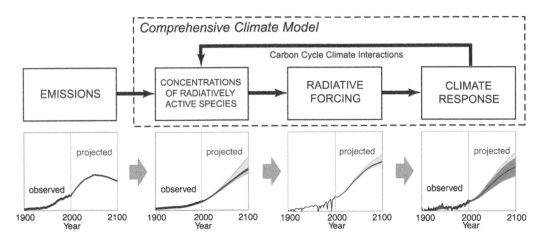

Fig. 31.1 Flow diagram of the computation steps between emission projections defined in a given socio-economic scenario and the climate response generated by the models. The model calculates the change in composition of the atmosphere, then the change in energy balance (radiative forcing) and, finally, the climate impact (temperature, precipitation etc.). Uncertainties increase with each step. Note that the diagram does not contain any uncertainty for the emission, since this is determined by the choice of scenario.

Source: IPCC 2007. *Climate Change 2007: The Physical Science Basis.* Working Group I Contribution to the Fourth Assessment Report of the Intergovernmental Panel on Climate Change, Figure 10.1. Cambridge University Press.

The last two points in the list, [5] and [6], relate to the life sciences. These are even more complex, and require knowledge of ecosystems and how they operate. It must be assessed whether life, and all its ramifications (population development, species development, and ecosystem structure and interaction), may be jeopardized, and how, for example, our food resources may be affected in consequence.

31.2 Climate models

The way in which a model is constructed and how it operates was described in Chapter 11, together with the recent improvements in modelling. To assess the ability of a model to provide an accurate representation of reality, tests were carried out based on well-documented climate situations drawn both from the present and from the past. Two earlier periods were selected for climate conditions that differed from now: the last glaciation, 20,000 years ago, when the average temperature was 5°C lower, and 6000 years ago, during the Holocene, when the temperature was slightly higher than now but the Sahara was substantially wetter. A comparison of the results obtained by climate models with those of climate reconstructions based on proxy data shows that the simulations are closer to reality when several models are used together. This is how the Intergovernmental Panel on Climate Change (IPCC) community proceeds in modelling projected climate. Some 20 existing models are deployed to assess climate change in the various proposed scenarios. The model results are then averaged, and the deviation from the mean defines the uncertainty. These results are referred to as *multimodel* results.

What do these models teach us about future climate change? Which scenarios are selected for simulating the future climate?

We start by briefly describing the scenarios and simulations that were considered in IPCC 2007, since they provide a clear introduction to the subject. Afterwards, we specify the new scenarios and simulations that are treated in IPCC 2013.

Chapter 32
Simulations: a survey

32.1 Long-term scenarios

32.1.1 CLIMATE STABILIZATION: THE LAST STAGE OF THE SCENARIO

Whether over fairly short or much longer timescales, all the proposed scenarios lead to stabilization of the global mean temperature and of the climate. In this final stage, anthropogenic GHG emissions are stabilized at a level at which the natural sinks of the planet (land-based biosphere and oceans) can absorb them (Part VI, Note 3). Such emissions should not exceed 2–3 GtC/yr. As this goal cannot be attained immediately (the present rate of emission, approximately 10 GtC/yr, is still increasing steadily), all scenarios allow for an initial increase in emission, followed by a plateau (stabilization) and then a decrease. The 'greener' the scenario, the sooner the decrease comes about.

32.1.2 THE COMING CENTURIES

Here, we review the different categories of scenario that, after a few centuries, stabilize the global mean temperature. *What are the final temperatures in each scenario?*

Figure 32.1 displays, as a function of time, six categories of CO_2 emission scenario that are addressed in IPCC 2007 SR. In each scenario, CO_2 emissions (almost 8 GtC/yr, or ~30 Gt of CO_2 in 2000) start by increasing. In the scenario with the severest emission restrictions, for instance, this growth stops in 2020 at about 10 GtC/yr before falling to a few GtC/yr towards 2100, with the atmospheric CO_2 stabilizing at around 350–400 ppm. This corresponds to a

Climate Change: Past, Present and Future, First Edition. Marie-Antoinette Mélières and Chloé Maréchal.
© 2015 John Wiley & Sons, Ltd. Published 2015 by John Wiley & Sons, Ltd.
Companion website: www.wiley.com\go\melieres\climatechange

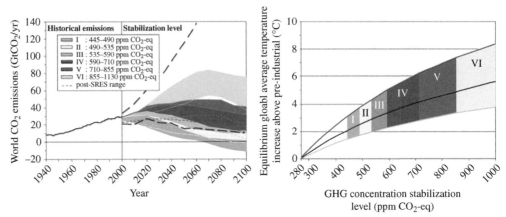

Fig. 32.1 Left-hand panel: global CO_2 emission from 1940 to 2000 and emission ranges for six categories of stabilization scenarios from 2000 to 2100. For each scenario the range of greenhouse gas (GHG) concentration at stabilization is indicated. Black dashed lines give the emission range of baseline scenarios published since the Special Report of Emission Scenarios (2000). Right-hand panel: the corresponding relationship between the stabilization target (indicated by the concentration in ppm CO_2-eq) and the probable increase in equilibrium global temperature with respect to pre-industrial values. Reaching equilibrium can take several centuries. In this scheme, doubling the CO_2 content in the atmosphere (560 ppm) raises the equilibrium temperature by +3°C (from ~2 to ~4.5°C), in agreement with the climate sensitivity parameter defined in Section 10.3.
Source: IPCC 2007. *Climate Change 2007: Synthesis Report*. Contribution of Working Groups I, II and III to the Fourth Assessment Report of the Intergovernmental Panel on Climate Change, Figure 5.1. IPCC, Geneva, Switzerland.

445–490 ppm CO_2-equiv when all anthropogenic greenhouse gases are taken into account (Part VI, Note 4). All six categories of scenario stabilize the atmospheric GHG levels between about 450 ppm and about 1000 ppm CO_2-equiv. The dark blue line in the figure shows the final projected warming after atmospheric GHG levels stabilize. Total uncertainty is indicated by the coloured curves lying above and below this line.

One characteristic of the curve is the final warming generated by doubling the atmospheric CO_2 from the pre-industrial level of 280 ppm to 560 ppm. This is the *climate sensitivity* (a concsept introduced in Chapter 10). It shows that, when all feedback mechanisms are taken into account, doubling the CO_2 content will ultimately warm the planet by 3°C (between 2.5 and 4.7°C, according to IPCC 2013).

In conclusion, Category I scenarios, with the most severe restrictions on emission, end up with a global warming of between 1.5 and 3°C (the green zone). In Category VI scenarios (the highest emission, the grey zone), the equilibrium global warming is about 5°C.

Even after atmospheric GHG levels have stabilized, it will in fact take several centuries for the Earth's energy systems to reach equilibrium, during which time the global temperatures will continue to rise by a few tenths of a degree per century. Global warming will affect sea levels over even longer timescales: thermal expansion will cause sea levels to rise for several centuries or millennia, and, over several thousands of years, melting ice sheets will gradually raise the sea level. In addition, the possible collapse of part of the West Antarctic Ice Sheet, which would rapidly accelerate sea-level rise, cannot be excluded. Figure 32.2 summarizes the possible changes at different timescales.

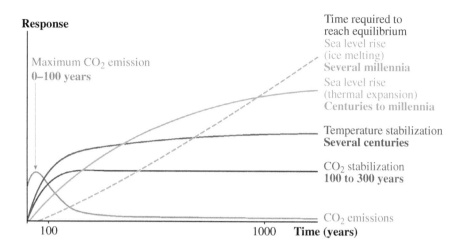

Fig. 32.2 A schematic diagram of the various response time constants of the Earth's system to CO₂ emission into the atmosphere. The CO₂ concentration, temperature and sea level continue to rise long after emissions are reduced.

32.1.3 OVER SEVERAL MILLENNIA

How will the atmospheric CO₂ vary in the long term, once anthropogenic emissions have stopped? With time, it will gradually decrease, at first by redistributing it into the land and the ocean, and later by reaction with igneous rocks. This decrease has been computed by coupled carbon-cycle–climate models, with the assumption of an excess of 5000 GtC emitted into the atmosphere at time zero (Fig. 32.3). In scenario RCP8.6, the order of magnitude of 1000 GtC emitted by human activity is plausible, as will be seen in Section 32.3.2. A decrease with time of comparable order of magnitude occurred during the climatic event of the *Palaeocene–Eocene Thermal Maximum*, 55 million years ago, after a sudden emission of GHG that raised the mean temperature by about 5°C. That emission was estimated to be a few thousand GtC (Chapter 24).

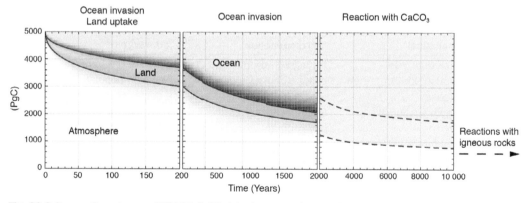

Fig. 32.3 Decay after release of 5000 PgC CO₂ into the atmosphere at time zero, and its subsequent redistribution into land and ocean as a function of time, computed by coupled carbon-cycle climate models. The width of the green and dark blue bands indicates carbon uptake by land reservoirs and ocean, respectively. The right-hand panel shows the longer-term redistribution including dissolution of carbonaceous sediments in the ocean (1 PgC = 1 GtC).

Source: IPCC 2013. *Climate Change 2013: The Physical Science Basis*. Working Group I Contribution to the Fifth Assessment Report of the Intergovernmental Panel on Climate Change, FAQ 6.2. Figure 2. Cambridge University Press.

32.2 IPCC 2007 scenarios for the 21st century

32.2.1 DIFFERENT ECONOMIC SCENARIOS

The scenarios studied by IPCC 2007 consider possible changes in demographics, global economics and new technologies. The 40 or so scenarios may be divided into three categories: a 'technological' solution (A1), a 'heterogeneous' solution (A2) and a 'convergent' solution (B1). The A1 family is subdivided into three scenarios (A1FI, A1B and A1T). All major economic trends are thus covered. Table 32.1 summarizes the main differences between these scenarios.

32.2.2 MODELLING AND SCENARIOS

The climate models referred to here include changes in emission of CO_2, GHGs and other atmospheric compounds that affect the climate, such as aerosols. They also take into account the role of the ocean and terrestrial biosphere as CO_2 sinks. The results below, excerpted from IPCC 2007, relate to scenarios B1, A2 and the intermediate scenario, A1B. These scenarios are often referred to as SRES scenarios, from the Special Report on Emission Scenarios. Figure 32.4 indicates the changes during the 21st century, in the following:

- CO_2, CH_4, and SO_2 emissions;
- atmospheric concentrations of the three greenhouse gases CO_2, CH_4, and N_2O; and
- radiative forcing and the resulting warming, expressed as the increase in the global mean temperature with respect to the period from 1980 to 2000.

TABLE 32.1 CHARACTERISTICS OF THE DIFFERENT FAMILIES A1, A2 AND B1 OF SCENARIOS.

Assumptions	Scenario A1: 'technological'	Scenario A2: 'heterogeneous'	Scenario B1: 'convergent'
Population change	Increase until 2050 (population 9 billion), then a decrease	Strong steady increase (15 billion population by 2100)	Increase until 2050 (9 billion population), then a decrease
Economic change	Very rapid economic growth; fewer differences between countries	Regional economic development extremely variable, depending on the country	Rapid change in economic structures; development of services and reduced flux of materials
Advance and spread of new technologies	Very rapid use of fossil fuels (A1FI), clean energy (A1T) or both (A1B)	Very slow and extremely variable, depending on the country	Rapid development of clean technologies; more efficient in controlling emissions

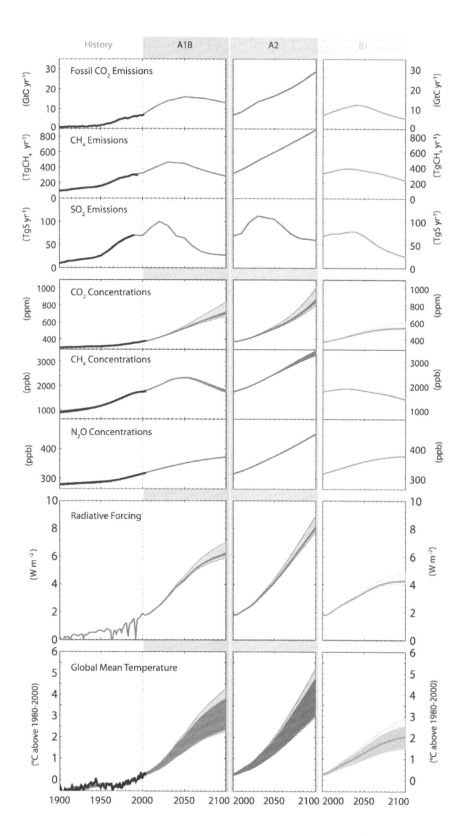

Fig. 32.4 Projections for SRES scenarios A1B, B1 and A2 of the emissions of fossil fuel CO_2, CH_4 and SO_2, together with the concentrations of CO_2, CH_4 and N_2O, the radiative forcing and the global mean temperature. The Special Report on Emissions Scenarios (SRES) is a report by the Intergovernmental Panel on Climate Change (IPCC) that was published in 2000. These projections are based on the set of models considered in the IPCC report of 2007. The shaded areas in the lower panels represent the different uncertainty levels.

Source: IPCC 2007. *Climate Change 2007: The Physical Science Basis*. Working Group I Contribution to the Fourth Assessment Report of the Intergovernmental Panel on Climate Change, adapted from Figure 10.26. Cambridge University Press.

32.2.3 GLOBAL MEAN TEMPERATURE CHANGE

Figure 32.5 summarizes the changes in CO_2 emission and global mean temperature for all three scenarios. In scenario A2, CO_2 and GHG emissions increase by a factor of almost four over the 21st century. Atmospheric CO_2 will attain a level of 820 ppm by 2100 and the global mean temperature will rise by 3.5°C (between 2 and 5.5°C). In 2100, emissions will still be increasing, as will also the atmospheric GHG levels and the global mean temperature.

In scenario A1B, CO_2 emissions increase by a factor of two over the coming century, raising the CO_2 level to 700 ppm, and causing a warming of about 3°C by 2100 (between 1.8 and 4.4°C). Emissions do not stabilize in this scenario either, and GHG levels and temperature both continue to rise.

In scenario B1, the temperature stabilizes towards the end of the 21st century. CO_2 emissions peak at about 12 GtC/yr around 2040, before decreasing over the rest of the century. The climate is practically in balance. The atmospheric CO_2 content levels out at 550 ppm and mean global warming gradually stabilizes at about 2°C (between 1.2 and 2.8°C). The

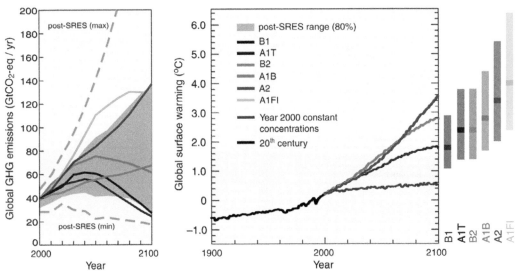

Fig. 32.5 Left: Global GHG emissions (in GtCO₂-eq/year) in the absence of climate policies: six illustrative SRES (Special Report on Emissions Scenarios by the Intergovernmental Panel on Climate Change (IPCC) published in 2000) marker scenarios (coloured lines) and the 80th percentile range of recent scenarios published since SRES (post-SRES) (grey shaded area). Dashed lines show the full range of post-SRES scenarios. The emissions include CO_2, CH_4, N_2O and F-gases. Right: Solid lines are multi-model global averages of surface warming for scenarios A2, A1B and B1, shown as continuations of the 20th-century simulation. These projections also take into account emissions of short-lived GHGs and aerosols. The pink line is not a scenario, but is for Atmosphere–Ocean General Circulations Model (AOGCM) simulations where atmospheric concentrations are held constant at year 2000 values. The bars at the right of the figure indicate the best estimate (solid line within each bar) and the likely range assessed for the six SRES markers scenarios at 2090–2099. All temperatures are relative to the period 1980–1999. 1 Gt CO_2/yr = 0.27 GtC/yr.

Source: IPCC 2007. *Climate Change 2007: Synthesis Report*. Contribution of Working Groups I, II and III to the Fourth Assessment Report of the Intergovernmental Panel on Climate Change, Figure SPM.5. IPCC, Geneva, Switzerland.

equilibrium temperature is determined principally by the total amount of CO_2 remaining in the atmosphere. This amount is the same if emissions are initially high and then reduced drastically, or if they are not as high initially but remain unrestricted for longer. From these two options, two opposing political approaches were developed in the early 21st century: one recommends prompt action to reduce emissions; the other would postpone restrictions to avoid disrupting the world economy, at the same time as developing techniques for rapid control of emissions at a later date.

Comparison between the scenarios

We now compare the temperature estimates for the three scenarios. In the first decades of the 21st century, all three scenarios exhibit a very similar temperature increase, because the GHG emissions hardly differ.

In these simulations, the global warming in the coming decades is about 1°C, which is comparable to that over the past few decades: the three scenarios project a continuation of the recent warming. Towards the middle of the century, differences in the temperature rise start to emerge clearly, increasing over time. At the end of the century, the average temperature increase in scenario B1 is close to 2°C, and about 3.5°C in scenario A2.

32.3 IPCC 2013 scenarios for the 21st century

Since IPCC 2007, observations of the planet have been greatly enriched, in particular with the increased number of satellite measurements, and models have been considerably improved. *In what way do the projected changes differ from those of IPCC 2007?*

32.3.1 MODELLING IN IPCC 2007 AND 2013

The scenarios used to calculate future climate changes are analysed in a Climate Model Inter-comparison Project: CMIP3 for IPCC 2007, and CMIP5 for IPCC 2013. The models used for climate simulation in CMIP5 are an improvement on those of CMIP3, but the relationship between the projected distribution of temperature and precipitation over the surface of the Earth and the degree of global warming nevertheless remains very similar. This finding is independent of the scenario considered. *The conclusions of IPCC 2013 thus clearly confirm those of IPCC 2007 as to the trends in temperature and precipitation over the Earth's surface.*

How well does CMIP5 simulate the recent climate change?

The simulations are illustrated in Fig. 32.6. The upper part of the figure shows the observed linear trend of the mean temperature for 1950–2011, and that simulated by CMIP5. Inspection of the two maps shows that the simulation is very close to reality. The same comparison can be made for precipitation (the bottom panel). Here the match is less good, and in some regions the observed and simulated trends are opposite. *Does this imply that modelling of precipitation in the 21st century is meaningless?*

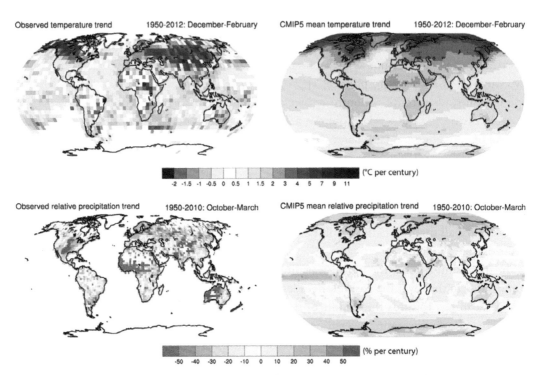

Fig. 32.6 Comparison between observed linear trend (left) and the mean trend simulated by the Climate Model Intercomparison Project CMIP5 (IPCC 2013) (right) for the period 1950–2010. Top: trend in temperature from December to February, in °C per century; bottom: trend in precipitation from October to March, in % per century. Grid boxes left white where fewer than 50% of the years have observations.

Source: IPCC 2013. *Climate Change 2013: The Physical Science Basis*. Working Group I Contribution to the Fifth Assessment Report of the Intergovernmental Panel on Climate Change, Box 11.2 Figure 1. Cambridge University Press.

It is now well established that an average warming of approximately 1°C has occurred over the past six decades. Average precipitation, however, does not show any clear trend: any average trend remains obscured by substantial year-to-year variability in precipitation, particularly in the equatorial regions. It will take a greater climate change – that is, more warming – before regional trends prevail over yearly regional fluctuations. However, the trends forecast by these simulations are slowly starting to emerge; for example, in the decreasing rainfall in the Mediterranean region.

32.3.2 SCENARIOS

In IPCC 2007, the emission scenarios represent various routes for the different emitted compounds on the basis of well-defined assumptions. In IPCC 2013, four new scenarios, referred to as Representative Concentration Pathways (RCPs), were developed. Instead of tracking changes in emissions over time, these track changes in the radiative forcing generated by the emissions. This approach separates the evaluation of the different economic choices from the calculation of climate change (which is due to radiative forcing alone).

Three of the four scenarios, RCP4.5, RCP6.8 and RCP8.5, are comparable to the previous SRES scenarios B1, A1B and A2 (Fig. 32.7). The atmospheric content of each GHG is displayed as a function of time in IPCC 2013. Here, we focus on the principal greenhouse gas,

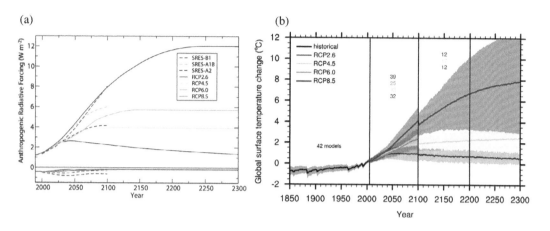

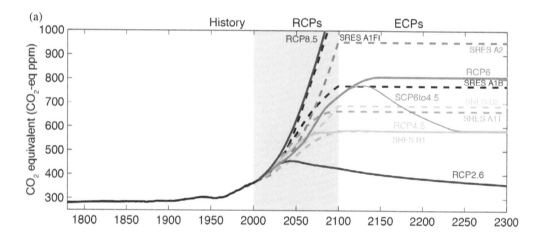

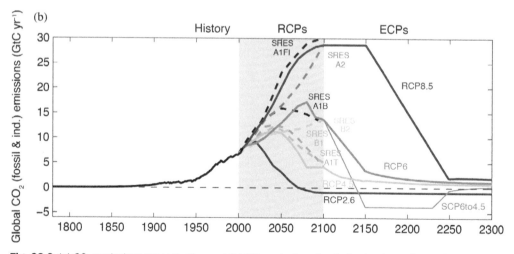

Fig. 32.8 (a) CO_2-equivalent concentration and (b) CO_2 emissions (excluding land use) for the four RCPs (including transition scenario from RCP6 to RCP4.5) as well as some SRES scenarios (see Fig. 32.5). In scenario RCP8.5 the CO_2 concentration continues to increase after 2100, stabilizing at 2000 ppm after 2250, nearly seven times the pre-industrial levels.

Source: IPCC 2013. *Climate Change 2013: The Physical Science Basis*. Working Group I Contribution to the Fifth Assessment Report of the Intergovernmental Panel on Climate Change, Box 1.1 Fig. 3. Cambridge University Press.

Fig. 32.7 (a) Comparison among the three SRES scenarios used in CMIP3 and the four RCP scenarios used in CMIP5 (RCP2.6, 4.5, 6.0, 8.5). The total anthropogenic (positive) and anthropogenic aerosol (negative) radiative forcing relative to preindustrial times (about 1765) for each scenario is indicated as a function of time. (b) Projected change in global annual mean temperature relative to 1986–2005 for the four RCP scenarios. For each scenario the temperature is the average of the simulated temperatures of all models (shaded band: ±1.6 standard deviation). Numbers indicate the number of models contributing to the different time periods. Discontinuities at 2100 are due to the difference in the number of models performing the calculations after 2100, and have no physical meaning.

Source: IPCC 2013. *Climate Change 2013: The Physical Science Basis*. Working Group I Contribution to the Fifth Assessment Report of the Intergovernmental Panel on Climate Change, Figures 12.3 and 12.5. Cambridge University Press.

CO_2. Figure 32.8 shows the resulting changes in CO_2 emission and the CO_2 concentration for each RCP scenario. Note that for RCP8.5, the emissions stabilize between 2100 and 2200 at 30 GtC/yr, before starting to decrease. The CO_2 concentration stabilizes at 2000 ppm only in 2250.

32.3.3 GLOBAL MEAN TEMPERATURE CHANGE

The increase in the global mean temperature (compared to 1986–2005) is plotted for the four scenarios in Fig. 32.7 until the year 2300. By the end of the present century, temperatures increase by $4 \pm 1°C$ in RCP8.5, which is comparable to the projected increase of about 3.5°C (between 2 and 5.4°C) of scenario A2 (Fig. 32.5).

Chapter 33
Future warming and its consequences

In this chapter, the description of the impact of future global warming on the different environmental components is based on the recent IPCC 2013 report.

33.1 Global warming

The increase in global mean temperature until the year 2300 in scenarios RCP2.6 to RCP8.5 is displayed in Fig. 32.7. Here, we discuss only the projected changes at the end of the 21st century. In these two scenarios, the rise in global mean temperature by 2100 is about 1°C and 4°C, respectively. *How will the warming affect the different regions of the world?*

   The colour-coded maps in the upper part of Fig. 33.1 illustrate these projected changes with respect to the period 1986–2005 for temperature and precipitation. In the two scenarios, two geographical patterns stand out:

- Warming is more noticeable on land than on the ocean.
- Warming increases towards the high latitudes of the Northern Hemisphere. In scenario RCP8.5, the increase in average temperature in the Arctic regions will exceed 10°C by the end of this century, while in some regions in the tropics the warming may be as much as 5–7°C.

Climate Change: Past, Present and Future, First Edition. Marie-Antoinette Mélières and Chloé Maréchal.
© 2015 John Wiley & Sons, Ltd. Published 2015 by John Wiley & Sons, Ltd.
Companion website: www.wiley.com\go\melieres\climatechange

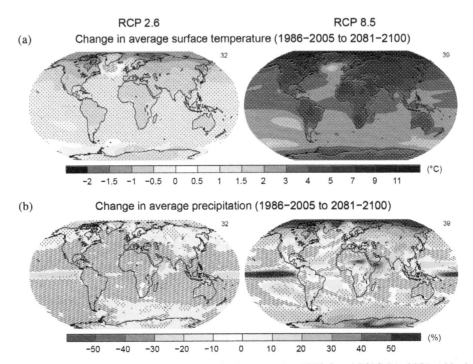

Fig. 33.1 Maps of CMIP5 multi-model mean results for the scenarios RCP2.6 and RCP8.5 in 2081–100 of (a) annual mean surface temperature change and (b) average per cent change in annual mean precipitation. Changes are shown relative to 1986–2005. The number of CMIP5 models used to calculate the multi-model mean is indicated in the upper right corner of each panel. Hatching indicates regions where the multi-model mean is small compared to natural internal variability (i.e. less than one standard deviation of natural internal variability in 20-year means). Stippling indicates regions where the multi-model mean is large compared to natural internal variability (i.e. greater than two standard deviations of natural internal variability in 20-year means). **Source:** IPCC 2013. *Climate Change 2013: The Physical Science Basis.* Working Group I Contribution to the Fifth Assessment Report of the Intergovernmental Panel on Climate Change, Figure SPM 8. Cambridge University Press.

These features are consistent with our observations of past climate variations, such as the glacial–interglacial oscillations during the Quaternary, or the milder fluctuations of the Holocene (the present warm period that started 12,000 years ago). It is notable that these two features have already been clearly detected in the recent warming (Part V).

33.2 The water cycle and precipitation

The water cycle pattern and latitudinal climate belts – a poleward shift?

The projected impact of global warming on the water cycle is shown in Fig. 33.2. It will:

- Intensify the water cycle, and accentuate the present characteristics of the different latitude climate bands. The Equator and the middle and high latitudes will receive more precipitation, and the tropics will receive less.
- Move the latitudinal climate belts towards the poles.

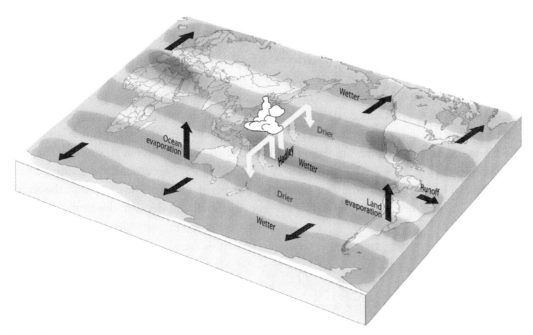

Fig. 33.2 A schematic diagram of projected changes in principal components of the water cycle. Blue arrows indicate major movements of water through the Earth's climate: water transport towards the poles by extra-tropical winds, evaporation from the surface and runoff from the land to the ocean. Darker regions denote areas likely to become either drier (red) or wetter (blue). Yellow arrows indicate a large change in atmospheric circulation by the Hadley cell. Its upward motion increases tropical rainfall, at the same time as suppressing subtropical rainfall. Model projections indicate that the Hadley circulation will move its down branch towards the poles in both hemispheres, accompanied by drying. At high latitudes, wetter conditions are projected because a warmer atmosphere will allow greater precipitation with greater movement of water into these regions.
Source: IPCC 2013. *Climate Change 2013: The Physical Science Basis.* Working Group I Contribution to the Fifth Assessment Report of the Intergovernmental Panel on Climate Change, FAQ 12.2, Figure 1. Cambridge University Press.

Both tendencies contain important clues to the questions raised in Chapter 3 about the changes that global warming can bring to the latitude climate belts. As the climate warms, it is likely that *the Hadley cell will widen, extending the subtropical dry zones towards the poles.* In the African–European region, this means that the Sahara desert zone will expand further north, and Northern Africa and Southern Europe will both become drier.

What will happen to the storms that regularly cross the mid-latitudes? They travel from west to east along a storm track at latitudes 40–50° of the Northern and Southern Hemispheres. In scenario RCP8.5 (IPCC 2013), *poleward shifts of 1–2 degrees in the mid-latitude jets* are likely by the end of the 21st century, together with a *poleward shift of several degrees of the storm tracks in the Southern Hemisphere.* But in the NH, substantial uncertainty remains about projected changes in the storm track, especially for the North Atlantic basin.

Precipitation

The principal changes in precipitation at the end of 21st century in scenarios RCP2.6 and RCP8.5 are illustrated in more detail in Fig. 33.1: increased rainfall in well-watered regions, and reduced rainfall where rain is already scarce. The trends, which appear clearly in RCP8.5, can be summarized as follows. In the equatorial belt and neighbouring tropical regions, where the rainfall is at present highest, precipitation will tend to increase further. By contrast, in the desert regions of the tropics and surrounding areas, rainfall will tend to decrease. In the monsoon region subtropics, the results of the model do not converge clearly; these are regions where rainfall gradients are strong and where the high uncertainty makes them difficult to model. The impact on the monsoon therefore remains indeterminate at present. For the high-rainfall regions at high latitudes in the Northern Hemisphere, the simulations yield increased precipitation in all seasons. The most pronounced changes projected in the water cycle (Fig. 33.2) appear around the Mediterranean basin (from North Africa to Southern Europe) bordering the Sahara Desert, and in the southern part of North America. In these regions, annual rainfall will decrease drastically. Southern Europe in particular will be affected by this trend, which, with the increase of temperature, will bring more marked episodes of drought.

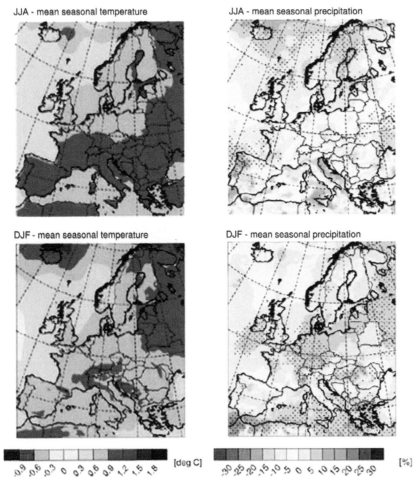

Fig. 33.3 Europe-scale projections from the ENSEMBLES regional climate modelling project in the SRES scenario A1B, for 2016–35 with respect to 1986–2005. Change in the mean seasonal temperature (left) and precipitation (right), for summer (top) and winter (bottom). The stippling in (c) and (d) highlights regions where 80% of the models agree in the sign of the change of precipitation (for temperature all models agree on the sign of the change).

Source: IPCC 2013. *Climate Change 2013: The Physical Science Basis.* Working Group I Contribution to the Fifth Assessment Report of the Intergovernmental Panel on Climate Change, Figure 11.18. Cambridge University Press.

The results of the simulations at a regional level are of more immediate interest. Regional simulations are shown in IPCC 2013 (see IPCC 2013, Chapter 14 and Annex I) for the different continental regions of the planet and the polar regions. Maps of changes in seasonal precipitation and seasonal temperature are given for different time slices in the 21st century, and for the different scenarios. These maps complement the global-scale annual maps (Fig. 33.1) and contain additional details about the trends. We take as an example the projections of the SRES scenario A1B (Fig. 32.4). They refer to Europe and are taken from the ENSEMBLES regional climate modelling project for changes in the period 2016–35 with respect to 1986–2005 (Fig. 33.3). For summer temperatures, the projected warming is between 0.6 and 1.5°C, with the greatest changes over the land portion of the Mediterranean basin. For winter temperatures, the warming is from 0.3 to 1.8°C, with the greatest intensity in north-eastern Europe. This pattern tends to persist until the end of the century. For precipitation, the trend clearly shows less precipitation in winter in the Mediterranean basin, and in summer over Europe.

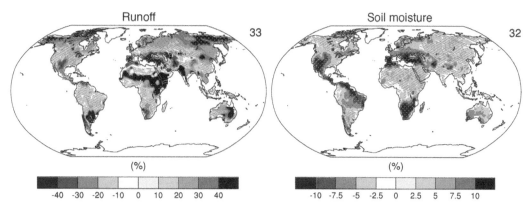

Fig. 33.4 Changes in the water cycle in scenario RCP8.5 projected for 2081–100 relative to the reference period (1986–2005) (~+4°C increase in the global mean temperature). Left: change in annual mean runoff (%). Right: change in annual mean moisture (mass of water in all phases in the uppermost 10 cm of the soil) (%). The number of CMIP5 models to calculate the multimodel means is indicated in the upper right corner of each panel. Hatched, unmarked and stippled regions respectively represent increasing levels of confidence in the simulations. **Source:** IPCC 2013. *Climate Change 2013: The Physical Science Basis.* Working Group I Contribution to the Fifth Assessment Report of the Intergovernmental Panel on Climate Change, Technical Summary – TFE.1 Fig 3. Cambridge University Press.

Soil moisture and runoff

What matters for vegetation is not so much the amount of rainfall but the level of moisture in the soil, which depends both on precipitation and evaporation at ground level. Figure 33.4 shows how soil moisture and runoff will change at the end of the 21st century in scenario RCP8.5. Although rainfall increases due to warming, soil moisture levels in many continental regions hardly change, because warming accelerates evaporation. Changes in rainfall and soil conditions also affect water drainage on land (runoff).

33.3 Extreme events

The simulations show that *tropical cyclones* will intensify, bringing stronger winds and higher precipitation. Some studies suggest that these changes have already begun. The number of category 4 and 5 cyclones has increased in the past 30 years (Chapter 25).

And what about heat waves and cold spells? As average temperatures rise, the temperature distribution curve will shift to higher values (Fig. 25.7), bringing more frequent and more pronounced heat waves. This is the trend that has been observed over the past few decades (Chapter 25). By contrast, cold spells will become less frequent.

33.4 Snow and ice

Global warming will inevitably lead to milder winters and earlier melting of snow cover. *How will this affect glaciers and snow cover in the mountains?* The studies for the mid-latitudes in Europe provide an example. Recent models are based on the assumption of a doubling of CO_2 by the 2050s. While they still need further refinement, the preliminary results indicate

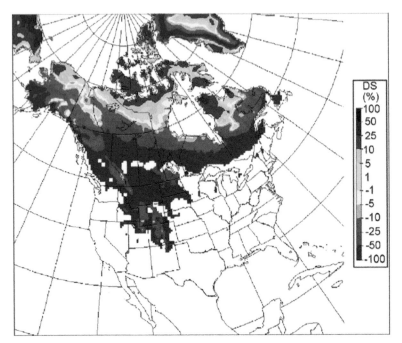

Fig. 33.5 Percentage changes in snow depth in March for 2041–70 with respect to 1961–90 in scenario A2. Projection by the Canadian Regional Climate Model. **Source:** IPCC 2007. *Climate Change 2007: The Physical Science Basis.* Working Group I Contribution to the Fourth Assessment Report of the Intergovernmental Panel on Climate Change, Figure 11.13. Cambridge University Press.

that global warming will outweigh the impact of the increased snowfall, and that snow cover will decrease. The mid-altitudes will be the most affected, as they will lose up to 50 days of snowfall at about 1600 m altitude. In North America, Fig. 33.5 shows that, for the major part of the land between 40°N and 60°N, the loss in snowfall in scenario A2 may be as high as 50–100% (projections of the Canadian Regional Climate Mode for the middle of the 21st century). The area covered by spring snow in the Northern Hemisphere will shrink by the end of the 21st century. Projected reductions vary between 7% (RCP2.6) and 25% (RCP8.5), and near-surface permafrost area is projected to decrease by between 37% and 81%, respectively.

What about ice floes? In the Arctic in summer, the warming of the 20th century has already caused it to lose 75% of its mass, and melting progresses year by year. A question then arises that has important economic implications: *when will the surface of the Arctic Ocean be entirely ice-free in late summer?* The answer is 'in the near future'. Depending on the approach, it may occur in the next 50 years (IPCC 2013), or earlier (extrapolation of the recent trend).

33.5 The sea level

One of the issues that has caught the public eye is the rise in sea level. Satellite observations between 1993 and 2011 show that the average sea level has risen slowly but steadily by 3.3 mm/yr, far more than the 20th-century average of 1.7 mm/yr (Fig. 27.1). *Will this rise accelerate in the 21st century?*

In IPCC 2013, between the lowest (RCP2.6) and the highest (RCP8.5) scenario, the projected increase in sea level at the end of the 21st century ranges from about 0.45 m (0.3–0.6 m) to about 0.7 m (0.5–0.9 m), respectively. In the latter scenario, the rise will be as great as 11 mm/yr. In high-emission scenarios with stabilization levels between 700 and 1100 ppm, the projections for the year 2300 indicate an increase of 1–3 m. These models include thermal expansion of the ocean and melting from glaciers and ice sheets. There is *medium confidence* in these estimations, but larger values cannot be excluded (IPCC 2013). Various mechanisms that could increase the ice melting rate, such as that responsible for the recent acceleration of outlet glacier flow, or the possible collapse of part of the ice sheet, are still not fully understood.

What happened to the sea level in the past when climates were a few degrees warmer than in the recent pre-industrial period? Two periods are relevant: (i) the last interglacial (130–116 kyr), which was 2°C warmer, when the sea level was about 7 m higher than now; and (ii) the Mid-Pliocene (about 3 million years ago), which was 2–3°C warmer, when the sea level was about 10 m higher (with a greater uncertainty). Such increases in sea level require substantial melting from the Greenland and the Antarctic Ice Sheets, and could involve some collapse. It therefore remains an open question as to whether an abrupt increase from such a collapse could overtake the projected regular increase in sea level in the 21st century.

33.6 Ocean acidification

Like sea-level rise, acidification of the ocean surface is already under way, as a consequence of the increase in atmospheric CO_2 dissolving in the ocean. Observations over recent decades show that the surface-water pH has already dropped by 0.1, from pH 8.2 to 8.1, and that this phenomenon has already affected the deeper waters in the North Atlantic, where the surface water sinks into the deeper ocean.

For the pH of the surface water at the end of the 21st century, the two scenarios RCP2.6 and RCP8.5 project a further decrease in pH (acidification) by about 0.1 and 0.5, respectively (Fig. 33.6). The latter value corresponds to an atmospheric CO_2 content of 1000 ppm being reached in 2100. Such a level of acidification has not been attained for at least several million years, since which time atmospheric CO_2 levels have never exceeded 400 ppm (Fig. 18.1). This degree of acidification could have grave repercussions on key marine organisms (such as corals and certain plankton), that build their shells and skeletons from calcium carbonate, given that the stability of calcium carbonate diminishes when pH levels drop. Orr et al. (2005) used 13 models of the ocean-carbon cycle to assess calcium carbonate saturation under the IS92a scenario, (a scenario close to the RCP8.5 scenario for the 21st century) for

Change in ocean surface pH (1986–2005 to 2081–2100)

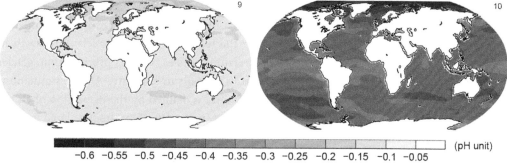

Fig. 33.6 Maps of CMIP5 multi-model mean results of change in ocean surface pH for the scenarios RCP2.6 (left) and RCP8.5 (right) in 2081–100. Changes are shown relative to 1986–2005. The number of CMIP5 models used to calculate the multi-model mean is indicated in the upper right corner of each panel.

Source: IPCC 2013. *Climate Change 2013: The Physical Science Basis*. Working Group I Contribution to the Fifth Assessment Report of the Intergovernmental Panel on Climate Change, Figure SPM 8, Cambridge University Press.

future emissions of anthropogenic carbon dioxide. They found that Southern Ocean surface waters will begin to become undersaturated with respect to aragonite, a metastable form of calcium carbonate, by the year 2050; by 2100, this undersaturation could extend throughout the entire Southern Ocean and into the subarctic Pacific Ocean. When live pteropods (which frequently belong to dominant zooplankton species in polar regions) were exposed to the predicted level of undersaturation during a two-day experiment, their aragonite shells showed notable dissolution.

Ongoing laboratory experiments suggest that different organisms react in different, complex ways to a reduction in pH, some showing decreased calcification while others react by increasing calcification. Fabry et al. (2008) presented arguments showing that ocean acidification and the synergistic impacts of other anthropogenic stressors provide great potential for widespread changes to marine ecosystems. Recent *in situ* observations in the coral reefs of Papua New Guinea (Fabricius et al. 2011), where the pH fell from 8.1 to close to 7.4, however, cast light on this issue (Chapter 28). For a pH value between 8.1 and 7.8, the taxonomic richness of hard corals was reduced by 39%, and below pH 7.7 (> 1000 ppm of CO_2) the reef simply stops developing; no reefs have been found at a lower pH level.

33.7 Climate predictions: what degree of confidence?

33.7.1 SOURCES OF UNCERTAINTY IN THE PROJECTED GLOBAL MEAN TEMPERATURE

Three sources of uncertainty are involved in the projections of the global mean temperature. In IPCC 2013, these are listed as follows:

- *The internal variability* of the climate system. These uncertainties were listed earlier in this book (Chapters 22 and 23). In the coming decades, they will not amount to more than a few tenths of a degree Celsius.
- *The inherent uncertainty of the model.* The simulations of climate change are performed by several groups (as many as 40, depending on the scenario), whose models differ both in their structure and their parameterization. Certain processes, such as precipitation, glacier dynamics and ocean circulation, are notoriously difficult to simulate. All the simulations are grouped into a final curve, where the uncertainty is defined by the dispersion of all the simulations. Thus, for the highest GHG emission scenario, RCP8.5, the simulated increase in mean temperature in the year 2100 is +4°C, with an uncertainty of ±1°C (Fig. 32.7). What new information have the improved models brought in the most recent decades? This point is discussed below.
- *Differences between the projected scenarios.* These depend on the choices made by humankind. This is by far the greatest uncertainty. In the year 2100, the projected global mean warming with respect to the beginning of the 21st century ranges from ~+1°C in scenario RCP2.6 to ~+4°C in RCP8.5. But are these scenarios still relevant? This point is discussed below.

Taking into account these three sources of uncertainty, and according to the choice of scenario made by human beings, warming in the year 2100 with respect to the beginning of the 21st century could either fall almost to zero (scenario RCP2.6, compensated by natural variability) or could be as high as about +6°C (scenario RCP8.5, reinforced by natural variability) and could continue to increase thereafter.

33.7.2 IMPROVEMENTS TO THE MODELS

Climate models are constantly being improved. The latest generation of models does a better job of incorporating the physics of the different mechanisms involved and representing interactions between the mechanisms. Current models take into account warming-related feedback, such as the decreased flux of the land carbon sinks and the acceleration of ice-sheet melting. Progress is also being made in understanding the formation and development of clouds and aerosols, and in improving the spatial resolution of the models.

Despite improvements since the first IPCC report in 1992, basic climate predictions for the 21st century have not changed significantly. It is notable, however, that feedbacks are now better understood (overall, they still remain positive) and that, in comparable scenarios, the warming estimates have shifted to higher temperatures. Generally, our improved understanding of past climates confirms the magnitude of the climate crisis that is projected for the near future. *The crisis appears as one of the major threats to the planet.*

33.7.3 ARE THE ECONOMIC SCENARIOS OUTDATED?

To what extent do the various economic scenarios reflect future economic trends? In an attempt to answer this question, we compare the range of anthropogenic CO_2 emission scenarios with observational data from recent years (Fig. 33.7). Since 2000, fossil fuel–related CO_2 emissions have reached the upper range of these scenarios. The growth in emissions is due principally to the greater use of coal in emerging countries. Moreover, recent developments in unconventional fossil-fuel reserves (shale gas, oil sands etc.) do not point towards reduced CO_2 emissions. *At present, we are moving towards the highest of the scenarios. This scenario is tracking the real situation.*

33.7.4 TIPPING POINTS AND THE POSSIBILITY OF AN ABRUPT CLIMATE EVENT

A further uncertainty comes from the existence of critical thresholds within the climate system that could give rise to an abrupt climate change. Such changes stem from positive feedbacks that accelerate the response of the system. An abrupt climate change can be defined as *'large-scale change in the climate system that takes place over a few decades or less, persists (or is anticipated to persist) for at least a few decades, and causes substantial disruptions in human and natural systems … A number of components or phenomena within the Earth system have been proposed as potentially possessing critical thresholds (sometimes referred to as tipping points, (Lenton et al. 2008)), beyond which abrupt or non-linear transitions to a different state ensue'* (IPCC 2013).

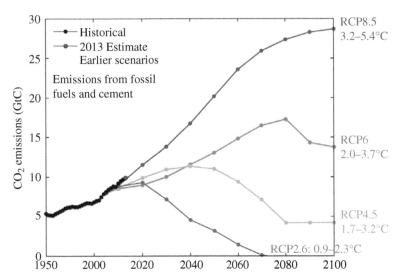

Fig. 33.7 Observed fossil fuel and cement CO_2 emissions compared with projected emissions in the different IPCC scenarios. The most recent trend follows scenario RCP8.5. Emissions are on track for a 3.2–5.4°C 'likely' increase in temperature.

Sources: Global Carbon Budget and trends 2013, based on Peters et al. (2012) and CDIAC data. Reproduced by permission of the Global Carbon Project.

Abrupt climate events occurred during the last ice age, where the climate changed violently within a few decades (Chapter 19). During the warm interglacial periods, by contrast, the climate remained much more stable. Could similar abrupt changes in the Earth's climate nevertheless happen under future warming? The main features within the Earth system that have been proposed as potentially susceptible to abrupt change are: the collapse of the Atlantic Meridional Oceanic Circulation, the collapse of ice sheets, the release of permafrost carbon and clathrate methane, tropical/boreal forest dieback, the disappearance of summer Arctic sea ice and so on. These mechanisms are reviewed in, for example, Lenton et al. (2008), and in IPCC 2013.

Can we rely on models to anticipate these rapid events? Unfortunately, the mechanisms involved are in many cases still poorly determined. An example is the melting of the Arctic sea ice in summer: the models predict its disappearance around 2050, but extrapolations from observations place this date earlier. Similarly, the mechanisms that are involved in destabilizing the West Antarctic Ice Sheet, the base of which rests on the ocean floor, are poorly understood.

Below are three frequently asked questions about abrupt changes.

Arctic warming and methane/carbon dioxide release

Could warming of the Arctic cause a sudden increase in methane emission? These regions harbour large volumes of methane in the form of clathrates, stable solid ice-like structures that release water and methane when they melt. Such gas hydrates form at high pressure and low temperature, both in seabed sediments and in the Arctic permafrost. Warming could cause the clathrates to melt and rapidly release large volumes of methane. A release of this type could have happened during the Palaeocene–Eocene Thermal Maximum (PETM), 55 million years ago, when the temperature rose abruptly by 5°C. That event, however, took place in a climate context that was very different from ours, when the mean temperature on Earth was some 10°C warmer than now (see Section 24.1). The study of past climates offers some clues. The amount of methane in the atmosphere over the past 800,000 years has now been measured. Methane levels peaked during interglacial periods, but even in the warmest interglacials, when average temperatures were about 2°C higher than in the present Holocene (Chapter 19), the levels remained below 800 ppb. This suggests that atmospheric methane levels may not be significantly affected by global warming of 2°C. If, however, future warming exceeds 2°C, we have no example with which a comparison can be made. What do recent estimates point to (Lenton 2012)? The most likely scenario is a long-term chronic methane source made up of many small events (Archer et al. 2009). According to IPCC 2013, it is 'very unlikely' that methane from clathrates will undergo catastrophic release during the 21st century.

What about CO_2 release from permafrost? As permafrost thaws, microbes degrade the carbon within the soil, releasing carbon dioxide and methane into the atmosphere. *To what extent could future warming melt the permafrost and bring on a tipping point?* To date, no studies have convincingly demonstrated that permafrost warming is a tipping element (Lenton 2012). This judgement is shared in the IPCC 2013 report, which states that *'existing modelling studies of permafrost carbon balance under future warming … do not yield consistent results beyond the fact that present-day permafrost will become a net emitter of carbon during the 21st century under plausible future warming scenarios. They also reflect an insufficient understanding of the relevant soil processes during and after permafrost thaw … and preclude any quantitative assessment of the amplitude of irreversible changes in the climate system potentially related to permafrost degassing and associated feedbacks.'*

Ice-sheet collapse

When warming occurs, an ice sheet can in certain circumstances partially collapse due to instability, causing a rapid discharge of ice and an abrupt rise in sea level. In Greenland, loss of ice other than by melting is limited to well-defined outlets. Discharge of icebergs is self-limiting because, as the ice sheet shrinks, contact with the ocean, and hence iceberg calving, both decrease. By contrast, when the bedrock of the ice sheet is below sea level, rapid loss of ice mass can occur. This essentially applies to the West Antarctic Ice Sheet. Moreover, decay of the ice shelf (the thick floating platform of ice attached to the coastline) due to warming can lead to abruptly accelerated ice flows further inland.

The IPCC 2013 report states that 'the available evidence suggests that it is exceptionally unlikely that either Greenland or West Antarctica will suffer a catastrophic abrupt and irreversible near-complete disintegration during the 21st century'. If we recall that such a disintegration would raise the sea level by either 6 or 3 m, this is good news! We must nevertheless bear in mind that partial disintegration, or a rapid increase in sea level, cannot be excluded. It should be recalled that this occurred during the Meltwater pulse 1A in the last deglaciation (Chapter 19), when the sea level suddenly rose by 4 m per century between 14,700 and 14,300 years ago. Moreover, the fact that the West Antarctic Ice Sheet almost certainly melted during the last interglacial (which was 2°C warmer, with a sea level about 7 m higher), together with the particularly unstable nature of this ice sheet, are a warning that this risk cannot be excluded.

Collapse of the AMOC

Collapse of the Atlantic Meridional Overturning Circulation could strongly reduce the deep ocean circulation and give rise to a colder climate in the North Atlantic region. This happened over the Atlantic both during the last ice age, when Heinrich events led to cold snaps, and also at the beginning of the current interglacial period, during the Younger Dryas cold snap, 12,800 years ago (Chapter 19). *Could the present warming cause similar disruption of the AMOC in the next few decades?* Models that prefigure a temporary slowing down of the AMOC nevertheless show no collapse in the coming century, and the IPCC 2013 report states 'it is unlikely that the AMOC will collapse beyond the end of the 21st century'.

33.8 In summary, the future is already with us

The projected changes in the 21st century are in line with what has been observed in recent decades. Recent changes in the distribution of average warming (land versus ocean, Equator versus poles) and changes in rainfall, ice cover and ocean conditions (sea-level rise and increasing acidification) are all consistent with the simulations. These ongoing changes, which have become more marked over the past 30 years, will, in most scenarios, become increasingly pronounced during the 21st century.

How will the new climate compare to those that preceded it? Unless GHG emissions are drastically controlled and reduced, climate warming will bring major changes to the conditions on the Earth by the end of the 21st century. An entirely new climate will set in, one that the Earth has not known for at least several million years.

Chapter 34
The choice

34.1 Can future warming be counteracted naturally?

Can natural climate changes in the near future aggravate, attenuate or even offset the global warming caused by human activity? To answer this question, we review the principal causes of natural climate variation and assess their effects.

The impact of orbital parameters

On the timescale of several millennia, the climate of the Earth is affected by its position relative to the Sun, which gives rise to the glacial–interglacial alternations (Parts III and IV). While the position changes cyclically, with periods of 22,000, 41,000 and 100,000 years, the consequences can be accelerated by feedback. As we have already seen, they can occur quite abruptly (during deglaciation, for instance, sea levels rose by as much as 4 m per century over about five centuries). Past records show that the end of a warm period (such as the present interglacial) can rapidly lead into glaciation. Since warm periods last some 10,000 years on average, it could well be that we are at the end of such a period and on the threshold of a glaciation. Such a situation would offset future warming. However, we saw earlier (Chapter 21) that the present unusual configuration of the Earth's orbit, which is becoming more and more circular, is unfavourable for glaciation in the next several tens of thousands of years. In fact, *over the coming 10,000 years, the Northern Hemisphere will naturally become slightly warmer.*

Climate Change: Past, Present and Future, First Edition. Marie-Antoinette Mélières and Chloé Maréchal.
© 2015 John Wiley & Sons, Ltd. Published 2015 by John Wiley & Sons, Ltd.
Companion website: www.wiley.com\go\melieres\climatechange

Volcanic eruptions

We saw in Chapter 23 that explosive volcanic eruptions, by injecting large amounts of aerosols into the stratosphere, lower the average temperature on Earth by a few tenths of a degree. The cooling effect lasts for two or three years, until the volcanic aerosols settle back to the Earth's surface. To have a lasting cooling effect on the Earth's climate, volcanic activity would need to continue. However, reconstructions of past volcanic eruptions (Fig. 23.3) do not indicate intense, continuous volcanic activity over recent millennia. Nevertheless, we cannot exclude the possibility of a slight increase in volcanic activity, but *it would cause the climate to cool slightly for only a couple of years after each eruption.*

Solar activity

Solar activity varies slightly for periods lasting decades or centuries, causing alternating warmer and cooler climates. The present decrease of solar activity in the current solar cycle (Fig. 13.1) is an illustration of these fluctuations. These, however, can be more pronounced, as they were during the Little Ice Age (LIA). The LIA, which lasted from the 16th century to the 19th century, is the most recent illustration of a low-activity phase. Climate reconstructions indicate that several fluctuations of comparable amplitude occurred during the Holocene, lasting from a few centuries to a millennium (Part IV). This implies that solar fluctuations produce temperature variations that in no case exceed one degree. At present, the Sun is in a more active phase than during the LIA. *Its activity could therefore decrease again in the near future. As in the past, however, cooling would be only temporary.* In contrast to the potential anthropogenic global warming of several degrees, such cooling would not exceed 1°C.

In summary

We now have a relatively good understanding of past climate change at various timescales, ranging from decades to millennia. Over this time range, no natural climate change emerges that could sustainably offset an anthropogenic global warming of several degrees. At best, natural causes can produce only a temporary global cooling of about 1°C. *Global anthropogenic warming will therefore be the principal factor that governs climate change over the coming centuries.*

34.2 Which choice of scenario?

34.2.1 THE IMPACT ON LIFE: THE FUNDAMENTAL ISSUE

We have just seen the impact of warming on the physical world in the different scenarios of human activity during the 21st century. Ultimately, however, the essential question is not how warm the climate will become or how much rain will fall. Rather, it is *what consequences will climate change have on life and on biodiversity? What impact will climate change have on the different relationships that permit ecosystems to function? How will it disrupt the balance between the various pieces of this enormous puzzle?* We have not referred to this question, which lies outside the scope of this book, although it is the most sensitive issue by far. It may nevertheless be recalled that the past 30 years have already given us a foretaste of some of the consequences that can be expected:

- relocation of wildlife and plants to cooler areas (northward, in the Northern Hemisphere);
- disruption of seasonal cycles, with an advancing warm season that disturbs the balance of ecosystems and causes failure of synchronization between food availability and species reproduction (thereby jeopardizing species that are unable to adapt quickly); and
- changes in the physiology and behaviour of animal species.

These mechanisms, already observed in recent decades, are detrimental to biodiversity, and bring certain species to the brink of extinction. Up to now, average temperatures have risen by about 1°C. In future scenarios, however, not only will much higher temperatures be attained, but they will also be accompanied in certain regions by a deficit in rainfall (e.g. the Mediterranean and Southern Europe, Mexico etc.), with much more severe consequences. In such a context, vegetation and agriculture will be particularly affected.

As an illustration, we consider the changes projected for France, a mid-latitude country with a maritime climate. Figure 34.1 shows, for scenarios B2 and A2 (about +2.2°C and

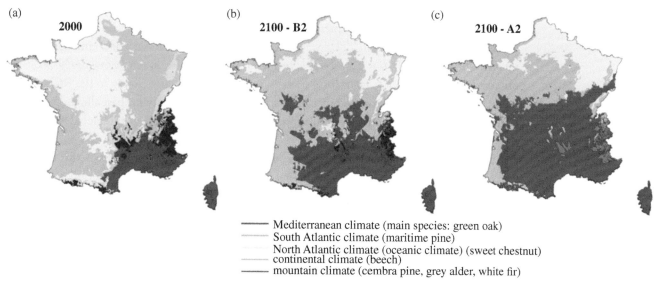

Mediterranean climate (main species: green oak)
South Atlantic climate (maritime pine)
North Atlantic climate (oceanic climate) (sweet chestnut)
continental climate (beech)
mountain climate (cembra pine, grey alder, white fir)

Fig. 34.1 The geographical distribution of seven chorological groups estimated by discriminant analysis of 70 tree species (a) as a function of current climate, and then extrapolated to future climates in 2100 for (b) a B2 SRES scenario and (c) an A2 SRES scenario. These scenarios correspond to an increase in global mean temperature with respect to 2000 of approximately +2.5°C and approximately +3.5°C, respectively. The map for 2000 (a) can be read as that of the current climate as represented by trees. Red, Mediterranean climate (main species green oak); orange, South Atlantic climate (maritime pine); blue, mountain climates (Swiss pine (*Pinus cembra*), grey alder and white fir); green, continental climate (beech); yellow, North Atlantic climate (oceanic climate) (sweet chestnut). The maps for 2100 show the possible evolution of potential climate niches of French forest species (see text).
Source: Vincent Badeau/INRA Nancy, France. Reproduced with permission.

+3.5°C with respect to the year 2000), how the distribution of potential climatic niches of tree species would change at the end of the 21st century (Badeau et al. 2010). These models, however, tell us nothing about the real potential relocation capability of the tree species. Taking into account constraints such as the dispersion of seeds, or diaspora (i.e. the physiological and mechanical constraints of forest species and constraints resulting from fragmentation of the forest environment), the germination capability of seeds on new sites, on-site competition among species, the development of (or exposure to) new pathogens and so on, the present picture based only on future migration of the climate zones could be deeply modified. On the one hand, it is assumed that the natural rates of dispersal of tree species are slow, and probably incompatible with the expected rapid migration of climate zones in the 21st century. On the other hand, the 'plasticity' of tree species (i.e. their ability to withstand new climatic stresses) is still poorly known. The worst consequences would be too slow northward migration and dieback in the southern limits.

If we consider the projections for the 21st century, a consensus emerges: global warming could easily reach +3 to +4°C, seriously disrupting the biosphere as we know it. *What is the basis for this statement?* Past climates can help us to appreciate the potential impact.

34.2.2 THE SIZE OF THE CHANGE COMPARED TO THE PAST

The recent anthropogenic warming (about +1°C) lies within the bounds of the natural fluctuations that have punctuated the present interglacial period. In the upper IPCC 2013 scenario (RCP8.6), the projected mean global warming for the year 2100 is about +5°C (and continues to increase after that date). *This is the scenario that is now being enacted, as can be seen in Fig. 33.7:* the observed emissions of CO_2 from fossil fuels and cement are holding closely to the RCP8.5 scenario, well above the track of the other three scenarios. Comparison with past climates, outlined in the three points below, enables us to appreciate the sheer magnitude of this global warming:

- *What does a change of +5°C in average temperature imply?* On the global level, this is the difference in global mean temperature between a glacial and an interglacial period. At the end of a glacial period, this increase implies warming by about 10°C in Western Europe, and by about 20°C on the Greenland Ice Sheet. In the middle and high latitudes of the Northern Hemisphere, the plant cover is completely transformed.
- *How fast did such climate changes take place in the course of the past million years?* Glacial–interglacial transitions (Part IV) typically occur over several thousand years, giving time for animal and plant species to relocate. Those unable to migrate or adapt die out. The Quaternary was punctuated by species extinctions, particularly at the onset of strong glaciations. Faster climate changes have occurred in the past, but not over the whole of the Earth. The projected +5°C global warming at the end of the 21st century would be of the same magnitude as in a glacial–interglacial transition: it would occur globally, but in the time interval of a century rather than several millennia.
- Finally, and *most importantly, the future warming will take place in an interglacial, rather than a glacial period.* Since 2.6 million years ago, the planet has alternated between two

fundamentally different states, a warm climate as we have now and a glacial climate. In the course of several dozen oscillations, the huge swings in temperature at the middle and high latitudes of the Northern Hemisphere have given rise to species selection. The sole survivors were those able to adapt or migrate fast enough to follow these transitions, or to find refuge in some special region. In other words, the flora and fauna of the middle and high latitudes have adapted to this type of oscillation: from cold to the present type of warm climate, and again from the present warm type back to cold. In future scenarios, however, since the warming will take place in an interglacial (and not in a glacial period), the flora and fauna of the middle and high latitudes will have to adapt to a totally new environment. This new climate, roughly +5°C warmer, would be even warmer than in the Mid-Pliocene (about 3 million years ago), when global mean temperatures were substantially warmer, about 2–3°C above pre-industrial temperatures. *This new climate would be comparable to what prevailed more than 10 million years ago (Fig. 18.1), since which time species have evolved considerably. The genus* **Homo**, *which is estimated to be between 2.3 and 2.4 million years old, would be propelled into an entirely new context (Fig. 34.2).*

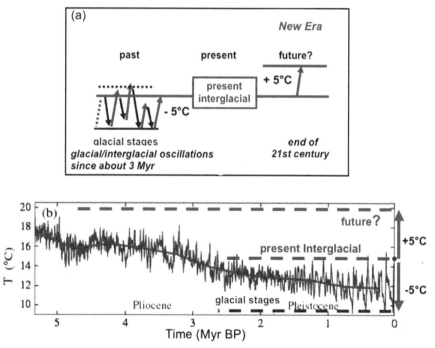

Fig. 34.2 (a) A comparison between the change in mean surface temperature during the glacial–interglacial oscillations of the past 2.6 million years and the projected warming by +5°C in scenario RCP8.5 at the end of the 21st century. An increase of +5°C would usher in a new climatic era such as the Earth has not known for millions of years, since before the existence of the genus *Homo*. (b) The reconstructed mean surface temperature for the past 5 million years: in the coolest glacial stages (dark blue dashed line) and at the present interglacial value (pink dashed line). Glacial–interglacial oscillations have been taking place for the past 2.6 million years. In the last five oscillations, the temperature drop between interglacials and glacials was 5°C (from about +15° to +10°C). A 5°C increase (red-line) from the current interglacial climate would project the Earth into a new climatic condition (mean temperature +20°C) that has not existed for at least 5 million years. At that time, the average temperature of the Earth's surface did not exceed +18°C.

Source: (b) Modified from Hansen et al. (2013b). Reproduced with permission of Makiko Sato.

The impact on life of the projected warming is a legitimate cause for concern, since it surpasses anything that has occurred over the past several million years. It will assuredly bring significant changes to the fragile balance among living species and to the functioning of ecosystems. Significant loss of biodiversity will occur, because adaptation is limited by:

- the rapidity of change of the climate;
- its magnitude;
- its sign (warming in an already warm interglacial period); and
- loss, due to appropriation by humans, of available living space for migration.

If we consider the basic interests of the human race alone, it is immediately obvious that the greatest risk is to agricultural resources. *How can we evaluate the reduction in GHG emissions that human society must enforce in order to limit the consequences of future warming?*

34.3 Global warming: no more than 2°C

The aim of the European Union is to limit global warming to 2°C, with temperatures stabilizing at the end of the 21st century. As it involves global emissions, this goal obviously concerns the whole of human society. The European Union has defined its own contribution: in October 2014 the EU leaders agreed to cut greenhouse gas emissions by at least 40% below 1990 levels by 2030. The reason for this limit is that although warming by 2°C is certain to affect biodiversity in many regions, its effect on the balance of resources that are essential to humankind would be manageable.

What is the basis for this value? Instead of specifying the considerations that lead to this estimate, we employ an argument that was introduced at the 2005 Exeter conference on the impact of climate change. The ecosystems of the planet can be classified into 36 groups that span the diversity of life on Earth. For each group, the impact of global warming (in °C) is then evaluated in terms of a percentage loss (either of area or of population) to that ecosystem. The magnitude of the loss depends on the type of ecosystem. For the complete set of ecosystems on Earth, it is shown that global warming of more than 2°C within such a short time lapse would have devastating effects on viability. When the geographical shift of favourable habitat for a species is too fast, or when the change in local climate conditions is too abrupt, the diversity in many types of ecosystem suffers severely.

34.4 The 'Triple Zero' challenge

For the temperature to stabilize, GHG emissions must first stabilize, and then decrease to well below the levels of 2000. We concentrate on the first stage; that is, stabilization of GHG emissions. Restricting ourselves to CO_2, we examine its implications for human society. These can be expressed through a qualitative relationship based on eqn. IV.1 of Part IV, Note 1, according to which CO_2 emissions are a function of the population and its structure.

To highlight the three important societal factors that govern emissions, the following simplified scenario is adopted: emissions from 'developed' or 'Northern' countries, and emissions from 'developing' or 'Southern' countries. For simplicity, the emerging countries are excluded from this scenario. The emissions are reduced to per capita averages. *What conditions must be met for global CO$_2$ emissions to remain constant?* One *simplistic solution* is to maintain each of the three following variables:

- constant per capita 'North' emissions;
- constant per capita 'South' emissions; and
- constant world population.

In this scenario, global CO$_2$ emissions remain constant indefinitely. This implies that:

- CO$_2$ emissions be stabilized in the 'North'. We observe that in these 'Northern' countries, consumption, energy production and CO$_2$ emissions are at present increasing by about 2% per year. Population increase, however, is modest.
- CO$_2$ emissions be stabilized in the global South; that is, per capita emissions, which constitute one fifth of those in the global North, remain constant and do not rise towards Northern levels. This is clearly not the case, especially as the emerging countries are developing.
- World population be stabilized. At present, world population is increasing at a rate of 1.1% per year. It is expected that by the middle of the 21st century, the world population will reach 9 billion human beings (about 50% more than at the beginning of the century).

The above assumptions are encapsulated in the following relationship, in which the left-hand side expresses the changes in the relevant factors with respect to the present, and the right-hand side is the resulting change in atmospheric CO$_2$. This is the *'Triple Zero' relationship*:

$$
\underset{\substack{\text{Change in CO}_2 \\ \text{emissions per} \\ \text{inhabitant of} \\ \text{the 'North'}}}{0} + \underset{\substack{\text{Change in CO}_2 \\ \text{emissions per} \\ \text{inhabitant of} \\ \text{the 'South'}}}{0} + \underset{\substack{\text{Change} \\ \text{in world} \\ \text{population}}}{0} \Rightarrow \underset{\substack{\text{Change in} \\ \text{atmospheric} \\ \text{CO}_2}}{+2\,\text{ppm/yr}}
$$

If the three factors on the left-hand side remain constant, emissions will stabilize at their current levels. Atmospheric CO$_2$ will increase regularly at its present rate of ~2 ppm/yr, since half of all emitted CO$_2$ accumulates in the atmosphere (Chapter 29), and leads to a continual increase in temperature. Emissions could also remain constant under other types of scenario – for instance, if one of the terms increases and the other two decrease. Thus, if the population doubles, per capita emissions must be cut by half, and so on.

This somewhat caricatural example starkly illustrates what are the three challenges that society must face up to in order to control GHG emissions: *demography, CO$_2$ production and 'North/South' cooperation.*

Chapter 35
Climate change in the present state of the planet

This book on climate change cannot be concluded without mentioning the global context. Climate change is but one component of the current changes with which human societies are confronted. In this global context, other interrelated forces are at work, whether environmental (continuing widespread degradation), economic (increasing demand for resources and raw materials, and their possible depletion) or social (changes in demographics and sharing of resources). All these considerations compel us to devise a new mode of development for our societies, a mode that is not only sustainable but also preserves the present environmental balance of the planet. The main points of this global context are summarized below.

35.1 Environmental degradation

One means of taking stock of the impact of human beings on the environment is to consider changes in each of the three compartments: air, water and soil. These form the setting for the fourth compartment, that in which life, in all its forms, develops. Quantitatively and qualitatively, the deterioration of these elements has reached alarming proportions (the quality and availability of fresh water, degradation of coastal and open ocean waters, air pollution in urban and industrial areas, planet-wide background noise, exhaustion of cultivated soils, increasing aridity, desertification and local pollution of soils, etc.). The biosphere is already severely affected, especially in the animal kingdom: habitat destruction, through the extension of agricultural land, deforestation or pollution, and intensive exploitation of animal resources (fishing and hunting), have led to a decrease in the number of individuals and species and have destabilized certain ecosystems. The reports by the Millennium Ecosystem Assessment (2005) review the various instance of degradation observed in the living world.

35.2 Depletion of energy resources

The second approach considers the energy requirements of the present model of the world economy. At present, 90% of all energy needs are met by fossil fuels, excluding biomass (non-fossil fuel), and 80% if biomass is included. Figure 29.6 shows the fraction of fossil fuel resources consumed each year. This raises the following fundamental question: *Will the available fossil-fuel resources run out in the near future?* To assess when the current rates of consumption will exhaust fossil fuels is a difficult undertaking, since the answer depends heavily on estimates of reserves. An order-of-magnitude approximation suggests that coal and oil could be exhausted within about one century, and gas within a few centuries. These orders of magnitude are modulated by two factors: if consumption increases, reserves will not last as long; however, improved extraction processes and the discovery of new deposits could increase the known reserves. At present, shale gas, bituminous sand deposits and deep offshore oilfields add to the current fossil-fuel resources. Therefore, we cannot rely on the imminent disappearance of fossil fuels to halt their emission. But the profound damage that these new resources inflict on the environment is a powerful argument for halting them.

35.3 Inexorable world population growth?

Demographics are a key component of the current global changes, as also is individual consumption. The Industrial Revolution, the cause of the main changes in CO_2 emissions, was accompanied by a demographic revolution during which the world's population exploded from 1.5 billion in 1900 to more than 7 billion in 2013. This growth automatically increased demand for food resources (agricultural and other) and for consumer goods. The world's population is now growing at a rate of about 1.1% per year. Current forecasts place this figure at about 9 billion people by the middle of the 21st century.

35.4 A new type of development?

Numerous studies conclude with the daunting observation that the way in which humanity is developing at present leads to an impasse. This is based on the fact that resources (fossil energy, raw materials and living resources) are finite, and that demand is ever-growing. If these paradigms do not change (expenditure of living resources without regard for their preservation, habitat destruction, production and use of material goods heedless of the damage to the environment, and population growth), in the coming decades, life functions and their vital links will be seriously endangered.

From whichever angle we view the current situation, reconsideration of our current mode of development is imperative. The following three points constitute the basis of the necessary changes:

- To contain future climate change within levels that the present biosphere can tolerate, greenhouse gas emissions must rapidly be stabilized, and then reduced in the next few decades. As these emissions stem essentially from the use of fossil fuels, our society must drastically reduce its consumption of fossil energy.
- As the population rises and as more people accede to Western lifestyles, the demand for raw materials, animal and plant resources, and land is increasing. It is obvious that here, too, resources are limited. Moreover, our use of resources largely ignores the degradation that it causes to the physical and living world. The aim of our society must be to reduce consumption, to recycle and to preserve resources.
- We are already forfeiting the resources of future generations. The present increase in population further aggravates this debt. Conditions for stabilizing the world's population must therefore be implemented without delay.

All these considerations demand a revolution both in our mentality and in our technology.

PART VI SUMMARY

Over the future decades, climate change will be determined in part by natural variability and in part by human agency. Climate projections therefore need each of these effects to be defined. Human activity mainly involves emission into the atmosphere of greenhouse gases (GHGs), which amplify the natural greenhouse effect. In the coming years, these emissions will be governed by the two key factors, emission per head of population and the number of inhabitants on Earth. Climate models simulate future climate change under various socio-economic scenarios. These reflect different global approaches: (i) a proactive attitude, in which GHG emissions are reduced to their lowest levels and average warming is limited to +2°C – this would stabilize the climate by the end of the century; or (ii) the present *laissez-faire* attitude, leading to warming of the order of +5°C by the end of the century, and no stabilization. The models are still being refined, but the order of magnitude of the warming predictions has not changed between the first report (IPCC 1992) and the current one (2013).

The climate projections are extensions of the changes observed in the warming of recent decades, and possess the same features as climate changes in the past. They range from low to high greenhouse gas emissions, with the following characteristics:

- Non-uniform warming that depends on latitude, more pronounced at high latitudes in the Northern Hemisphere, and more pronounced on land masses rather than in the oceans. At the same time, heat waves will become more frequent, and extreme cold snaps less frequent. The changes will lead to a generalized retreat of the cryosphere (contraction of the snow cover, glaciers, permafrost, sea ice etc.).
- Changes in the water cycle that exacerbate present disparities: increased precipitation in current high-rainfall areas (the equatorial belt and the mid-latitudes), and reduced rainfall in arid tropical regions. At the same time, the Hadley cell will expand towards the poles, bringing greater aridity to, for example, the Mediterranean region (North Africa and Southern Europe), southern North America (the Southern USA and Northern Mexico), South Africa and so on.

Can anthropogenic warming be countered by natural changes in the climate? Our knowledge of past climates and our understanding of climate processes tell us that there are no means by which natural fluctuations can bring about global cooling of 5°C, either in a few decades or even in a few millennia. Only the onset of a new glacial climate could do so, but we can rule out an ice age for the next several tens of thousands of years. Other natural events that could attenuate current anthropogenic warming include: an explosive volcanic eruption that would cool the planet by a few tenths or even one degree for several years only; or a decrease in solar activity that might yield the same result for several decades or centuries. However, the global impact of natural climate variability is limited to less than one degree over several decades. It is clear that natural causes can mitigate the predicted global warming at most by only one degree, and for a limited period of time.

The worst-case scenarios, warming by +5°C, could be reached by the end of the century (uninterrupted warming). These projections are of deep concern to the scientific community for three reasons:

- *The magnitude of the change*, which is of the same order as that during glacial–interglacial transitions. Historically, such a degree of warming on the continents has brought about near-total replacement of the vegetation at middle and high latitudes.
- *The rate of change:* historically, during glacial–interglacial transitions, warming took place over several millennia, which allowed species to migrate and adapt. This warming will be an order of magnitude faster, taking place over a single century.
- *Finally and most crucially, the warming will occur in an interglacial stage, when the climate is already warm.* Since the beginning of the Quaternary, warming has always occurred between glacial and interglacial climates. Those interglacial stages were, within one or two degrees, comparable to ours. Over roughly 2 million years, our ecosystems have been selected on the basis of their capacity to adapt and resist changes between glacial climates (with an average temperature of about +10°C) and interglacial climates (with average temperatures of about +15°). They have never had to adapt to the transition between an interglacial climate (+15°C) and a warmer climate of average temperature +20°C. Such warm climates have existed in the past, but many millions of years before the appearance of the genus *Homo*.

The Palaeocene–Eocene Thermal Maximum episode, which occurred 55 million years ago, provides an interesting illustration of the impact of such sudden warming. At the planetary level, biodiversity decreased dramatically in the oceanic benthic community, and underwent large changes on the continents. A new geological epoch began, the Eocene.

While it is difficult to assess the impact of climate change on the Earth's vegetation and ecosystems, everything indicates that, if the biosphere is to adapt without having to pay too heavy a price, global warming must be limited to +2°C. To meet this recommendation, greenhouse gas emissions must be reduced drastically. This will require a revolution in energy production.

PART VI NOTES

1. By 'primary energy', we mean the total energy required to produce useable energy (such as electric power). Only a fraction of the primary energy can be transformed into electricity. At present, global electricity production is ensured by thermal power stations fuelled principally by fossil fuels (coal, oil and natural gas). In these power stations, only 35–46% of the energy produced by the fossil fuels (primary energy) is transformed into electricity (35% in plants commissioned in the 1970s; 46% in supercritical processes that became available in the early 2000s).

2. The Annex B countries are the 39 most industrialized countries, which are subject to the binding obligations contained in the Kyoto Protocol controlling their GHG emissions.

3. In this context, the notion of atmospheric CO_2 stability refers to periods of several centuries. This stability is, however, relative: once the composition of the atmosphere has stabilized, atmospheric CO_2 levels will decrease over several tens of thousands of years towards their initial value.

4. The correspondence between CO_2 concentration and total GHG concentration (expressed in CO_2 equivalents) is as follows:

I : $350 - 400$ ppm CO_2 or $445 - 490$ ppm CO_2 – equiv.

II : $400 - 440$ ppm CO_2 or $490 - 535$ ppm CO_2 – equiv.

III : $440 - 480$ ppm CO_2 or $535 - 590$ ppm CO_2 – equiv.

IV : $480 - 570$ ppm CO_2 or $590 - 710$ ppm CO_2 – equiv.

V : $570 - 660$ ppm CO_2 or $710 - 855$ ppm CO_2 – equiv.

VI : $660 - 790$ ppm CO_2 or $855 - 1130$ ppm CO_2 – equiv.

PART VI FURTHER READING

Archer, D. (2012) *Global Warming – Understanding the Forecast*. John Wiley & Sons, Hoboken, NJ; see Part 3.

Climate Change 2013: The Physical Science Basis. Working Group I Contribution to the Fifth Assessment Report of the Intergovernmental Panel on Climate Change (IPCC), Cambridge University Press; available at http://www.ipcc.ch/report/ar5/wg1/-; see Chapters 11–14.

Fletcher, C. (2013) *Climate Change – What the Science Tells Us*. John Wiley & Sons, Hoboken, NJ; see Chapters 4–7.

Houghton, J. (2009) *Global Change – The Complete Briefing*, 4th edn. Cambridge University Press, Cambridge, UK; see Chapters 6 and 7.

Neelin, J.D. (2010) *Climate Change and Climate Modeling*. Cambridge University Press, Cambridge, UK; see Chapter 7.

Ruddiman, W.F. (2008) *Earth's Climate – Past and Future*, 2nd edn. W.H. Freeman, New York; see Part 5.

Conclusion

We have just laid out the basic concepts that permit an overall view of climate change and may serve to guide choices that must be made by society. We have concentrated in turn on the following:

- the fundamental role of the climate in sustaining life and allowing it to flourish on Earth, and how any change in the climate affects the equilibrium of the tissue of life;
- the existence of an average climate on Earth (mean global climate), which can be used as an indicator of the present changes;
- how the climate works in terms of energy, and the essential role of the greenhouse effect;
- natural climate changes at different timescales and their supposed causes;
- the present warming and its repercussions on the environment;
- computer simulations of climate change in the various possible socio-economic scenarios of the 21st century; and
- the sheer size, unprecedented over millions of years, of the climate change that is anticipated if greenhouse gas emissions are not regulated.

Now that we have come to the end of this study, what can we say to the question that, as the climate engine is so complex and its workings may still conceal numerous unknowns and surprises, is it not pretentious to predict its future changes? Is it not pretentious to confront

Climate Change: Past, Present and Future, First Edition. Marie-Antoinette Mélières and Chloé Maréchal.
© 2015 John Wiley & Sons, Ltd. Published 2015 by John Wiley & Sons, Ltd.
Companion website: www.wiley.com\go\melieres\climatechange

society with a choice that has such grave consequences, when we may have grasped is only a part of reality?

Certainly, our knowledge of the workings of the climate is far from complete, and numerous questions are still unanswered. Nevertheless, we have grasped enough of the main traits to understand the current climate change, and to attribute it, with very high probability, to the increase in greenhouse gases produced by human activity. On this basis, we can foresee in rough outline the climate changes during the 21st century as a function of the different societal choices that could be made. If emissions continue at their current level, then in all probability the climate will evolve towards *a situation that is totally new with respect to the past several million years that went before us*. We have also shown that it is totally illusory to rely on a natural change that can significantly modify the order of magnitude of the projected warming.

Since about 3 million years ago, the world climate has indeed oscillated between two profoundly different climates, an interglacial stage with a warm climate that is comparable to our own, and a glacial stage in which the climate is cold, with a drop in the average temperature of the order of 5°C and the presence of additional large icecaps. These, during the past million years, lowered the sea level by about 120 m. According to available palaeoclimate data, the mean temperature of the warmest interglacial stages since about 3 million years ago was never warmer than that of the present warm stage (which has been in place for 12,000 years) by more than about 2°C. Between these two climates, glacial and interglacial, the vegetation and the fauna differed greatly in the middle and high latitudes: in the ebb and flow of all those climate oscillations (around 50) over these past few million years, the flora and fauna that survived were selected by adaption and migration.

Towards the end of the 21st century, uncontrolled emission of greenhouse gases could plunge the planet into an entirely different context: an average warming of 5°C could be attained, which no natural fluctuation can successfully offset. It would involve a return to climate conditions that reigned many million years ago, a world in which the species *Homo* had not yet emerged, a world that was much warmer than now, and in which the flora and fauna were quite different. The change in climate that will take place in a few decades is enormous, the stuff of a science fiction scenario. And if this turns out to be (the choice depends always on society), then flora and fauna will adapt only with great difficulty. The magnitude and the speed of the change will inevitably harm the wealth of the biosphere, since life, whether defined in terms of the operating efficiency of ecosystems or in terms of the number of species, depends entirely on the climate that prevails in each place. The climate change would, in particular, affect our resources, and in the first place our food resources.

This is the reason why the scientific community has mobilized so strongly to share the current state of knowledge with the rest of society. All the observations, analyses and current previsions force us to concentrate our minds on the imperious need to reduce greenhouse gas emissions drastically.

The required reduction will take place in a changing world environment that is clearly marked by degradation of the natural surroundings. The necessary change in direction of society throughout the world is consistent with the wider aim of management that preserves the environment and makes better use of the resources offered by the planet; that is, 'sustainable' development. In this project, which is much wider than the climate question, a new paradigm emerges: another conception of the development of society planet-wide, where the wealth of the natural environment and the resources that it offers can be maintained. It is only under these conditions that the capital that we have inherited can be transmitted.

In this context, we must remember that *the principal uncertainty in future changes of the climate and the environment is without question the ability of man to evolve.* This evolution, which is essential, will take place through the ability of scientists and teachers to explain the risks that are involved on a worldwide scale, and the ability of all the members of society in the world (citizens, politicians, artists etc.) to seek solutions for a new deal. In this way, a vast construction site is opened, where the creativity of each individual is not only necessary but indispensable.

References

Arrhenius, S. (1896) On the influence of carbonic acid in the air upon the temperature of the ground. *The London, Edinburgh and Dublin Philosophical Magazine and Journal of Science,* Series 5, **41** (251), 237–276.

Archer, D., Buffett, B. and Brovkin, V. (2009) Ocean methane hydrates as a slow tipping point in the global carbon cycle. *Proceedings of the National Academy of Sciences of the United States of America,* **106**, 20,596–20,601.

Badeau, V., Dupouey, J.-L., Cluzeau, C., Drapier, J. and Le Bas, C. (2010) Climate change and the biogeography of French tree species: first results and perspectives, in *Forests, Carbon Cycle and Climate Change* (ed D.Loustau), Editions Quae, Update Sciences et Technologies, Versailles, Chapter 11, pp. 231–252.

Bard, E. (2006) Variations climatiques naturelles et anthropiques. *BRGM Geosciences,* **3**, 30–35.

Baronni, C. and Orombelli, G. (1996) The Alpine 'Ice man' and Holocene climatic change. Quaternary Research, **46**, 78–83.

Barthlott, W., Lauer, W. and Placke, A. (1996) Global distribution of species diversity in vascular plants: towards a world map of phytodiversity. *Erdkunde,* **50**, 317–328.

Barthlott, W., Hostert, A., Kier, G. et al. (2007) Geographic patterns of vascular plant diversity at continental to global scales. *Erdkunde,* **61**, 305–315.

Climate Change: Past, Present and Future, First Edition. Marie-Antoinette Mélières and Chloé Maréchal.
© 2015 John Wiley & Sons, Ltd. Published 2015 by John Wiley & Sons, Ltd.
Companion website: www.wiley.com\go\melieres\climatechange

Battisti, A., Stastny, M., Netherer, S. et al. (2005) Expansion of geographic range in the pine processionary moth caused by increased winter temperatures. *Ecological Applications*, **15** (6), 2084–2096.

Beaugrand, G., Luczak, C., and Edwards, M. (2009) Rapid biogeographical plankton shifts in the North Atlantic Ocean. *Global Change Biology*, **15**, 1790–1803.

Beaugrand, G., Reid, P.C., Ibanez, F., Lindley, J.A. and Edwards, M. (2002) Reorganisation of North Atlantic marine copepod biodiversity and climate. *Science*, **296**, 1692–1694.

Becker, M., Meyssignac, B., Letetrel, C., Llovel, W., Cazenave, A. and Delcroix, T. (2012) Sea level variation in Pacific Islands since 1950. *Global and Planetary Change*, **80–81**, 85–86.

Bender, M., Labeyrie, L.D., Raynaud, D. and Lorius, C. (1985) Isotopic composition of atmospheric O_2 in ice linked with deglaciation and global primary productivity. *Nature*, **318**, 349–352.

Berger, A. and Loutre, M.F. (2002) An exceptionally long interglacial ahead? *Science*, **297**, 1287–1288; doi: 10.1126/science.1076120

Blunier, T. and Brook, E. (2001) Timing of millennial-scale climate change in Antarctica and Greenland during the last glacial period. *Science*, **291**, 1090112.

Boden, T.A., Marland, G. and Andres, R.J. (2011) *Global, Regional, and National Fossil-Fuel CO_2 Emissions*. Carbon Dioxide Information Analysis Center, Oak Ridge National Laboratory, US Department of Energy, Oak Ridge, TN; doi: 10.3334/CDIAC/00001_V2011

Bond, G.C. and Lotti, R. (1995) Iceberg discharges into the North Atlantic on millennial time scales during the last glaciation. *Science*, **267**, 1005–1010.

Bond, G.C., Kromer, B., Beer, J. et al. (2001) Persistent solar influence on North Atlantic climate during the Holocene. *Science*, **278**, 1257–1266.

Bond, G.C., Showers, W., Cheseby, C. et al. (1997) A pervasive millenial-scale cycle in the North Atlantic Holocene and glacial climates. *Science*, **294**, 2130–2136.

Bradley, R. (2000) Past global changes and their significance for the future. *Quaternary Science Review*, **1**, 391–402.

Budyko, M.I. (1969) The effect of solar radiation variations on the climate of the Earth. *Tellus*, **XXI**, 611–619.

Capron, E., Landais, A., Lemieux-Dudon, B. et al. (2010) Synchronizing EDML and North GRIP ice cores using $\delta^{18}O$ of atmospheric oxygen ($\delta^{18}O_{atm}$) and CH_4 measurements over MIS5 (80–123 kyr). *Quaternary Science Review*, **29**, 222–234.

Cita, M.B., Vergnaud-Grazzini, C., Robert, C., Chamley, H., Ciaranfi, N. and D'Onofrio, S. (1977) Paleoclimatic record of a long deep-sea core from the eastern Mediterranean. *Quaternary Research*, **8**, 205–235.

Costa, S.M.S. and Shine, K.P. (2012) Outgoing longwave radiation due to directly transmitted surface emission. *Journal of the Atmospheric Sciences*, **69** (6), 1865–1870.

Cullen, H.M. and deMenocal, P.B. (2000) North Atlantic influence on Tigris–Euphrates streamflow. *International Journal of Climatology*, **20**, 853–863.

Dahl, S.O. and Nesje, A. (1996) A new approach to calculating Holocene winter precipitation by combining glacier equilibrium-line altitudes and pine-tree limits: a case study from Hardangerjøkulen, central southern Norway. *The Holocene*, **6**, 381–398.

de Beaulieu, J.-L. (2006) Apport des longues séquences lacustres à la connaissance des variations des climats et des paysages pléistocène. *Comptes Rendus Palevol*, **5**, 65–72.

de Beaulieu, J.-L., Richard, H., Clerc, J. and Ruffaldi, P. (1994) History of vegetation, climate and human action in the French Alps and the Jura over the last 15,000 years. *Dissertationes Botanicaea*, **234**, 253–275.

deMenocal, P.B., Ortiz, J., Guilderson, T. et al. (2000) Abrupt onset and termination of the African Humid Period: rapid climate response to gradual insolation forcing. *Quaternary Science Reviews*, **19**, 347–361.

Denton, G.H. and Karlen, W. (1973) Holocene climatic variations – their pattern and possible cause. *Quaternary Research*, **3**, 155–205.

Deschamps, P., Durand, D., Bard, E. et al. (2012) Ice sheet collapse and sea level rise at the Bölling warming 14,600 yr ago. *Nature*, **483**, 559–564.

Devictor, V., van Swaay, C., Brereton, T. et al. (2012) Differences in the climatic debts of birds and butterflies at a continental scale. *Nature Climate Change*, **2**, 121–124; doi: 10.1038/NCLIMATE1347

Dufresne, J.-L. and Bony, S. (2008) An assessment of the primary sources of spread of global warming estimates from coupled atmosphere–ocean models. *Journal of Climate*, **21**, 5135–5144.

Dykosky, C.A., Edwards, R.L., Hai Cheng et al. (2005) A high-resolution absolute-dated Holocene and deglacial Asian monsoon record from Dongge Cave, China. *Earth and Planetary Science Letters*, **233**, 71–86.

Elsig, J., Schmitt, J., Daiana Leuenberger, D. et al. (2009) Stable isotope constraints on Holocene carbon cycle changes from an Antarctic ice core. *Nature*, **461**, 507–510.

Emeis, K.-C., Schulz, H., Struck, U. et al. (2003) Eastern Mediterranean surface water temperature and $\delta^{18}O$ composition during deposition during deposition of sapropels in the late Quaternary. *Paleooceanography*, 18; doi: 10.1029/2000PAOOO617

EPICA Community Members (2004) Eight glacial cycles from an Antarctic ice core. *Nature*, **429**, 623–628.

Fabricius, K., Langdon, C., Uthicke, S. et al. (2011) Losers and winners in coral reefs acclimatized to elevated carbon dioxide concentrations. *Nature Climate Change*, **1**, 165–169.

Fabry, V., Seibel, B.A., Feely, R.A. and Orr, J.C. (2008) Impacts of ocean acidification on marine fauna and ecosystem processes. *ICES Journal of Marine Science*, **65**, 414–432.

Fasullo, J.T. and Trenberth, K.E. (2008) The annual cycle of the energy budget. Part II: meridional structures and poleward transports. *Journal of Climate*, **21**, 2313–2325.

Fettweis, X., Tedesco, M., van den Broeke, M. and Ettema, J. (2011) Melting trends over the Greenland ice sheet (1958–2009) from spaceborne microwave data and regional climate models. *The Cryosphere*, **5**, 359–375.

Fleitmann, D., Burns, S.J., Mudelsee, M. et al. (2003) Holocene forcing of the Indian monsoon recorded in a stalagmite from southern Oman. *Science*, **300**, 1737–1739.

Fleming, K. (2000). Glacial rebound and sea-level change constraints on the Greenland ice sheet. PhD thesis. Australian National University.

Fleming, K., Johnston, P., Zwartz, D., Yokoyama, Y., Lambeck, K. and Chappell, J. (1998) Refining the eustatic sea-level curve since the Last Glacial Maximum using far- and intermediate-field sites. *Earth and Planetary Science Letters*, **163** (1–4), 327–342.

Foster, G. and Rahmstorf, S. (2011) Global temperature evolution 1979–2010. *Environmental Research Letters*, **6**; doi: 10.1088/1748-9326/6/4/044022

Fourier, J.-B. (1824) On the temperature of the terrestrial sphere and interplanetary space. *Mémoire de l'académie royale des sciences*, **7**, 569–604.

Francou, B. and Vincent, C. (2007) *Les glaciers à l'épreuve du climat*. Ed. Belin, Paris.

Gasse, F. (2000) Hydrological changes in the African tropics since the Last Glacial Maximum. *Quaternary Science Reviews*, **19**, 189–211.

Genty, D., Blamart, D., Ghaleb, B. et al. (2006) Timing and dynamics of the last deglaciation from European and North Africa $\delta^{13}C$ stalagmite profiles – comparison with Chinese and South Hemisphere stalagmites. *Quaternary Science Review*, **25**, 2118–2142.

Global Carbon Project (GCP) (n.d.) http://www.globalcarbonproject.org/carbonbudget

Gordley, L.L., Marshall, B.T. and Chu, D.A. (1994) LINEPAK: algorithms for modeling spectral transmittance and radiance. *Journal of Quantitative Spectroscopy & Radiative Transfer*, **52** (5), 563–580.

GRIP Project Members (1993) Climate instability during the last interglacial periods recorded in the GRIP ice core. *Nature*, **364**, 203–207.

Groombridge, B. and Jenkins, M.D. (2000) *Global Biodiversity, Earth's Living Resources in the 21st Century*. World Conservation Press, Cambridge, UK.

Grootes, P.M., Stuiver, M., White, J.W.C., Johnsen, S. and Jouzel, J. (1993) Comparison of oxygen isotope records from the GISP2 and GRIP Greenland ice cores. *Nature*, **366**, 552–554.

Grosjean, M. , Suter, P.J., Trachsel, M. and Wanner, H. (2007) Ice-borne prehistoric finds in the Swiss Alps reflect Holocene glacier fluctuations. *Journal of Quaternary Science*, **22** (3), 203–207.

Grove, J.M. (2004) *Little Ice Ages: Ancient and Modern*, **2** vols. Routledge, London.

Haeberli, W., Hoelzle, M., Paul, F. and Zemp, M. (2007) *Integrated monitoring of mountain glaciers as key indicators of global climate change: the European Alps. Annals of Glaciology*, **46**, 150–160.

Hanel, R.A., Schlachman, B., Rogers, D. and Vanous, D. (1971) Nimbus 4 Michelson Interferometer. *Applied Optics*, **10**, 1376–1382.

Hansen, J., Sato, M. and Ruedy, R. (2012) Perception of climate change. *Proceedings of the National Academy of Sciences of the United States of America*, **109** (37), E2415–E2423.

Hansen, J., Sato, M. and Ruedy, R. (2013a) Reply to Rhines and Huybers: changes in the frequency of extreme summer heat. *Proceedings of the National Academy of Sciences of the United States of America*, **110** (7), E547–E548.

Hansen, J., Sato, M., Kharecha, P. and von Schuckmann, K. (2011) Earth's energy imbalance and implications. *Atmospheric Chemistry and Physics*, **11**, 13,421–13,449.

Hansen, J., Sato, M., Russell, G. and Kharecha, P. (2013b) Climate sensitivity, sea level, and atmospheric carbon dioxide. *Philosophical Transactions of the Royal Society A*, **371**; doi: 10.1098/rsta.2012.0294

Haug, G.H., Hughen, K.A., Sigman, D.M., Peterson, L.C. and Röhl, U. (2001) Southward migration of the intertropical convergence zone through the Holocene. *Science*, **293**, 1304–1308.

Hays, J.D., Imbrie, J. and Shackleton, N.J. (1976) Variations in the Earth's orbit: pacemaker of the ice ages. *Science*, **235**, 1156–1167.

Holzhauser, H. (2009) Auf dem Holzweg zur Gletschergeschichte, in *Hallers Landschaften und Gletscher. Beiträge zu den Veranstaltungen der Akademien der Wissenschaften Schweiz*

2008 zum Jubiläumsjahr 'Haller300'. Sonderdruck aus den Mitteilungen der Naturforschenden Gesellschaft in Bern, Neue Folge, **66**, 173–208.

Holzhauser, H., Magny, M. and Zumbühl, H.J. (2005) Glacier and lake-level variations in west-central Europe over the last 3500 years. *The Holocene,* **15** (6), 789–801.

IPCC 2007: *Climate Change 2007. The Physical Science Basis,* Working Group I (WGI); *Impacts, Adaptation and Vulnerability,* Working Group II (WGII); *Mitigation of Climate Change,* Working Group III (WGIII). Contribution to the Fourth Assessment Report of the Intergovernmental Panel on Climate Change, Cambridge University Press, Cambridge, UK.

IPCC 2007 SR: *Climate Change 2007: Synthesis Report,* Contribution of Working Groups I, II and III to the Fourth Assessment Report of the Intergovernmental Panel on Climate Change, IPCC, Geneva, Switzerland.

IPPC 2013: *Climate Change 2013. The Physical Science Basis,* Working Group I (WGI). Contribution to the Fifth Assessment Report of the Intergovernmental Panel on Climate Change, Cambridge University Press, Cambridge, UK.

Jones, G.V. (2006) Climate and terroir: impacts of climate variability and change on wine, in *Fine Wine and Terroir – the Geoscience Perspective* (eds R.W.Macqueen and L.D. Meinert), Geoscience Canada Reprint Series Number 9, Geological Association of Canada, St. John's, Newfoundland, pp. 1–14.

Jones, G.V., Reid, R. and Vilks, A. (2012) Climate, grapes, and wine: structure and suitability in a variable and changing climate, in *The Geography of Wine: Regions, Terroir, and Techniques* (ed. P. Dougherty), Springer, Dordrecht, pp. 109–133.

Jorgenson, M.T., Racine, C.H., Walters, J.C. and Osterkamp, T.E. (2001) Permafrost degradation and ecological changes associated with a warming climate in central Alaska. *Climatic Change,* **48**, 551–579.

Jouzel, J., Masson-Delmotte, V., Cattani, O. et al. (2007) Orbital and millennial Antarctic climate variability over the past 800,000 years. *Science,* **317**, 793–797.

Kalnay, E., Kanamitsu, M., Kistler, R. et al. (1996) The NCPE/NCAR 40-year Reanalysis Project. *Bulletin of the American Meteorological Society,* **77** (3), 437–471.

Kiehl, J.T. and Trenberth, K E. (1997) Earth's annual global mean energy budget. *Bulletin of the American Meteorological Association,* **78**, 197–208.

Koerner, R.M. and Fisher, D.A. (1990) A record of Holocene summer climate from a Canadian high-Arctic ice-core. *Nature,* **343**, 630–631.

Kopp, G. and Lean, J.L. (2011) A new, lower value of total solar irradiance: evidence and climate significance. *Geophysical Research Letters,* **38** (1); doi: 10.1029/2010GL045777

Kottek, M., Grieser, J., Beck, C., Rudolf, B. and Rubel, F. (2006) World map of the Köppen–Geiger climate classification updated. *Meteorologische Zeitschrift,* **15**, 259–263.

Kuper, R. and Kroeplin, S. (2006) Climate-controlled Holocene occupation in the Sahara: motor of Africa's evolution. *Science,* **313**, 803–807.

Latham, R.E. and Ricklefs, R.E. (1993) Continental comparisons of temperate-zone tree species diversity, in *Species Diversity: Historical and Geographical Perspectives* (eds R. E. Ricklefs and D. Schluter), University of Chicago Press, Chicago, pp. 294–314.

Lauritzen, S.-E. (1996) Calibration of stable isotopes against historical records: a Holocene temperature curve from North Norway? In *Climatic Change: the Karst Record,* Vol. **2** (ed S.-E. Lauritzen), Karst Waters Institute special publication, Charles Town, WV, pp. 78–80.

Lauritzen, S.E. and Lundberg, J. (1998) Rapid temperature variations and volcanic events during the Holocene from a Norwegian speleothem record, in *Past Global Changes and Their Significance for the Future*, Volume of Abstracts, IGBP–PAGES, Bern, p. 88.

Lenton, T.M. (2012) Arctic climate tipping points. *Ambio*, **41**, 10–22.

Lenton, T.M., Held, H., Kriegler, E. et al. (2008) Tipping elements in the Earth's climate system. *Proceedings of the National Academy of Sciences of the United States of America*, **105**, 1786–1793.

Le Quéré, C., Peters, G.P., Andres, R.J. et al. (2014) Global carbon budget 2013. *Earth System Science Data Discussions*, **6**, 235–263.

Le Roy Ladurie, D., Rousseau, D. and Vasak, A. (2011) *Les fluctuations du climat: de l'an mil à aujourd'hui*. Fayard, Paris.

Li, C., Battisti, D.S., Schrag, D.P. and Tziperman, E. (2005) Abrupt climate shifts in Greenland due to displacements of the sea ice edge. *Geophysical Research Letters*, **32**, L19702; doi: 10.1029/2005GL023492, 2005

Lisiecki, L. and Raymo, M. (2005) A Pliocene–Pleistocene stack of 57 globally distributed benthic $\delta^{18}O$ records. *Paleoceanography*, **20**; doi: 10.1029/2004PA001071

Liu, Z., Wang, Y., Gallimore, R. et al. (2007) Simulating the transient evolution and abrupt change of North Africa atmosphere–ocean–terrestrial ecosystem in the Holocene. *Journal of Quaternary Science Review*, **26**, 1818–1837.

Loeb, N.G., Lyman, J.M., Johnson, G.C. et al. (2012) Observed changes in top-of-the-atmosphere radiation and upper-ocean heating consistent within uncertainty. *Nature Geoscience*, **5**, 110–113.

Magny, M. (2013) Orbital, ice-sheet, and possible solar forcing of Holocene lake-level fluctuations in west-central Europe. A comment on Bleicher. *The Holocene*, **23**, 1202–1212.

Manabe, S. and Wetherald R.T. (1967) Thermal equilibrium of atmosphere with a given distribution of relative humidity. *Journal of the Atmospheric Sciences*, **24** (3), 241–258.

Maréchal, C., Télouk, P. and Albarède, F. (1999) Precise analysis of copper and zinc isotopic compositions by plasma-source mass spectroscopy. *Chemical Geology*, **156**, 251–273.

Maréchal, C., Nicolas, E., Douchet, C. and Albarède, F. (2000) Abundance of zinc isotopes as a marine biogeochemical tracer. *Geochemistry, Geophysics, Geosystems*, **1**, doi:10.1029/1999GC000029.

Mark, N., Lister, D., Hulme, M. and Makin, I. (2000) A high-resolution data set of surface climate over global land areas. *Climate Research*, **21**, 1–25.

Maslanik, J., Stroeve, J., Fowler, C. and Emery, W. (2011) Distribution and trends in Arctic sea ice age through spring 2011. *Geophysical Research Letters*, **38** (12); doi: 10.1029/2011GL047735

Mayewsky, P.A. and Bender, M. (1995) The GISP2 ice core record – paleoclimate highlights. *Review of Geophysics*, (**33**), 1287–1296.

Mélières, M.-A., Rossignol-Strick, M. and Malaizé, B. (1997) Relation between low latitude insolation and $\delta^{18}O$ change of atmospheric oxygen for the last 200 kyrs, as revealed by Mediterranean sapropels. *Geophysical Research Letters*, **24**, 1235–1238.

Milankovitch, M.M. (1941) *Kanon der erd bestrahlung und seine Anwendung auf des Eizeitenprobleme*, Special Publication 132, Section on Mathematical and Natural Sciences, 33, Köninglich Serbische Akademie, Beograd.

Millennium Ecosystem Assessment (2005) *Ecosystems and Human Well-Being: Synthesis*, Island Press, Washington, DC.

Milne, G.A., Long, A.J. and Bassett, S.E. (2005) Modelling Holocene relative sea-level observations from the Caribbean and South America. *Quaternary Science Reviews*, **24** (10–11), 1183–1202.

Mourik, A.A., Bijkerk, J.F. and Cascella, A. (2010) Astronomical tuning of the La Vedova High Cliff section (Ancona, Italy) – implications of the Middle Miocene Climate Transition for Mediterranean sapropel formation. *Earth and Planetary Science Letters*, **297**, 249–261.

North Greenland Ice Core Project Members (2004) High-resolution record of Northern Hemisphere climate extending into the last interglacial period. *Nature*, **431**, 147–151.

Oerlemanns, J. (2005) Extracting a climate signal from 169 glacier records. *Science*, **308**, 765–767.

Olson, D.M., Dinerstein, E., Wikramanayake, E.D. et al. (2001) Terrestrial ecoregions of the world: a new map of life on Earth. *BioScience*, **51** (11), 933–938 (map by L. Spurrier, WWF, 2013).

Ohmura, A. (2004) Cryosphere during the twentieth century, in *The State of the Planet: Frontiers and Challenges in Geophysics* (eds J.Y. Sparling and C.J. Hawkesworth), American Geophysical Union, Washington, DC, pp. 239–257.

Orgil, D. and Gribbin, J. (1979) *The Sixth Winter*, Simon and Schuster, New York.

Orr, J., Fabry, V.J., Aumont, O. et al. (2005) Anthropogenic ocean acidification over the twenty-first century and its impact on calcifying organisms. *Nature*, **437**, 681–686.

Ortlieb, L. (2000) The documentary historical record of El Niño events in Peru: an update of the Quinn record (sixteenth through nineteenth centuries), in *El Niño and the Southern Oscillation: Variability, Global and Regional Impacts* (eds H. Diaz and V. Markgraf), Cambridge University Press, Cambridge, UK, pp. 207–295.

PAGES 2k Consortium (2013) Continental-scale temperature variability during the past two millennia. *Nature Geoscience*, **6**, 339–346.

PALEOSENS Project Members (2012) Making sense of paleoclimate sensitivity. *Nature*, **491**, 683–691.

Peixoto, J.P. and Oort, A.H. (1992) *Physics of Climate*. American Institute of Physics, New York.

Peters, G., Andrew, R.M., Boden, T. et al. (2012) The challenge to keep global warming below 2°C. *Nature Climate Change*, **3**, 4–6.

Peters, G., Minx, J.C., Weber, C.L. and Edenhofer, O. (2011) Growth in emission transfers via international trade from 1990 to 2008. *Proceedings of the National Academy of Sciences of the United States of America*, **108** (21), 8903–8908.

Petit, J.R., Jouzel, J. Raynaud, D. et al. (1999) Climate and atmospheric history of the past 420,000 years from the Vostok ice core, Antarctica. *Nature*, **399**, 429–436.

Queney, P. (1974) *Eléments de météorologie*, Masson, Paris.

Quinn, W.H., Neal, V.T. and Antunez de Mayolo, S.E. (1987) El Niño occurrences over the past four and a half centuries. *Journal of Geophysical Research*, **92**, 14,449–14,461.

Rabatel, A., Vincent, J., Naveau, P., Bernard, F. and Grancher, D. (2005) Dating of Little Ice Age glacier fluctuations in the tropical Andes: Charquini glaciers, Bolivia, 16°S. *Comptes Rendus Geoscience*, **337**, 1311–1322.

Rabatel, A., Machaca, A., Francou, B. and Jomelli, V. (2006) Glacier recession on Cerro Charquini (16 degrees S), Bolivia, since the maximum of the Little Ice Age (17th century). *Journal of Glaciology*, **52**, 110–118.

Ravanel, L. and Deline, P. (2010) Climate Influence on rockfalls in high-Alpine steep rock-falls: the north side of the aiguilles de Chamonix (Mont Blanc Massif) since the end of the 'Little Ice Age'. *The Holocene*, **21**, 357–365.

Reynolds, R.W., Rayner, N.A., Smith, T.M., Stokes, D.C. and Wanqiu Wang (2002) An improved *in situ* and satellite SST analysis for climate. *Journal of Climate*, **15**, 1609–1625.

Roberts, N. (2014) *The Holocene, an Environmental History*, 3rd edn, John Wiley & Sons, Ltd, Chichester.

Robinet, C., Imbert, C.-E., Rousselet, J. et al. (2012) Human mediated long-distance jumps of the pine processionary moth in Europe. *Biological Invasion*, **14**, 1557–1569.

Rossignol-Strick, M. (1985) Mediterranean quaternary sapropels, an immediate response of the African monsoon to variation of insolation. *Paleogeography, Paleoclimatology, Paleoecology*, **49**, 237–263.

Rothman, L.S., Jacquemart, D., Barbe, A. et al. (2004) The HITRAN 2004 molecular spectroscopic database. *Journal of Quantitative Spectroscopy & Radiative Transfer*, **96**, 139–204.

Rousseau, D. (2006) Surmortalité des étés caniculaires et surmortalités hivernales en France. *Climatologie*, **3**, 43–54.

Ruddiman, W.F. (2003) The anthropogenic greenhouse era began thousands of years ago. *Climatic Change*, **61**, 261–293.

Ruddiman, W.F. (2005) *Plows, Plagues and Petroleum: How Humans Took Control of Climate*. Princeton University Press, Princeton, NJ.

Ruddiman, W.F. (2007) The early anthropogenic hypothesis: challenges and responses. *Reviews of Geophysics*, **45**, RG4001.

Schär, C., Vidale, P.L., Lüthi, D. et al. (2004) The role of increasing temperature variability in European summer heat waves. *Nature*, **427**, 332–336.

Sellers, W.D. (1969) A global climatic model based on the energy balance of the Earth–atmosphere system. *Journal of Applied Meteorology*, **8**, 392–400.

Severinghaus, J., Beaudette, R., Headly, M.A., Taylor, K. and Brook, E.J. (2009) Oxygen-18 of O_2 records the impact of abrupt climate change on the terrestrial biosphere. *Science*, **324**, 1431–1434.

Siegenthaler, U., Stocker, T.F., Monnin, E. et al. (2005) Stable carbon cycle–climate relationship during the Late Pleistocene. *Science*, **310**, 1313–1317.

Singarayer, J.S., Valdes, P.J., Friedlingstein, P., Nelson, S. and Beerling, D.J. (2011) Late Holocene methane rise caused by orbitally controlled increase in tropical sources. *Nature*, **470**, 82–85.

Stocker, T.F. and Johnsen, S.J. (2003) A minimum thermodynamic model for the bipolar seesaw. *Paleoceanography*, **18** (4), art. no. 1087; doi: 10.1029/2003PA000920

Strasser, A., Hilgen, F.J. and Heckel, P.H. (2006) Cyclostratigraphy – concepts, definitions and applications. *Newsletters on Stratigraphy*, **42**, 75–114.

Stuiver, M. and Grootes, P.M. (2000) GISP2 oxygen isotope ratios. *Quaternary Research*, **53**, 277–284.

Tyndall, J. (1861) On the absorption and radiation of heat by gases and vapors, and on the physical connexion of radiation, absorption, and conduction. *Philosophical Magazine*, series 4, **22**, 169–194, 273–285.

Tzedakis, C., Channell, J.E.T., Hodell, D.A., Kleiven, H.F. and Skinner, L.C. (2012) Determining the natural length of the current interglacial. *Nature Geoscience*, **5**, 138–141.

Van Vliet-Lanoë, B. and Lisitsyna, O. (2001) Permafrost extent at the Last Glacial Maximum and at the Holocene Optimum: the Climex map, in *Permafrost Response on Economic Development: Environmental Security and Natural Resources* (eds. R. Paepe, V.P. Melnikov, E. Van Overloop and V.D. Gorokhov), Kluwer, Dordrecht, pp. 215–225.

Vincent, C., Le Meur, E. and Six, D. (2005) Solving the paradox of the end of the Little Ice Age in the Alps. *Geophysical Research Letters*, **32**, L09706; doi: 10.1029/2005GL022552

Vincent, C., Kappenberger, G., Valla, F., Bauder, A., Funk, M. and Le Meur, E. (2004) Ice ablation as evidence of climate change in the Alps over the 20th century. *Journal of Geophysical Research*, **109** (D10), D10104; doi: 10.1029/2003JD003857

Waelbroeck, C., L. Labeyrie, L., Michel, E. et al. (2002) Sea-level and deep water temperature changes derived from benthic foraminifera isotopic records. *Quaternary Science Reviews*, **21**, 295–305.

Wang, P., Wang, B. and Kiefer, T. (2009) Global monsoon in observations, simulations and geological records. *PAGES News*, **17** (2), 82–83.

Wang, Y., Cheng, H., Edwards, L.R. et al. (2005) The Holocene Asian monsoon: links to solar changes and North Atlantic climate. *Science*, **308**, 874–875.

Wang, Y.J., Cheng, H., Edwards, L.R. et al. (2001) A high-resolution absolute-dated late Pleistocene monsoon record from Hulu Cave, China. *Science*, **294**, 2345–2348.

Wanner, H., Beer, J., Bütikofer, J., et al. (2008) Mid- to late Holocene climate change: an overview. *Quaternary Science Reviews*, **27**, 1791–1828.

Webster, P.J., Holland, G.J., Curry, J.A. and Chang, H.-R. (2005) Changes in tropical cyclone number, duration, and intensity in a warming environment. *Science*, **309**, 1844–1846.

Whittaker, R. (1975) *Communities and Ecosystems*. Macmillan, New York.

Wild, M., Folini, D., Schär, C., Loeb, N., Dutton, E.G. and König-Langlo, G. (2013) The global energy balance from a surface perspective. *Climate Dynamics*, **40**, 3107–3134.

Zachos, J.C., Dickens, G.R. and Zeebe, R.E. (2008) An Early Cenozoic perspective on greenhouse warming and carbon-cycle dynamics. *Nature*, **451**, 279–283.

Zachos, J., Bohaty, S.M., John, C.M., McCarren, H., Kelly, D.C. and Nielsen, T. (2003) A transient rise in tropical sea surface temperature during the Paleocene–Eocene Thermal Maximum. *Science*, **302**, 1551–1554.

Zachos, J., Röhl, U., Schellenberg, S.A. et al. (2005) Rapid acidification of the ocean during the Paleocene–Eocene Thermal Maximum. *Science*, **308**, 1611–1615.

Index

Page numbers in *italics* refer to illustrations

Climate Change: Past, Present and Future, First Edition. Marie-Antoinette Mélières and Chloé Maréchal.
© 2015 John Wiley & Sons, Ltd. Published 2015 by John Wiley & Sons, Ltd.
Companion website: www.wiley.com\go\melieres\climatechange

Printed and bound by CPI Group (UK) Ltd, Croydon, CR0 4YY